The New Politi
and Pleasure

Edited By

Peter Bramham
Leeds Metropolitan University, UK

and

Stephen Wagg
Leeds Metropolitan University, UK

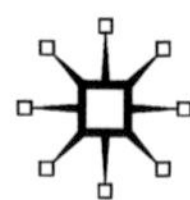

First published 2011 by
PALGRAVE MACMILLAN

Palgrave Macmillan in the UK is an imprint of Macmillan Publishers Limited, registered in England, company number 785998, of Houndmills, Basingstoke, Hampshire RG21 6XS.

Palgrave Macmillan in the US is a division of St Martin's Press LLC, 175 Fifth Avenue, New York, NY 10010.

Palgrave Macmillan is the global academic imprint of the above companies and has companies and representatives throughout the world.

Palgrave® and Macmillan® are registered trademarks in the United States, the United Kingdom, Europe and other countries

ISBN 978-0-230-21683-9 hardback
ISBN 978-0-230-21684-6 paperback

This book is printed on paper suitable for recycling and made from fully managed and sustained forest sources. Logging, pulping and manufacturing processes are expected to conform to the environmental regulations of the country of origin.

A catalogue record for this book is available from the British Library.

Library of Congress Cataloging-in-Publication Data

The new politics of leisure and pleasure / edited by Peter Bramham and Stephen Wagg.
p. cm.
ISBN 978-0-230-21684-6
1. Leisure. 2. Leisure–Social aspects. 3. Pleasure. I. Bramham, Peter. II. Wagg, Stephen.
GV14.N49 2010
306.4'812–dc22 2010032360

10 9 8 7 6 5 4 3 2 1
20 19 18 17 16 15 14 13 12 11

Printed and bound in Great Britain by
CPI Antony Rowe, Chippenham and Eastbourne

The New Politics of Leisure and Pleasure

Also by Peter Bramham

HOW STAFF RULE: Structures of Authority in Two Community Schools

UNDERSTANDING LEISURE (*co-authored*)

LEISURE AND URBAN PROCESSES (*co-edited*)

LEISURE POLICIES IN EUROPE (*co-edited*)

LEISURE RESEARCH IN EUROPE (*co-edited*)

SOCIOLOGY OF LEISURE: A Reader (*co-edited*)

SPORTS DEVELOPMENT: Policy, Process and Practice (*co-edited*)

SPORT, LEISURE AND CULTURE IN THE POSTMODERN CITY (*co-edited*)

Also by Stephen Wagg

THE SOCIAL FACES OF HUMOUR (*co-edited*)

BECAUSE I TELL A JOKE OR TWO: Comedy, Politics and Social Difference (*edited*)

BRITISH FOOTBALL AND SOCIAL EXCLUSION (*edited*)

CRICKET AND NATIONAL IDENTITY IN THE POST-COLONIAL ERA: Following On (*edited*)

EAST PLAYS WEST: Essays on Sport and the Cold War (*co-edited*)

AMATEURISM IN BRITISH SPORT: It Matters Not Who Won or Lost? (*co-edited*)

KEY CONCEPTS IN SPORT STUDIES (*co-authored*)

SPORT, LEISURE AND CULTURE IN THE POSTMODERN CITY (*co-edited*)

Contents

Notes on Contributors

Feona Attwood is Principal Lecturer in Communication at Sheffield Hallam University in the UK. Her research is in the general area of gender, sex, technology, the body and media. She is the editor of *Mainstreaming Sex: The Sexualization of Western Culture* (2009) and *porn.com: Making Sense of Online Pornography* (2010).

Terry Austrin is Associate Professor of Sociology at the University of Canterbury in Christchurch, New Zealand. He has also taught in universities in Britain and the United States. His research interests include gambling and sexuality and he is the author, with Huw Beynon, of *Masters and Servants: Class and Patronage in the Making of a Labour Organisation* (1997).

Shane Blackman is Reader in Cultural Studies at Canterbury Christ Church University in the UK and has written widely on young people's culture. He is the author of *Youth: Positions and Oppositions – Style, Sexuality and Schooling* (1995) and *"Chilling Out": The Cultural Politics of Substance Consumption, Youth and Drug Policy* (2004).

Peter Bramham is a Visiting Research Fellow, and former Reader, in the Carnegie Faculty at Leeds Metropolitan University, in the UK. His publications include *Understanding Leisure* (with John Capenerhurst, Les Haywood, I. P. Henry, Frank Kew and John Spink, 1995), A *Sociology of Leisure: A Reader* (edited with Chas Critcher and Alan Tomlinson), *Leisure and the Urban Process* (1989) and *Leisure Policies in Europe* (1993) (both edited with Ian Henry, Hans Mommaas and Hugo van der Poel). His most recent book is *Sport, Leisure and Culture in the Postmodern City*, a study of Leeds, edited with Stephen Wagg (2009).

Chas Critcher is a Professor in the Department of Media and Communication at the University of Swansea. He is currently making comparisons of moral panics across nations and deploys and tests models of moral panics against a range of examples including AIDS, rave/ecstasy, video nasties, child abuse and paedophilia. His latest book on the subject is *Critical Readings: Moral Panics and the Media* (2006). With John Clarke, he also wrote the influential study of leisure *The Devil Makes Work: Leisure in Capitalist Britain* (1995).

Philip Drake is a Lecturer in the Department of Film, Media and Journalism at the University of Stirling in Scotland. He conducts research on the cultural politics of celebrity and is completing a book on the political economy of the Hollywood film industry.

Paul Gilchrist is a Research Fellow in the Chelsea School at the University of Brighton in the UK. He is completing a doctorate on the cultural politics of heroism in British mountaineering between 1921 and 1995.

Richard Haynes is a Senior Lecturer in the Department of Film, Media and Journalism at Stirling University and the Director of Stirling Media Research Institute. He has a special interest in the relationship between sport and the media. His principal publications are *The Football Imagination: The Rise of Football Fanzine Culture* (1995); *Power Play: Sport, the Media and Popular Culture* (2000) with Raymond Boyle; *Football in the New Media Age* (2004) also with Raymond Boyle and *Media Rights and Intellectual Property* (2005).

John Horne is Professor of Sport and Sociology at the University of Central Lancashire in the UK. His most recent book is: *Sport in Consumer Culture* (2006). With Wolfram Manzenreiter, he also edited *Japan, Korea and the 2002 World Cup* (2002), *Football Goes East: Business, Culture and the People's Game in China, Japan and Korea* (2004) and *Sports Mega-Events: Social Scientific Analyses of a Global Phenomenon* (2006).

Brett Lashua lectures in Leisure Studies at Leeds Metropolitan University. He previously worked at the Institute of Popular Music at Liverpool University in the UK. He holds degrees from Kent State University in the United States and a doctorate from the University of Alberta in Canada. His research explores the intersections of youth, popular music and music-making technologies, space and place, and leisure.

Nicole Matthews lectures in Media and Critical and Cultural Studies at Macquarie University in Australia, having previously taught at Liverpool John Moores University in the UK. Her research is on the relation between media, practices of the self and formations of citizenship in neo-liberal political cultures. Her books include *Comic Politics: Gender in Hollywood Comedy after the New Right* (2000) and (as editor, with Nickianne Moody) *Judging a Book by its Cover: Fans, Publishers, Designers and the Marketing of Books* (2007).

Andy Miah is Professor of Ethics and Emerging Technologies in the Faculty of Business and Creative Industries at the University of the West of Scotland. He is also a Fellow of the Institute for Ethics and Emerging Technologies, USA and a Fellow at FACT, the Foundation for Art and Creative Technology, UK. He is author of *Genetically Modified Athletes*

(2004) and co-author with Dr Emma Rich of *The Medicalization of Cyberspace* (2008) and editor of *Human Futures: Art in an Age of Uncertainty* (2008). His research concerns the intersections of art, ethics, technology and culture and he has published broadly in areas of emerging technologies, particularly related to human enhancement.

Nigel Morgan is Professor of Tourism Studies at the University of Wales Institute Cardiff. He is editor, with Annette Pritchard and Roger Pride of *Destination Branding. Building the Unique Place Proposition* (2009).

Julian Petley is Professor of Film and Television at Brunel University in London, UK. He is the author of many books on the media, the most recent of which are *Culture Wars: the Media and the British Left* (with James Curran and Ivor Gaber) (2005), *Freedom of the Word* (2007) and *Freedom of the Moving Image* (with Philip French) (2008). With Martin Barker, he also edited *Ill Effects: The Media/Violence Debate* (2001).

Annette Pritchard is Professor and Director of Welsh Centre for Tourism Research at the University of Wales Institute Cardiff. She is also a Visiting Professor at the University of Languages and Communication in Milan and a Member of the Executive Committee of the UK's Association of Tourism in Higher Education.

Neil Ravenscroft is Professor of Land Economy at the University of Brighton. He researches and writes on sport, recreation and the environment. He has contributed to policy development in the area of recreational access to private land and is a member of the editorial board of *Leisure Studies*.

Stephen Wagg is a Professor in the Carnegie Faculty of Leeds Metropolitan University in the UK. He has written widely on the politics of comedy, as well as the politics of sport and of childhood. He edited *Because I Tell a Joke or Two: Comedy, Politics and Social Difference* in 1998.

Jackie West teaches in the Department of Sociology at Bristol University in the UK, where her research interests include the gambling and sex industries. With Minghua Zhao, she edited *Women of China: Economic and Social Transformation* (1999).

Introduction: Unforbidden Fruit: From Leisure to Pleasure

Peter Bramham and Stephen Wagg

> Leisure's role in people's lives is not purely economic. Leisure has important social, psychological and cultural dimensions. As leisure's share of the economy grows, so does its role in people's everyday lives. So the balance tilts from life being work- and production-centred to becoming leisure- or consumption-centred Roberts (2004: 2).

Ken Roberts has consistently argued[1] that the growth of leisure is an important feature of modern industrial societies. Modernity is made up of four distinctive facets – the **economic**, the **social**, the **political** and the **cultural**. Such formations have their own history, institutional trajectory and momentum. Each has a separate domain: the economic entails material distribution, price mechanisms, supply and demand in markets; the social is organised around face-to-face relations in families, neighbourhoods and local communities; politics is about the nature of authority and the distribution of power and finally; culture concerns communication and material and intellectual signification. Some academic traditions and disciplines have focused on one or other of these spheres. Indeed, one would hardly expect the same theoretical debates and issues to shape the different disciplines of economics, political science, sociology, and anthropology. Equally, each discipline has developed its paradigms, exemplars of good research methodology and distinctive techniques of framing and collecting data. In turn, each has been drawn into separate debates about policy and decision-making, so as to inform managerial issues of intervention and control. Therefore, the roots of each discipline have been shaped in some way by their own technical legacies – whether it is Keynesian economics, social engineering inside the welfare state or state-socialist collectivism, participation in the planning process or postcolonial anthropological narratives.

This book likewise has a distinctive intellectual heritage and purpose. It seeks to develop the contribution of critical cultural studies to the understanding of free time and leisure in the period of what some now term 'postmodernity'. Its focus is the new politics of freedom and constraint, of choice and regulation, in diverse leisure worlds. Contributions have been sought which discuss traditional free-time practices involving alcohol, reading, and sex, along with more orthodox leisure topics such as sport, outdoor recreation, gambling and tourism, and including newer mediated forms involving TV and broadcasting, film and the Internet.

From the 1970s the Centre for Contemporary Cultural Studies (CCCS) at Birmingham University was at the heart of new debates about the nature of class, work, community and changing patterns of consumption. Rather than accepting leisure simply as free time, Marxist writers challenged concepts both of freedom and of time under industrial capitalism. They questioned the conventional wisdoms, dominant academic disciplines and recreational managerialism that cast leisure simply as free time – as that precious time left over from work, which must be filled with serious leisure and gainful pursuit. Leisure and leisure studies were seen as ideological; both spoke to the individual as a subject who was exhorted to choose active leisure in the arts, sports and exercise. By way of contrast, CCCS writers were more interested in mass leisure forms and working-class consumption of film, television, gambling, smoking, drinking and sex.

It is not for the first time that Marxist theory has expressed concern about leisure and culture: the Frankfurt School, temporarily exiled in the USA in the 1930s and 1940s, feared the soporific and depoliticising effects of a hedonistic consumerism, fuelled by standardised mass entertainments in music, film and broadcasting. Advertising and corporate public relations were the new engines – or, as Adorno neatly phrased it, 'the cultural industries' – of modern capitalism. Their ideological task for modernity was one of both mediation and diversion: to encourage workers to acquiesce in capitalist work discipline and exploitation and to convince them that their future happiness would best be secured as willing consumers rather than as dissatisfied socialists or radical class warriors. For some Marxists leisure was dismissed as a space in people's lives where false needs were imposed by national elites and where entertainment and recuperation would serve to compensate for alienating work. Why should Marxists worry about sport, comedy, literature, or the mass media when there was revolutionary struggle amongst students and workers to organise and, of course, to theorise?

It was primarily in the late 1960s and early 1970s that the works of these 'new' German Marxists – Max Horkheimer and Theodor Adorno (2002), and later Jürgen Habermas (1971, 1979) became translated into English and widely read and debated in UK universities. The writings of the Marxists such as Herbert Marcuse (1991), Louis Althusser (1970) and Nicos Poulantzas (1978), also proved to be very influential. As the global economy shuddered with the 1973 Oil Crisis and experienced seismic redistributions in the international division of labour around manufacturing, Marxist analysis took a double dose of theory as it tried to swallow unpalatable concepts of 'overdetermination' and 'hegemony' – both germane to issues of control rather than to human freedom and choice.

It is more a question for the sociology of knowledge how key writers, texts or even concepts become icons of particular disciplines and how groups of writers become so influential as to establish schools of thought, or, in Max Weber's terms, offer distinctive *Weltanschauungen* on key questions that plague the social sciences. These epistemological and methodological problems, often stated in Kuhnian terms as 'paradigmatic',[2] become ever more complex when key texts or writers have transnational voices which are often delayed historically because of issues around publication, translation and copyright. Although synthesising the work of some continental writers, the CCCS also provided a theoretical space for feminists, Marxist feminists in particular, to question the existing sexual division of labour because Marxist writers sought to study leisure relations in the wider context of the economic, political and cultural relationships of contemporary capitalism. Nevertheless Marxism itself faced sustained criticism from black writers[3] as it stood accused of rendering 'race' and racism invisible.

British cultural studies

Inspired by writers such as Raymond Williams and E. P. Thompson, British Cultural Studies sought to distance itself from the determinism in the work of Louis Althusser. In the 1970s French structuralism was influential in a variety of disciplines, particularly anthropology, linguistics and psychoanalysis. However, it was Althusser's reading of the late Marx, his demand for scientific Marxist praxis and his analysis of ideology, that proved most corrosive not only for sociology but also for more humanist and empiricist versions of Marxist analysis which stressed human agency and choice. Althusser and others influenced by his approach[4] launched a theoretical assault on social science. Although maintaining the Marxist shibboleth of the economic base determining the superstructure 'in the last instance', Althusser suggested that ideological superstructures could be 'relatively autonomous' from the economic base. The concept of overdetermination permitted the class contradictions of capitalism to find their expression in the ideological superstructure for example in the Ideological State Apparatuses (ISAs) such as the family, education or politics. Althusser therefore challenged ideas of history and human agency. English Marxism and Italian Marxism, indebted to Antonio Gramsci, were deemed equally guilty of historicism and of theoretical naivety by engaging with a false problematic. By focusing on the agency of human subjects, such analyses were deemed to be ideological. Both British and Italian theoretical traditions failed, the Althusser school argued, to develop scientific categories of Marxism which could locate and explain the structural contradictions of capitalism.

In contrast to Althusser, British cultural studies were keen to embrace Italian Marxism and, in particular, Gramsci's newly translated work on hegemony and its purchase on struggles over and cultural consent in modernity. Indeed, as Stuart Hall's[5] introduction to the theoretical problematic of cultural studies acknowledges, Gramsci was one crucial strand of thinking that gave shape to subsequent diverse epistemological, theoretical, methodological and practical interventions. Researchers working in the Centre of Contemporary Cultural Studies (CCCS) at Birmingham University were keen to map out a new field of study, a new territory or space for engagement so as to flesh out sociology, with its skeletal focus on the social, and to challenge its sub-discipline, the sociology of literature, in order to broaden out its conservative Leavisite criticism of the 'Great Tradition' of English texts. In a variety of ways, the CCCS sought to raise new questions. It sought to refocus attention on class, on the mass media and on popular culture both as sites for resistance and identity. By drawing on Gramsci's concept of hegemony, the CCCS defined capitalism holistically, as a totality structured in dominance, with social, political and cultural ramifications. This shift in perspective away from traditional British sociology and the emergence of new problematics appeared in a plethora of multi-authored publications but was most graphically illustrated in *Policing the Crisis* (**1978**),[6] a critical study of the media panic about black street crime in the UK, orchestrated by the authoritarian state response by the police and the courts.

So when some writers[7] chose to celebrate the 'leisure revolution' at the beginning of the 1970s, Marxism was alive to inequality *both* at work *and* in leisure. This was not the result of ideological blindness as in the 1980s growing numbers, particularly young adults, experienced the oxymoron of 'enforced leisure' or unemployment as globalisation tightened its grip on national economies and local lives. Marxist writers suggested that it was not simply an empirical question of mapping what free-time activities were but more a theoretical question of why people spent their non-work time as they did.

In the field of Leisure Studies, these themes found their best expression in 1985 with the publication of *The Devil Makes Work* by John Clarke and Chas Critcher.[8] The book provided a coherent theoretical riposte to the major leisure studies of that decade, namely the textbooks of Stan Parker (1971),[9] Ken Roberts (1978)[10] and Robert and Rhona Rapoport (1975).[11] All three key texts were deemed guilty of a weak functionalist analysis which posited that market capitalism provided freedom of choice for individuals at different stages of their working lives and family cycle. *The Devil Makes Work* brought a distinctive new approach to the study of leisure, moving it away from largely untheorised empirical studies of provision and consumption and linking it explicitly to class, culture and inequality. For Clarke and Critcher the 'conventional wisdom' of leisure research failed to acknowledge that politics and economics directly affected leisure: *People do make choices (and not only in leisure), but their choices are made within the structures of constraint which order their lives* (Clarke and Critcher, 1985: 46). Class membership was defined as essential to people's leisure experiences, and following C. Wright Mills' work, Clarke and Critcher demanded that social theory must bridge history and biography. So understanding the passage of historical time, how it both shaped and was shaped by individual and collective agency were defined as central to critical social theory.

This challenge remains as pressing now as it was then. As we noted, it is over 20 years now since the ground-breaking *The Devil Makes Work*[12] was first published. It is certainly not the intention of this book to act as a second edition of the *The Devil Makes Work,* as some of its themes have already been revisited by Critcher and Bramham (2004). Meanwhile, as leading writers such as Chris Rojek[13] have noted, the world of leisure has moved on. In what many commentators describe as a postmodern, Post-Fordist society, work is no longer felt to be central to the lives of individuals, local communities or social classes. Leisure has become an ever greater site of excess, escape, transgression, resistance and change. New technologies of digitalisation in computing, linked into satellite communication systems, offer diverse networks for personal information exchange as well as for data archive retrieval. These have resulted in the rapid reconfiguration of terrestrial national broadcasting alongside innovation on the Internet to realise global platforms of news and entertainment.

The leisure project of modernity, orchestrated by local and national policy-makers in pursuit of rational recreation, has been superseded by new agendas, policy alliances and corporate forces. Rojek has spoken at one point of 'dark' or 'wild' leisure[14] and more recently[15] of the need for a 'action analysis' of leisure, one which should be ideologically committed to a new 'greenish' politics, which values the care of the individual self, as well as the care of others and care of the environment. Alongside this there has been a pattern of deregulation in many nation states and growing globalisation experienced through the Internet, travel and new patterns of communication. In numerous cases what once was marginal and deviant has now become the basis for widely available commercial entertainment. We have seen the emergence of complex new politics of pleasure, choice and desire. The distinctive focus of all of the following chapters in this book is to explore both change and continuity in the diverse fields of individual taste, community leisure and postmodern culture.

The chapters

The book is made up as follows. Peter Bramham's opening chapter provides a backdrop against which the remainder of the book can be read. It has three central purposes: to discuss the concepts of *leisure* and *generation* in relation to modernity and the development of social theory; to review the important sociological work on leisure in the British social sciences between the 1960s and the present day; and to chart the ways in which a generation of 'grasshoppers', steeped, ironically, in the playful philosophies of the 1960s, has now agreed to live by the dictates of neo-liberalism and to shun the state-sponsored security of life and leisure under managed capitalism.[16] Chapter 2, by Chas Critcher, is about the politics of *drinking* and it offers two further important illustrations of the book's themes. First, by drawing clear historical parallels between the so-called 'gin crazes' of the early eighteenth century and the current political anxiety over 'binge-drinking', it shows how the concept of the moral panic is still central to any understanding of the politics of leisure. Second, in its examination of the notion of 'moral regulation', it highlights a widespread tension in the contemporary politics of leisure between a free market in access to some activity and the perceived need, usually on moral grounds, for restricting that access.

In Chapter 3, Neil Ravenscroft and Paul Gilchrist examine what has, historically, been one of the most vital and vexed questions in the politics of leisure: *access to the countryside*. Their thoroughgoing analysis examines whether recent legislation by the 'New' Labour government in Britain has actually procured the 'right to roam' for the general public or the historic privileges of the 'landed interest' have, once again, been safeguarded.

Chapter 4, by Philip Drake and Richard Haynes looks at *television* – another key leisure site and one which, across the world, has undergone a series of deregulating measures since the 1980s. Drake and Haynes analyse a world, which, for many commentators, has moved 'beyond broadcasting' and now embraces a multiplicity of channels. This world has different regulatory regimes from the television of the 1970s and its programming – dominated as it is by 'reality' and celebrity-based formats – likewise bears little resemblance to the terrestrial television of 30 or 40 years ago. Through a close inspection of changes in the fields of technology, markets and policy, the writers assess the extent to which the promise of the 'information society' has been fulfilled.

In Chapter 5, Feona Attwood produces a detailed and insightful account of the leisure industry that has lately assembled around *sex*. Sex, as she points out, is an area wherein the moral judgement of private behaviour has been relaxed and variously commodified sex has correspondingly become seen as a means to pleasure, fulfilment and self-exploration. Our primary sexual relationship, she argues, is with ourselves. The chapter explores some of the ways in which, in the time of this 'late modern recreational sexuality', sex becomes a form of play.

When people observed that 'The devil makes work for idle hands' – a popular phrase since the early eighteenth century – the hands they had in mind invariably belonged to young people and youth is the subject of Chapter 6, written by Shane Blackman. Blackman focuses specifically on the matter of *youthful intoxication* and discusses the various moral panics that have built around this issue. Using history, social theory and policy analysis the chapter charts the ways in which youth and intoxication have been constructed as social and political problems. Intoxication, after all, was seen as a pleasurable and exploratory activity when engaged in by the upper classes. It only seemed to become the subject of political contention when young people became intoxicated. The chapter dissects the contradiction wherein youthful intoxication is regarded by governments and policymakers as pathological but by young people themselves simply as (albeit risky) leisure.

In Chapter 7 Jackie West and Terry Austrin outline recent developments in *gambling*, another previously guilty and criminalised pleasure, now sanctioned and widely available. As they show, governments across the world now invariably see gambling as a means to economic regeneration and its expansion has been facilitated by technological developments – particularly in the field on online betting. In this industry, in particular, there is tension between local regulatory regimes and globalising trends and, once again, an uneasy balance between permission and political restriction.

Chapter 8, by Andy Miah, analyses the *Internet*, a massive and comparatively new presence in most people's leisure lives. *YouTube*, *Facebook*, *Second Life*, *Wikipedia* and/or *MySpace* are now the common coin of daily existence for most people of, or below, a certain age. Crucial here has been the rise of Web 2.0 environments, the champions of which argue that it has involved 'the many wresting power from the few'. Chapter 8 assesses this claim.

Annette Pritchard and Nigel Morgan are the authors of Chapter 9, which is about *tourism*. Pritchard and Morgan argue that tourist experiences – experiences which, after all, have become more and more important to the leisure of most people in Western societies – are essentially about the body and that embodiment should have a more central place in the analysis of leisure.

Chapter 10, by Stephen Wagg, is a discussion of *comedy*, another leisure pursuit (and industry) to experience huge growth over the last 30 years. The chapter examines the cultural politics of comedy in Britain and the United States and, once again, the paradox of freedom and restriction that seems to have attended the emergence of 'alternative comedy' – usually surfacing in public discourse in the form of a pseudo-debate about 'political correctness'.

The topic of Chapter 11 is *books and reading*. Here Nicole Matthews discusses the apparent passing of the historic notion that books should uplift, educate and civilise their readers. She analyses contemporary reading in relation to Rojek's concept of 'edgework' (which widely informs this volume). Booksellers are now less inclined to see themselves as mediators of cultural taste and readers, she argues, will be more likely to read for pleasure, escape, arousal and the satisfaction of desire.

The following chapter deals with *sport and lifestyle*, two more burgeoning preoccupations in the leisure field. In this connection, John Horne takes up five themes: economic trends and their impact on the viability of sport sponsorship; the changing format of sport and its relationship to the mass media; the relationship between athletes and the institutions and organisations that enable sports to operate; developments in policy – both of government and of sports bodies; and the social relations and the associated lifestyles and identities which frame sport.

The penultimate chapter, by Brett Lashua, concerns *music* as leisure and tackles two contemporary issues: the politics respectively of *X Factor*-style television talent contests (and attempts to subvert them) and of the file-sharing or 'pirating' of music from the Internet. His well-informed and nuanced argument counsels against the notion that either trying to subvert the *X Factor* by campaigning to stop a competition winner topping the charts or downloading music free from the Internet might, in some simple sense, constitute 'fighting the power'.

Finally, Julian Petley produces a powerful polemic against the current, contradictory politics of *screen censorship* in Britain. In an essay fully consonant with the central concerns of the book he analyses the paradox whereby neo-liberal politics and philosophies have ostensibly produced a free market in viewing material in the UK but have failed to remove a range of regulations and restrictions on this same material. The attendant political rhetoric of 'balancing rights with responsibilities' and invocations of 'the public interest' demonstrate the confused nature of the contemporary politics of leisure and, once again, aptly illustrate the defining theme and purpose of this book.

Notes

1. See for example his two seminal textbooks on leisure, Roberts, K. (1978) *Contemporary Society and the Growth of Leisure* (London: Longman); Roberts, K. (2004) *The Leisure Industries* (London: Palgrave Macmillan).
2. The term 'problematic' is more favoured by the CCCS as it sought to raise different questions from those prevalent in traditional social science disciplines. See Kuhn, T. S. (1970) *The Structure of Scientific Revolutions* (Chicago: Chicago University Press).
3. See for example Gilroy, P. (1987). *"There ain't no black in the Union Jack": The Cultural Politics of Race and Nation* (London: Routledge); Gilroy, P. (1993) *The Black Atlantic: Modernity and Double Consciousness* (London: Verso).
4. See debate between Ralph Miliband and Nicos Poulantzas on the nature of the state in late capitalism.
5. Hall, S. (1980) 'Cultural studies and the centre: Some problematics and problems', in S. Hall, D. Hobson and P. Willis, *Culture, Media, Language*, 15–47,(London: Heinemann) 15–47.
6. Hall, S., C. Critcher, et al. (1978) *Policing the Crisis: Mugging, the State, and Law and Order* (London: Macmillan).
7. See for example Bell, D. (1973) *The Coming of Post-industrial Society: A Venture in Social Forecasting* (London: Heinemann); Jenkins, C. and B. Sherman (1979) *Leisure Shock* (London: Eyre Methuen); Stonier, T. (1983) *Wealth of Information: Profile of the Post-industrial Society* (London: Methuen).

8 Clarke, J. and C. Critcher (1985) *Devil Makes Work: Leisure in Capitalist Britain* (Basingstoke: Macmillan Press).
9 Parker, S. (1971) *The Future of Work and Leisure* (London: MacGibbon and Kee).
10 Roberts, K. (1978) *Contemporary Society and the Growth of Leisure* (London: Longman).
11 Rapoport, R. and R. Rapoport (1975) *Leisure and the Family Life Cycle* (London: Routledge).
12 Clarke, J. and C. Critcher, (1985) ibid.
13 Rojek, C. (2005) *Leisure Theory: Principles and Practice* (Basingstoke: Palgrave).
14 See particularly Rojek, C. (1993) *Ways of Escape: Modern Transformations in Leisure and Travel* (London: Macmillan); Rojek, C. (1995) *Decentring Leisure: Rethinking Leisure Theory* (London: Sage); and Rojek, C. (2000) *Leisure and Culture* (Basingstoke: Palgrave).
15 In *Leisure and Culture* ibid.
16 See Scott Lash and John Urry (1987) *The End of Organised Capitalism* (Cambridge: Polity Press).

Bibliography

Adorno, T. and M. Horkheimer (2002) *Dialectic of Enlightenment* (San Francisco: Stanford University Press).
Althusser, L. and E. Balibar (1970) *Reading 'Capital'* (London: New Left Books).
Bell, D. (1973) *The Coming of Post-Industrial Society: A Venture in Social Forecasting* (London: Heinemann).
Clarke, J. and C. Critcher (1985) *Devil Makes Work: Leisure in Capitalist Britain* (Basingstoke: Macmillan Press).
Critcher, C. and P. Bramham (2004) 'The devil still makes work', in J. Haworth and A. Veal, *Work and Leisure*, 34–50 (London and New York: Routledge).
Gilroy, P. (1987) *"There ain't no black in the Union Jack": The Cultural Politics of Race and Nation* (London: Routledge).
Gilroy, P. (1993) *The Black Atlantic: Modernity and Double Consciousness* (London: Verso).
Habermas, J. (1971) *Toward a Rational Society: Student Protest, Science and Politics* (London: Heinemann Educational).
Habermas, J. (1979) *Communication and the Evolution of Society* (Boston, MA: Beacon Press).
Hall, S. (1980) 'Cultural studies and the centre: Some problematics and problems', in S. Hall, D. Hobson and P. Willis, *Culture, Media, Language*, 15–47 (London: Heinemann).
Hall, S., C. Critcher et al. (1978) *Policing the Crisis: Mugging, the State, and Law and Order* (London: Macmillan).
Jenkins, C. and B. Sherman (1979) *Leisure Shock* (London: Eyre Methuen).
Kuhn, T. S. (1970) *The Structure of Scientific Revolutions* (Chicago: Chicago University Press).
Lash, S. and J. Urry (1978) *The End of Organised Capitalism*(Cambridge: Polity Press).
Marcuse, H. (1991) *One Dimensional Man: Studies in the Ideology of Advanced Industrial Society* (Boston, MA: Beacon).
Milliband, R. (1970) 'The capitalist state – Reply to Poulantzas', *New Left Review*, 59: 53–60.
Poulantzas N. (1969) 'Problems of the capitalist state', *New Left Review*, 58: 67–78.
Parker, S. (1971) *The Future of Work and Leisure* (London: MacGibbon and Kee).
Poulantzas, N. (1978) *State, Power, Socialism* (London: New Left Books).

Rapoport, R. and R. Rapoport (1975) *Leisure and the Family Life Cycle* (London: Routledge).
Roberts, K. (1978) *Contemporary Society and the Growth of Leisure* (London: Longman).
Roberts, K. (2004) *The Leisure Industries* (London: Palgrave Macmillan).
Rojek, C. (1993) *Ways of Escape* (London: Macmillan).
Rojek, C. (1995) *Decentring Leisure* (London: Sage Publications).
Rojek, C. (2005) *Leisure Theory: Principles and Practice* (Basingstoke: Palgrave Macmillan).
Stonier, T. (1983) *Wealth of Information: Profile of the Post-Industrial Society* (London: Methuen).

1

Choosing Leisure: Social Theory, Class and Generations

Peter Bramham

> It was not so much the relevance of history that sociologists failed to see as the relevance of time.
>
> Philip Abrams (1982) *Historical Sociology* (1982) xvi

***Chapter 1** sets the scene, establishing a context for the book as a whole. It performs two basic tasks: first, it links **leisure** to the history and development of **social theory** and, second, it discusses the relationship between the politics of leisure and the so-called 'baby boomer' **generation**, born in the late 1940s and early 1950s in Western societies. In the latter context it asks whether this generation have, in effect, chosen private leisure and disposable income over the public services and state welfare that they grew up with.*

This chapter discusses the concept of leisure in relation to social theory – particularly the social theory of recent times. To this end it offers a brief introduction to historical sociology and so outlines its commitment to the study of the nature of time and process in social relations. The classical sociology of Marx, Durkheim and Weber was preoccupied with explaining social change; societies were seen as time machines – moving or struggling from eighteenth-century preindustrialism to twentieth-century modern industrial forms. By the late 1990s this two-step model of modernity was felt to be out of date as many writers now detected a third phase or **era.** Societies, for diverse technological, economic and cultural reasons, were placed in a new postindustrial or postmodern phase. Although acknowledging these **eras** of modernity – what others call *la longue durée* or the 'glacial' time of industrialism – historical sociology is also interested in **epochs**, a shorter time period, that is experienced by generations or age cohorts as they mature through childhood into adulthood and old age. Drawing on insights from Marxist cultural studies, this chapter argues that the 1960s generation or postwar 'baby boomers' came to live in a 'leisure society' and by the 1980s had laid the political foundations for postmodernity. The neo-liberal project of Thatcherism in the UK and Reaganomics in the US sought to encourage market forces, minimise state regulation and expenditure and reduce the power of working-class collectivism, embedded in trade unions. This confluence of globalisation in world economy and changing state policies meant a new landscape for leisure. It is here that the third notion of time becomes important – that of the individual's lifespan or ***Dasein***, one's experience of everyday life. New Right ideologies, with deep roots in nineteenth-century liberalism, spoke directly to individual freedoms. Leisure was no longer a cultural space for national improvement in arts, sports and recreation as postmodernity encouraged individualisation. Leisure and culture have become disconnected from the 'imaginary' community of the nation state (Anderson, 1991) and reformulated as a site for the exploration of self-identity, lifestyle and consumption. To quote Anthony Giddens, leading academic spokesperson for these new politics:

> One concerns the primacy of lifestyle – and its inevitability for the individual agent. The notion of lifestyle sounds somewhat trivial because it is often thought of solely in terms of a superficial consumerism: as suggested by glossy magazines and advertising images. But there is something much more fundamental going on than such a conception suggests: in conditions of high modernity, we all not only follow lifestyles, but in an important sense we are forced to do so – we have no choice but to choose. A lifestyle can be defined as a more or less integrated set of practices which an individual embraces, not only because such practices fulfil utilitarian needs, but because they give material form to a particular narrative of self identity (Giddens, 1991: 8).

The problem of sociology

Sociology has historically represented itself as 'the queen of the social sciences', as the one single discipline that could contextualise the narrower approaches of economics and politics and provide a contemporary version of the premodern, preliterate analysis of culture offered by anthropology. Sociologists have self-consciously placed modern industrial society at the heart of their investigations, as the discipline has focused on the social: networks of face-to-face relationships within families, in local communities, at work and in leisure, within the boundaries of nation states. In the immediate postwar period, there were unchallenged assumptions about a homogeneous functionalist consensus in society, sustained within nation states by particular institutional structures (constructed of tradition, normative rules, roles and obligations) and constituent social systems. Sociologists took it for granted that society was coextensive with the territorial and cultural parameters of the nation state. One was superimposed on the other and both were self-contained, permanent self-regulating recursive systems.

During the 1960s and 1970s conflict theorists with neo-Marxist perspectives subscribed to a similar holistic territorial view, but one which saw a class-divided society with economically dominant groups seeking to impose a shared ideology onto a national populace. The context for this was state capitalism or, in Zygmunt Bauman's initial conceptualisation, a stage of modernity (1987), wherein legislators, primarily political missionaries and policy planners, were keen to impose economic and cultural order on society. The ruling elite were confident law makers and shaped civil society through social welfare. Certain lifestyles, drugs and sexualities were both acceptable and legitimate, whereas others were not. There was perceived to be an objective concreteness, a solidity, to social structure – a sense that, just as Durkheim suggested, social facts ***did*** constrain and shape people's lives, or within the influential Althusserian version of this perspective, individuals were defined as swimmers, both waving and drowning in a sea of capitalist ideology.

This generic quest to uncover the nature of social facts had gathered momentum when faced with the recurrent question framed by C. Wright Mills (1970) in ***The Sociological Imagination*** – 'what kind of society do we live in?' – and has now been challenged by the construction of a new answer or discourse, the *postmodern*. The postmodern response seeks to subvert all traditional sociology, including Marxism, because it questions the creation of any universal meta-narrative, or any grand theory of long-term change in human history, any discourse which posits a single or dominant structuring force, such as class or gender or race in the shaping of human society. In one sense, it suggests the end of modern sociology itself because it assumes the end of the politics of the nation state, its legislators having given up on the leisure project of modernity[1] – that is, the all important task of socialising generations into a national self-referencing culture. The nation state is no longer the carapace of civil society as the lack of political tutelage leaves populations, no longer protected by high levels of welfare expenditure and collective consumption, exposed to the vagaries of global market forces. These are new or postmodern times in which the economic focus shifts from work and production towards leisure and consumption; classes of production fragment into classes for consumption.

It is beyond the scope of this single chapter to provide a comprehensive review of the different debates, positions and perspectives adopted in relation to postmodernity. The various chapters of this book may be seen as jigsaw pieces that help to piece together the puzzle of 'new' postmodern leisure. The postmodern or cultural turn in theory has challenged the domain assumptions of social scientific endeavour because it has struck at the roots of the conventional epistemological wisdom supporting social-scientific knowledge. But if the debate is about anything, it is about change, as sociologists struggle to develop better models, heuristic devices or theoretical perspectives to analyse the nature of social changes and predict their general direction. Indeed, as Foucault (1980) claims, knowledge is power and naming is everything. Many writers are reluctant to accept the term 'postmodernity' and are more comfortable with phrases such as 'high modernity', 'late modernity', 'Modernity II' or 'liquid modernity'. These terms are invariably set against earlier forms, deployed as yardsticks or templates, such as 'simple modernity', 'early modernity', 'Modernity I' or merely 'modernity'. As one would expect of writers working from within a distinctly Marxist perspective, there is a continuing ambivalence about the postmodern, the processes of globalisation and the significance of postmodern changes. Such changes do not necessarily undermine a materialist conception of history; what Marxists see in contemporary societies are the various continuities and contradictions generated between footloose international capital, 'hollowed-out' nation states and local working classes.

Describing and explaining change are central to social analysis and history but there is little academic consensus about how best they are achieved. One useful starting point in Marxist analysis is to be found in Stuart Hall (1992) view of modernity that posits four separate, distinct formations – the economy, the polity, society and culture. One should not however assume that all four formations move at the same pace or travel in the same direction. Consequently, one must adopt an historical approach in order to understand different institutional domains and the tensions and dislocations that may occur. To take the United Kingdom as an example, its economy in the nineteenth century was at the forefront of world production, it had developed a skilled workforce with an individualised culture of labour power, yet its political system remained fiercely anti-democratic, locked into a seventeenth century monarchic regime, without any constitutional Bill of Rights for individual citizens and offering far from universal suffrage, but still capable of developing an empire and a colonial mission overseas. So, one reading of nineteenth century UK history would map the way in which economic changes in the sexual division of labour and the growth in trade-union organisation connected with or was insulated from partisan politics, efforts at widening suffrage, changes in class consciousness and a nationalist and xenophobic culture.

The problem with sociology

One of the major fault lines of sociological theory relates to the question of its scope and perceived failure to appreciate the uneven dynamics of the four formations of modernity and/or the inability to theorise particular relationships between the four separate elements of market, state, civil society and mass communications. Postmodern theorists have usually celebrated change in social and cultural formations and have ignored contradictions and constraints in economic and political formations. The neo-liberal political project of Thatcherism in the 1980s, driven by a belief in global market forces, privatisation and flexible workfare could be considered to have resulted in the polarisation of classes, generating insecurity and fear in people's everyday lives. This corrosive individualisation and the simultaneous retreat of the nation state from collectivised politics have been central themes in seminal works by both Zygmunt Bauman (1999, 2005, 2006, 2007), Ulrich Beck (1992, 1999) and Beck and Beck-Gernheim (2002). The dark side of Thatcherism in the UK for traditional 'legislators' (Bauman, 1987) was that the marketisation of public services opened up a Pandora's box of individualised desire that politics and society failed to regulate comfortably. Rather than an established missionary elite in control of national culture, there was the babble of diverse 'experts' providing very different discourses on lifestyles and leisure. In the world of broadcasting it is that historical shift from a Reithian BBC of the 1930s, through the advertising-led commercial channels of the late 1950s and 1960s, towards new-millennium 'freeview boxes' and Internet platforms which synergise individual choice, consumption and commodification.

Beck (1992) has interpreted postmodern changes to indicate the family as the site of contradictions. A basic one is that the logic of the market, driving inexorably towards modernisation and individuation, undermines what it is posited on: the existence of a 'normal' nuclear family and its inherent gender divisions. The equality of the market is incompatible with the inequalities of family life. The normative constraints of gender dissolve in the face of increasing individuation so that the definition of roles in a heterosexual partnership becomes more fluid and problematic. The very proliferation of the apparent lifestyle choices of residence, work and household structure reveals the constraints on choice, opening up greater possibilities of conflict within the family. It becomes 'the setting but not the cause' (Beck, 1992: 103) of the disjunction wherein 'production and family work are subjected to contrary organisational principles' (Beck, 1992: 107). The freeing of women from their traditional roles is attributed to five changes: increasing life expectancy, the restructuring of housework, contraceptive and family planning, the fragility of marriage and educational opportunity. As women gain greater control over their biographies, so their traditional dependence on men becomes ever more irksome. The family is where these contradictions are acted out, with the result that 'the traditional unity of the family breaks apart in the face of decisions demanded of it' (Beck, 1992: 117). Since the clock cannot be turned back and equality is not achievable within the confines of the established family, the only solution is to reform employment and family structures so they can be made more compatible with each other. Beck concludes that sociology deals in 'zombie' categories – where traditional concepts such as 'the family' and 'leisure' bear little relationship to the individual's experience of the complexities of postmodern life.

Throughout the 1970s, mainstream sociology was confronted by proliferating Marxist analyses which emphasised the economic and material foundations to society, politics and culture. During the next decade feminist analysis also challenged the 'malestream' approaches of sociology and of Marxism and obliged sociologists to acknowledge and engage with the foundational divisions of gender, of 'public' and 'private' spheres and their consequences for the economy, society, politics and culture. What has further complicated theoretical debates has been a general reluctance to acknowledge that modernity as a concept has been rooted in a dichotomous conceptualisation of the 'West and the Rest'. Postcolonial theories demanded that there could be no single meta-narrative about modernity, no one definitive universal account, let alone one that ignored or marginalised the dependency and power relations of colonialism and the implications imperialist histories carried for the development of non-western nation states, national identities and the history of the 'Other'. (see Hall 1992). The task remains to understand modernity and change. Marxist, feminist, and postcolonial thinkers demand an awareness of the resilience of processes of class, gender and 'race' divides. For their part, postmodernist thinkers are keen to highlight weaknesses in these critical and radical approaches.

Inside Pandora's Box: Two steps forward?

Heidrun Friese and Peter Wagner in Featherstone and Lash's (1999) *Spaces of Culture* summarise a postmodern debate between structuralist and culturalist accounts of the nature of social life. Structuralist accounts, they argue, emphasise the determined nature of social life whereas culturalist accounts focus on meaning and interpretations. For Leisure Studies this debate has often been framed as one between leisure as 'control' and 'constraint', and leisure as 'freedom' (Coalter, 1989). This has not simply entailed a return to debates between the 'two sociologies' (Dawe, 1979) – functionalist systemic accounts pitted against demands by interactionists to *'bring people back in'*, and to break away from perspectives that treat humans as 'cultural dupes', those unconscious slavish bearers of structural relations and processes. That debate merely reproduced the historic dualism of the micro and the macro, the individual and society, agency and structure and of freedom and control. The postmodern debate is about human agency, individualisation and choice and leisure is an important but neglected site for theorising change.

One claim from Marxist cultural studies as harbinger[2] of the postmodern turn in social theory, is that the very fabric of social order has changed and is now stretched over time and space. The nature of social classes has changed and they have born witness to the seismic shift from a Fordist to a Post-Fordist society. To adopt Martin Jacques and Stuart Hall's ideas these are

'New Times', catalysed by the politics of Thatcherism.[3] This interest in ideology and political formations represented both a theoretical and political intervention, a break away from the concerns of mainstream sociology. Other writers, for example Anthony Giddens (Giddens, 1979, 1981, 1984) emphasised the need for the *spatialisation* of social theory. Historically space and locality had been taken for granted and consequently were neglected dimensions in sociological analysis. Such processes of *time-space distanciation,* particularly stimulated by innovation in computing and fibre optics, have resulted in distortions amongst institutions which are no longer recognisable, as they once were, as work, families, leisure and so on. Societies have been changed by technology, not least by the new technologies of information surveillance and telecommunications.

As perhaps one might expect, postmodern theorists argue that culture has become much more important in people's everyday lives. What Mike Featherstone (1990) celebrates in *Postmodernism and Consumer Culture* is the aestheticisation of *la vie quotidienne*: life must be lived with 'authenticity', secured and cherished through stylish consumption. It is assumed that the politics of modernity have disappeared and so has the relevance of the nation state, democratic politics and governance. Transnational or global culture has invaded society and simultaneously heightened the vulnerability of the host nation state to 'alien' forces. Social systems are now more penetrable by other societies or nation states. The nation state no longer has boundaries that are impermeable to 'outsiders' nor is it a secure and safe container for society and the social system. What theory needed in the light of these changes was to concentrate on culture at the boundaries, the crossovers and the margins rather than to focus on the centre or heart of the nation state. The postmodern gaze (Urry, 1990) turned towards the margins, those liminal spaces where individuals engaged in 'edgework', the testing and subversion of legal rules, norms and social conventions; in short, trying out the deviant imagination. Abnormal forms of leisure 'include drug abuse, alcoholism, dangerous sexualities, violence and murder' (Rojek, 2000). Leisure Studies has usually concentrated on 'serious leisure' or enthusiasms, voluntary leisure activity in which the individual develops a career, self identity and skilful performance. In contrast (Rojek, 2000: 17) argues that casual leisure: consists of desultory, opportunistic, circumstantial leisure activity which is motivated by a desire for automatic stimulation and immediate gratification. Examples of casual leisure include channel hopping, hanging around shopping malls, random drinking and smoking.

Another important metaphor of postmodern theory and change was one of cultural hybridity, – the idea that inside bourgeois nation states lay difference and diversity. Traditional boundaries between high and low culture became blurred. The centre could no longer hold and national culture and tastes imploded and fragmented alongside new patterns of labour migration and settlement. Postcolonial theories have developed because the occidental grip on modernity and its single meta-narrative has been challenged; the naive 'white' account of the West, celebrating scientific, technological and democratic progress, was only one strand of history. There was also a 'black' or 'double consciousness' about modernity (Gilroy, 1993); industrial society and the dominance of the 'West' over the 'Rest' (Hall, 1992) grew out of military power, economic exploitation and the horrors of the slave trade. Colonialism and imperialism had been crucial developments of the Enlightenment and industrialism rather than merely contingent sites where other causal structural processes were unfolding. Diverse voices from 'race', gender and environmental perspectives gained ground and grew louder above the 'white noise' of the mainstream. There was a new rainbow alliance of black/brown, pink and green pigmentation: all talking about identity politics and demanding to be heard and equally valued. 'Political correctness' established itself in policy and everyday discourses.

Traditional societies and nation states can no longer be seen as secure hermeneutic containers of our everyday worlds as economy and society become global. National elites can no longer control economic processes nor dictate media and cultural tastes. As Chris Rojek (1993) has argued, people are restless, mobile and desire change. There are now global 'flows' (Appadurai, 1990) of diverse peoples, ideas, technology, finance, information and so on. In modernity, people's lives were shaped by collectivities – their families, local communities, social classes and nation states, but these have been excoriated of locale and of tradition. They are disembedded by 'expert' systems of knowledge, communication and other processes of time-space distanciation. People must trust the expert knowledge systems that manage computers, medicine, transport systems and mass communications. Chains of interdependency and power become more complex and extended and no one seems to be in control of them. Other major consequences of this process are growing individualisation and changing class, 'race' and gender roles which further fragment nation states and so absolve individuals from collective rules, roles and obligations. As Bauman (2007) most recently argued, people now live in 'liquid modernity': there has emerged a global figuration in which all human activity is bonded together by a free market economic framework that is utterly beyond the control of individual citizens. For Bauman, this is seen to create the appearance of a 'new global disorder', which saturates individual lives with feelings of fear, anxiety and uncertainty, and which men and women are solely responsible for managing their everyday lives (Davis, 2008: 1238–9).

Marxist history and class cultures: One step backwards

Rather than focusing exclusively on how external processes of globalisation are opening up the nation state, Marxist writers have long stressed that nation states and national culture are differentiated internally. Culture is structured, hierarchical and varied. Abercrombie et al. (1980) suggest that the 'dominant ideology thesis' has more historical validity and accuracy than any consensual view of social order. Dominant ideologies bound the elite together, thereby providing political direction to an imperialist and colonial agenda at the turn of the century, whereas subordinate classes belonged to a distinctive class culture, one more concerned with the material necessities of living and the selling of labour power within the constraints of an unequal society. As Pierre Bourdieu argued, middle classes themselves could not change the consequences of industrialism; they could only try to distance themselves from naked materialism through high culture, the cultivation of taste, art and aesthetic lifestyles. Hence, fractions of the middle classes could detach themselves, at least perhaps in their leisure, from the bleak and divisive realities of class exploitation and poverty, ground out by capitalist commerce, manufacturing and finance. For some Marxist writers, and the *Devil Makes Work* is a good example, working-class culture was distinctive, collective and solidaristic. It had a socialist agenda or at the very least the radical democratic potential to resist, subvert or 'penetrate' conservative ideas and those corrosive illusions about the naturalness or inevitability of capitalist inequalities. For the working class, there was always the possibility of sharing meanings, collective communication networks and signification in the cause of articulating and securing public collective purposes and thus constructing 'community'.

During the late nineteenth century, as part of the Enlightenment project, most European nation states sought political solutions to working-class radicalism and public disorder by encouraging a shared inclusive national consensus around 'invented traditions' (Ranger and Hobsbawm, 1981). Wrinkles of local, ethnic and class differences were ironed out by missionaries of domestic metropolitan culture. Postmodern analysis argues that within late modernity these localised class and work cultures are no longer relevant or at least that they have become increasingly dedifferentiated, more flexible and less significant in everyday life.

There has been a rich tradition of Marxist cultural analysis that engaged with the nature of the relationship between economic and cultural formations in modernity or, in terms of Marxist discourse, between the economic base and the cultural superstructure. Throughout his academic career Raymond Williams sought to develop a Marxist analysis of culture and communications. He attempted to challenge a conventional sociology of art and culture by suggesting three major traditions of defining culture – the cultivation of the human mind; secondly, a selective tradition of high culture of literature, art and drama and, thirdly and more radically, as 'a lived experience', a 'structure of feeling' as a whole way of life, historically located as a critical response to industrial society (Williams, 1963).

The emphasis on collective history, on lived experience and human agency were central tenets of E. P. Thompson's life and work, particularly in his seminal study of emergent working-class consciousness, **The Making of the English Working Class** (Thompson, 1963). As he showed, by the 1840s the English working class had made itself. It was different from the middle classes. For example, in Oldham, John Foster (1974) suggested that working-class consciousness had a radical and even revolutionary edge and was challenging bourgeois control of the town. Throughout the nineteenth century there were fierce historical debates about the relative autonomy of working-class culture and its independent forms of organisation. The late nineteenth century, according to Gareth Stedman-Jones (1983) in **Languages of Class**, witnessed the incorporation of the working classes into commercialised leisure, forming a capitalist 'culture of consolation'. Rather than embarking on economic and political change for a radical and socialist future, the English working class had settled down within British culture. There was no radical working-class consciousness to make connections between economics and politics. In Britain, at least, these two formations were separated and were to remain so. Although there were frequent outbreaks of economic struggle and dispute between capitalist employers and workers, there was also sustained political consensus in support of the British state and not least of its imperial role, principally because empire supported industry. There may have been dissensus about relative rates of pay and economic rewards for labour power but such anomie did not extend to politics or to reshaping liberal democracy. Immiseration and alienation at work, the cornerstone predictions and motors of Marxist analysis and class struggle did not lead to radical political change in the UK.

The nature of time

The sociology of leisure has been built on two main foundations, the concepts of time and of freedom. Leisure researchers have argued that free time is a product of contemporary society and as hours of work have declined in industrial society there has been a concomitant growth in people's free time. However, conceptualising time in historical analysis is far from straightforward. Many writers, including Peter Burke (1980), Charles Tilly (1981) and Alex Callinicos (1989) have argued that historians and sociologists have much to learn from each other in this regard. Indeed, society can be seen both as a fixed and stable, a persistent structure, and also as an ongoing process of breakdown, renewal and development. Some historians look on the past as process, a perpetual movement of destruction and reconstruction. Consequently, different societies can be viewed as travellers in both time and space.

Historians such as Ferdinand Braudel (1975) have stressed the uneven nature of time with changes taking place at different speeds: first, the fast moving time of events and individuals (*histoire événementielle* or *courte durée*); secondly, the time of economic systems, nation states and so on with their slow rhythms (*histoire conjoncturelle*); finally, geographical time, the history of mankind and the environment – with its constant repetition, irresistible force and ever-recurring cycles (*histoire structurale*) (see Burke, 1980: 94). But for most individuals the only important currency in time is their own lifespan. Time can only be experienced within one's personal biography and everyone faces the core existential anxiety of having little idea of how long one will live. Even when faced on a daily basis with terminal illness, doctors and carers are usually reluctant to be specific and predict the time of the patient's death and, of course, there is always the hope of remission. So individuals must live their lives in the present, in the Here and Now, (or, as we saw earlier, what some German philosophers term the ***Dasein*** of the individual). Each individual not only has his or her own subjective experience but also has a personal history or biography with important moments or milestones in their life course – birth, childhood, adolescence, adulthood and so on. Major life events such as marriage or remarriage or a birth or a death in the family may lead individuals to claim that they are starting again, turning over a new leaf or starting a new chapter in their book of life. So we can measure time in the Here and Now, the immediacy of the individual's personal experience, grading where they are in the ageing process, within the life course. The concept of career is also significant in the narrative of self (Giddens, 1991) as individuals plan the length of their education, work experiences as well as commitments to family, leisure and the wider community.

The importance of generations

In contrast to the immediacy of the named individual's life stand the collective experiences of generations, those cohorts of individuals who are born in a particular decade, go to school together, work together and grow old together. Karl Mannheim (1952) argues that each generation shares a common location in the socio-historical process. Each generation grows up in, experiences and makes sense of its historical setting or context, as it is shaped by distinctive economic, political, social and cultural formations. Cultural tensions between the parent generation and the next (younger) generation are at the heart of Phil Cohen's (1972) analysis of youth subcultures: leisure and new leisure styles become a 'magical solution' to the problems of growing up, gaining adult identities and space, in a class-divided society. But generational analysis,[4] because of longevity and ageing demographics, must now not only look at parent–child cultural relations but also take on a three-generational dimension. For instance, the 1960s teenage 'baby-boomers' find themselves, 40 years on (now as pensioners in the noughties), caught between the demands of ageing infirm parents who require care and resources, and those of their own children who, now grown up, require capital and finance to help them through the present uncertain and unpredictable times. The second half of this chapter argues that cultural tensions between parent and youth generations should benefit from hindsight, or 'the condescension of human history'. It was not so much cultural struggle that distinguished the 1960s generation in its teenage years, though that was not unimportant, but it was rather the significance of the political choices made two decades later that opened the floodgates in the UK to the postmodern economy and change. The two are closely related.

Generational time is best measured in blocks of 30 years or so, the chronological difference or the 'generation gap' between parents and children. Geoff Pearson (1983) has argued that sustained debates about young people and 'law and order' reflect tensions that exist and persist between generations. Every generation has periodically voiced identical fears about its young people. There is a recurring moral panic about youth that worries about their lack of physical fitness or their moral degeneracy and, as a consequence, a possible social breakdown of 'the British way of life'. Current media anxieties over childhood obesity, fast foods, 'binge drinking', eating disorders and unhealthy lifestyles are simply the latest chapter in this history of concern.

Marxist analysis can help in the understanding of leisure lifestyles by exploring the changing collective class experiences of generational cohorts and how these are expressed over a lifetime. This requires, in addition, some Durkheimian analysis which focuses on the *'conscience collective'*, that social solidarity which crystallises at distinctive historical episodes. Durkheim's work on deviance stresses the functions of social rules and rituals, the effervescence and celebration of shared experience and memories. Peer pressure and shared historical experience are crucial contexts and resources for individuals growing up and growing old together, producing a distinctive 'spirit' or *Zeitgeist* for every age, never more so than in sets of leisure experiences.

Each generation inherits both material and cultural resources from the previous generation and in modernity each generation endeavours to win some cultural space from its parents and from the past. So the second level or measure of time can best be described by the word **epoch** which refers to a distinctive historical period or perhaps a decade, when a generational cohort sets about shaping their own institutions, class relations and history.

A plethora of Marxist authors sought to make sense of what Bernice Martin (1981) termed the 'expressive' cultural revolution of the 1960s and the spectacular subcultural styles of mods, rockers, hippies and others in both the UK and USA. Mike Brake (1980: 84) has argued that such research suggests a partial and distorted view of youth, a celebration of masculinity and one to be found primarily within the white English working class. This concentration reflects not only the biographies of male white researchers in the 1970s but the central Marxian 'problematic' of class structuration and working-class resistance to middle-class cultural hegemony.

The doyen of the Centre of Contemporary Cultural Studies, Phil Cohen (1984), argued that youth subcultural styles were not only based in material class conditions but were a reaction and response to parental cultural styles as well as to dominant ideologies. Each youth generation must come to terms with dominant middle-class discourses about the body, work and education, in addition to distancing itself from its parental culture. Ostentatiously grounded in leisure, youth subcultures provided 'magical solutions' to the problems and contradictions of growing up in 1960s class-divided society. These solutions were class based, mediated by distinctive commodities, mass entertainment and popular culture. The precise focus of subcultural concern varied with lifestyle, sexual relations, political radicalism, drug use, deviance and criminality. Brake (1980) identified three main strands in 1960s youth culture – the bohemian, the delinquent and the political. Not all writers were convinced that class relations were central to understanding youth and the experience of adolescence. Class has been problematised further by gender divisions and not least by 'race' and ethnic identity. Other writers developed black cultural studies and radical feminist approaches to understanding youth within the CCCS (see CCCS (1982) *The Empire Strikes Back).*

The racialisation of inner-city 'problem youth' in the 1970s was fully explored and theorised in Hall et al.'s (1978) definitive work on the amplification of black street crime in **Policing the Crisis: Mugging, the State and Law and Order**. This groundbreaking work synthesised discourses about moral panics, labelling perspectives on deviance and demonisation, with an empirical case study of the configuration of media, policing and criminal court outcomes. **Policing the Crisis** typified thinking in the CCCS; it was collaborative work offering a fresh Marxist commentary that embraced the economy, the state, the media and their contribution to the sustaining of a class-divided civil society in the UK. By the early 1980s Stuart Hall had grounded cultural studies in the politics of ideology, the legitimation crisis in the state and the emergence of 'authoritarian populism'. 'Thatcherism' became a central concept and was debated by Hall and Jacques (1983) and Bob Jessop et al. (1988) as Marxist writers set out to capture the novel agenda of the New Right in the UK during the 1980s and 1990s – namely, that of a 'free market, strong state, iron times'.

One of the purposes of this chapter is to explore the historical context in which generational identities and agency are forged and one key to understanding the emergence of the postmodern is to trace the history of the generation that are both its architects and beneficiaries. The generational cohort of postwar 'baby boomers' became teenagers in the growing affluence and consumerism of the UK in the 1960s. In sharp contrast, their parents' formative teenage years had been spent in the 1930s and 1940s, – times of austerity, deprivation and poverty. Naturally in a class-divided capitalist society levels of deprivation are relative to class position and power but generational analysis converges on long-term processes and negotiated outcomes of institutional change. So the adult generations of the 1940s and 1950s had, on balance, sufficient collectivist political will to lay the institutional foundations of the welfare state for their children. Their social reformist legacy was funded both by social insurance and progressive taxation policies, which resulted in the expansion of educational opportunities, pensions and healthcare. If the 1940s generation were Aesop's proverbial ants working hard and deferring gratification, the 1960s generation embraced the grasshopper lifestyle, playing hard and enjoying the present. If we stretch the metaphor, we shall see later, the grasshopper 1960s parents in their turn have produced a 1980s youth generation of butterflies, who have to survive the risks of a more precarious environment and uncertain future.

Much ink has been spilt about the 1960s, but most writers agree that the times were not so much revolutionary as assuredly radical – with the emergence of student power, shifting sexual and racial power relations and, not least, media and cultural revolutions around music and consumption. For the purposes of this chapter, it is useful to deploy Bernice Martin's concept of '*liminality*'. The 1960s witnessed a generation of youth that challenged and redrew cultural boundaries, which challenged cultural norms, rituals and pushed rules to their limits.

Bernice Martin (1981) referred to the cultural anthropology of Mary Douglas with its key concepts of grid and structure (a 'high grid' denotes a very structured and stratified society and a 'low grid' a more egalitarian one) to understand the expressive cultural revolution of the 1960s. Martin argued that the puritan cornerstones of working-class respectability were under direct attack from the mass media with its interest in 'drugs, sex and rock 'n' roll'. Deferred gratification and deference were simply dismissed as old fashioned and inappropriate by 'counter-culture' hedonists. In leisure, transgression and subversion have also been recurrent themes in Rojek's work (2000) on dark and deviant leisure. Despite occupying different class positions, teenagers growing up in the 1960s had the collective experience of shaping and being shaped by liminal processes in affluent times. As argued elsewhere (Bramham, 2005) this youth cohort has carried that existential propensity with them into later life. Traditional life stages and family cycles have been undermined by subversion and challenged by rejection. The 1960s generation became rebellious parents in the mid-1970s, restless middle-aged divorcees in the 1980s/1990s and atavistic born-again pensioners at the turn of the millennium. Sustained by the feminist movement, many have set about challenging the conventions of marriage, childcare, sexuality and gender roles. As one consequence of this, the constraints of 1950s respectable working-class culture have been diluted. Flexibility in work patterns has changed the sexual division of labour. Most importantly, white working-class culture had to come to terms with issues around 'race' and ethnicity, embodied in what Sivanandan (1990) terms the 'reserve army of labour'.

The 1960s cohort refused to grow up and pass gracefully into middle and old age in the manner of their parents. They have carried their rebellious edge forward, their thirst and taste for liminality, their deconstruction of stereotypes as they sought to rewrite traditional boundaries of middle age and explore longevity, becoming that oxymoron, the 'young old'. The central argument proposed here is the importance of experiential continuities in history and in generation. This position directly contrasts with that adopted by others. For example, Merleau-Ponty (1964) in **Sense and Non-Sense** argues that at different stages of our lives we are different persons that 'accidentally' inhabit the same body. Different selves get woven together retrospectively into a 'false' biographical unity' [quoted in Featherstone (1995) p. 60].

Grasshoppers, ants and butterflies

The 1960s generation had grown up in affluence and was restless to explore new tastes, new tourist destinations, new consumption styles and later on, new patterns of early retirement. By the 1980s this generation of 1960s baby boomers had in large numbers rejected the fiscal disciplines of social reformism and its growing burden of the welfare state. In keeping with this cohort's existential predilection for individualised consumption, they settled in politics for a radical New Right discourse of choice and consumerism and so opted for a less expensive and less intrusive residual state. Thatcherism spoke to the electorate as self-interested individuals and was anti-collectivist; it offered some of the electorate a chance to participate in 'a property–owning democracy' with low income tax thresholds, share-ownership in newly privatised nationalised industries, as well as encouraging private pensions, healthcare and the 'right to buy' social housing. If the 1960s baby boomers had grown up in a world of free dental, nursing and healthcare, fixed-salary, inflation-proof pensions and publicly funded comprehensive education, they were seemingly not prepared to make similar collective provision for the next generation. They had further benefited as property owners from rising house prices, and according to Hutton (1995), the top 40% of the population, were wealthy from the one-off windfall of 1980s privatisations. They had both the cultural and economic capital to be interested in post-modernity and engage with 'new times' as they took on the internet, mobile phones, globalised TV and fads in diet and lifestyles, along with a restlessness for make-overs, changing places, and holiday homes in the UK or abroad. But, in choosing the politics of the neo-liberal project and welcoming markets, deregulation and individualisation, they were absolved from the legislative legacy of their own parental generation, namely welfarism. With Thatcherism, traditional collective politics were dissolved and a 'hollowed out' state laid down the framework for new times, new formations and globalisation.

In the world of leisure and sports studies, figuration sociology makes much of the functional interdependence and growing democratisation of social and cultural networks. Divisions of class, gender and 'race' become more egalitarian and the distance between generations becomes less pronounced. Rather than 'dancing in the light', those who were teenagers in the 1980s now find themselves dancing in the shadows of their parents.[5] Stated simply, middle-aged groups now have the material and cultural capital along with the demographic weight to be leisure trendsetters and market heavyweights. Their tastes in clothes, music, holidays and so on provide an important backdrop to the choices available for young people growing up in the 2000s. Traditional icons of youthful rebellion and style are now subverted and consumed by the middle-aged. Paul Willis' analysis in Common Culture (1990) documented the ways in which 1960s teenagers customised mass-produced media, music and clothes as expressions of individualised tastes. His favourite example then was of powerful motorbikes; to buy, own and drive one of these now carries dangerous consequences in increasing road accidents for the elderly.[6] Gender and sexual stereotypes also face scrutiny. Critics challenge the gay press's preoccupation with male youth or advertising's disregard for the 'grey pound' and older role models in placing products.[7]

It is somewhat ironic that New Right politicians found so much fault with the permissive revolution of the 'swinging 1960s' and blamed progressive parenting and education policies for welfare dependency and a lack of enterprise in the 1980s workforce. For it is that very generation which formed an important component of the electorate that chose Thatcherism, best characterised as that 'odd couple' of ideologies: a fusion of liberalism and conservatism. Different generations warmed to this hybrid 'new' ideological politics, securing a platform for the neo-liberal project over many years. Nevertheless, the 'grass/shopper' approach to consumption and, in particular, to recreational drugs challenged the ethic of hard work, so cherished by past generations of parents and grandparents. So key politicians, sought out by the mass media, felt the need to confess to breathing in the drugs-laden air of the 1960s: some inhaled and others claimed not to have done so. But the fault lines of both welfare politics and national culture had been recast. The assault on tradition, primarily through media humour (the 'satire boom' of the 1960s, for instance), the subversive interest in liminality, in progressive education, in alternative lifestyles and sexual identities, in the deconstructing of British identity, in alternatives to the institution of marriage, whilst turning a blind eye to the changing nature of communities and to spiralling rates of crime were corrosive and hard to reverse. Conservative politicians in the 1980s and 1990s found this to their cost as they hypocritically advocated a return to 'family, community and Victorian values'.

This more individualised 'grasshopper' approach, of enjoying abundance in the present (provided essentially by the ant-like endeavour of their parents), together with a marked collective reluctance to invest long term in material infrastructure for following generations, has been highlighted most by environmentalist critiques. Green thinking (whether of light or darker, more radical hue) has pointed out that grasshopper lifestyles are unsustainable. Whether it is a question of mobility, lifestyle, landscape or biodiversity, things should change and change both globally and locally. Yet there is no sign that the 1960s generation are prepared to abandon their cars, holiday homes, cheap mass tourism by airplane or consumerist lifestyles.

The third and final version of time then is not measured lightly in individual lives or in generational decades but in a much weightier and long-lasting currency, which spans one or more centuries. These represent major phases of human history such as medieval times or modern industrial society. It is precisely this long-term structural change from one sort of society to another which created leisure. It is the shift from preindustrial society to industrial society that created leisure time. Giddens argues that we should think in terms of these times as **eras**, as periods of long-term seismic structural change. But it can be termed as glacial time when change, hardly noticed by individuals or generations, may eventually be quite radical in character. Environmentalists have this time scale in mind when they express concern about population growth, pollution and global warming. Riding around town in 'gas guzzling' 4x4 'all terrain sports' vehicles and enjoying cheap air travel suits our generation in the first decade of the new millennium but our grandchildren and their children may face a bleak environmentally depleted future.

Conclusion

This chapter has argued that social theory must be informed by generational analysis in explaining change and, ultimately, leisure. Rather than focusing solely on social relations, there is a need to acknowledge the economic, political and cultural formations of modernity. This too demands an historical understanding of three levels of time – ***Dasein***, **epochs** and **eras**. The second part of the chapter has introduced a brief generational analysis with a focus on the 'Sixties' cohort of 'baby-boomers'. This generation led the expressive revolution in youth culture in the 1960s, celebrated leisure lifestyles and saw neo-liberal politics in the 1980s as a magical resolution to the burdens of the welfare state. This hyperactive generation was simply not ready to settle into existing conventional patterns of class, gender and not least 'race'. Postcolonial patterns of immigration provided new and different historical challenges in the 1960s from those faced earlier in the nineteenth and twentieth century. All in all, the old institutional solutions conceived by past generations – the vision of self, life course, family and community, of work and of religion and leisure simply seemed no longer appropriate. The sexual revolution – for example, in the wide availability of contraception – challenged gender power relations and feminism raised questions about patriarchal processes in public and private spheres. The next generation in some ways became less important. They were suspended or delayed or reduced by smaller family size, as the large demographic bulge of the baby boom chronologically grew older. But socially and culturally, as we have seen, this was resisted. In some ways the 1960s cohort were reluctant parents...in the sense that they put their demands for self-expression and identity alongside the needs of their children rather than subordinating themselves into traditional roles as parents. It was not so much that the next generation of children was the cherished future, but that they as the parent generation were reluctant to be sidelined and felt they had a crucial role to play.

Understanding social relationships, both within and between generations, is only part of the intellectual journey as historical sociology demands a grasp of how relationships change over time. Yet each generation or age cohort grows up in, experiences and makes sense of its own historical setting and class position, shaped by distinctive economic, political, social and cultural formations. The present chapter acknowledges that there is growing individualisation in lifestyles, but class and generational analysis remain central to understanding the class, gender and racial divisions within leisure and culture. The 'baby boomer' 1960s generation developed spectacular youth cultures as 'a magical solution' to growing up in a class-divided society. It was in leisure worlds where youth negotiated distinctive cultural spaces away from and in relation both to dominant and to subordinate parent cultures. Sustained by the politics of affluence and welfarism, youth became a developing market, differentiated by tastes in music, dress, style and appearance and in activities. For the CCCS youth subcultures represented 'the magical recovery of community': 'skinhead culture selectively reaffirms certain core values of traditional working class...The reaffirmation is symbolic, rather than a 'real' attempt to recreate some aspects of the parent culture' (Clarke, 1976: 99). The focal concerns of territory, with football and fanship, and aggressive masculinity offer one response to class contradictions and experience of family, school, work, and neighbourhood. The traditional institutions of kinship, trade unions and the Labour Party developed in the parent culture of the 1940s and 1950s no longer draw support from the next generation. Twenty years on the 'grasshopper' generation came to endorse the neo-liberal project as 'a magical solution' to the politics of welfare and globalisation, since this best suited their context, historic experiences and generational aspirations.

Notes

1 For a full discussion of the nature and direction of the leisure project see Bramham, P. (2008) 'Rout(e)ing the leisure project', in P. Gilchrist and B. Wheaton, *Whatever Happened to the Leisure Society? Theory, Debate and Policy* (Eastbourne: Leisure Studies Association) **102**: 1–12.

2 This is not to suggest that all Marxists subscribed to the ideas of 'New Times' and postmodernity see Sivanandan, A. (1990) 'All that melts into air is solid', *Race and Class*, **31**: 1–31. Indeed, the 'postmodern turn' emerged out of Marxism and the new cultural studies, notably via the publication of Marxism Today, but many felt that these writers thereby ceased to be Marxists.
3 Martin Jacques was the influential editor of *Marxism Today* from 1977–1991. One edition, October 1988, was entitled 'New Times', defined as one of the three crucial debates in the journal's history. See http://www.amielandmelburn.org.uk/collections/mt/index_frame.htm (accessed 13 November, 2009).
4 The nature of the relationship between the parent and youth generation has been developed at length elsewhere see Bramham, P. (2005) 'Habits of a Lifetime: Age, generation and lifestyle', in P. Bramham and J. Caudwell *Sport, Active Leisure and Youth Cultures*, **86** (Eastbourne: LSA Publications).
5 Or more likely dancing on the same dance floor, as 50 year clubbers rekindle their interest in the rave and dance scene by dancing through the night fuelled by recreational drugs.
6 See article on 'born again bikers', *The Guardian/Observer*, 28.06.2004.
7 See Sandra Smith 'Is beauty youth, and youth beauty?', *The Guardian*, 31.03.2004.

Bibliography

Abercrombie, N., S. Hill and B. Turner (1980) *The Dominant Ideology Thesis* (London: Allen and Unwin).

Abrams, P. (1982) *Historical Sociology* (Shepton Mallet: Open Books).

Anderson, B. (1991) *Imagined Communities: Reflections on the Origins and Spread of Nationalism* (London and New York: Verso).

Appadurai, A. (1990) 'Disjuncture and difference in the global cultural economy', in M. Featherstone, *Global Culture, Nationalism, Globalization and Modernity* (London: Sage), 295–310.

Bauman, Z. (1987) *Legislators and Interpreters: On Modernity, Post-modernity, and Intellectuals* (Cambridge: Polity Press).

Bauman, Z. (1999) *Globalisation* (Cambridge: Polity Press).

Bauman, Z. (2005) *Liquid Life* (Cambridge: Polity).

Bauman, Z. (2006) *Liquid Fear* (Cambridge: Polity).

Bauman, Z. (2007) *Liquid Times: Living in an Age of Uncertainty* (Cambridge: Polity).

Beck, U. (1992) *Risk Society* (London: Sage).

Beck, U. (1999) *What is Globalisation?* (Cambridge: Polity).

Beck, U. and E. Beck-Gernheim (2002) *Individualization, Institutionalized Individualism and its Social and Political Consequences* (London: Thousand Oaks; New Dehli: Sage).

Brake, M. (1980) *The Sociology of Youth Culture and Youth Subcultures. Sex and Drugs and Rock 'n' Roll?* (London: Routledge and Kegan Paul).

Bramham, P. (2005) 'Habits of a lifetime: Age, generation and lifestyle', in P. Bramham and J. Caudwell, *Sport, Active Leisure and Youth Cultures*, **86** (Eastbourne: LSA Publications).

Bramham, P. (2008) 'Rout(e)ing the leisure project', in P. Gilchrist and B. Wheaton *Whatever Happened to the Leisure Society? Theory, Debate and Policy*, **102**: 1–12 (Eastbourne, Leisure Studies Association).

Braudel, F. (1975) *The Mediterranean and the Mediterranean World in the Age of Philip II*, Vol. 1 (London: Fontana). Edition, translated from French by Siân Reynolds. First published in 1949 in French, *La Méditerranée et le monde méditerranéen à l'époque de Philippe II*.

Burke, P. (1980) *Sociology and History* (London: George Allen & Unwin).
Callinicos, A. (1989) *Making History: Agency, Structure and Change in Social Theory* (Cambridge: Polity Press).
Center for Contemporary Cultural Studies (CCCS) (1982) *The Empire Strikes Back*, University of Birmingham (London: Routledge).
Clarke, J. (1976) 'The skinheads and the magical recovery of community', in S. Hall and T. Jefferson, *Resistance through Rituals*, 99–118 (London, Melbourne, Sydney, Auckland, Johannesburg: Hutchinson).
Coalter, F. (1989) *Freedom and Constraint: The Paradoxes of Leisure, Ten Years of the Leisure Studies Association* (London: Routledge).
Cohen, P. (1984) 'Subcultural change in a working class community', in E. a. W. Butterworth, D, *A New Sociology of Modern Britain* (Glasgow: Fontana).
Cohen, P. (1975) 'Subcultural conflict and working class community', in E. Butterworth and D. Weir, *The Sociology of Modern, Britain*, 92–102 (Glasgow: Fontana/Collins).
Davis, M. (2008) 'Liquid sociology', *Sociology*, **42** (6): 1237–43.
Dawe, A. (1979) 'Theories of social action', in T. Bottomore and R. Nisbet, *A History of Sociological Analysis*, 362–417 (London: Heinemann).
Featherstone, M. (1990) *Postmodernism and Consumer Culture* (London: Sage).
Featherstone, M. (1995) *Undoing Culture* (London: Thousand Oaks; New Delhi: Sage Publications).
Foster, J. (1974) *Class Struggle and the Industrial Revolution: Early Capitalism in Three English Towns* (London: Weidenfeld and Nicolson).
Foucault, M. (1980) *Power/knowledge* (New York: Patheon).
Friese, H. and P. Wagner (1999) 'Modernity and contingency moving culture', in M. Featherstone and S. Lash, *Spaces of Culture: City, Nation World* (London: Sage).
Giddens, A. (1979) *Central Problems of Social Theory Action, Structure and Contradiction in Social Analysis* (London: Macmillan).
Giddens, A. (1981) *A Contemporary Critique of Historical Materialism* (London: Macmillan).
Giddens, A. (1984) *Constitution of Society: Outline of the Theory of Structuration* (Cambridge: Polity Press).
Giddens, A. (1991) *Modernity and Self-identity* (Cambridge: Polity Press).
Gilroy, P. (1993) *The Black Atlantic: Modernity and Double Consciousness* (London: Verso).
Hall, S. (1992) 'The west and the rest: Discourse and power', in S. Hall and B. Gieben, *Formations of Modernity*, 275–320 (Cambridge: Polity Press in association with the Open University).
Hall, S., C. Critcher, T. Jefferson, J. Clarke and B. Roberts (1978) *Policing the Crisis: Mugging, the State, and Law and Order* (London: Macmillan).
Hall, S. and M. Jacques (1983) *The Politics of Thatcherism* (London: Lawrence Wishart in association with Marxism Today).
Hobsbawm, E. and T. Ranger, (eds) (1983) *The Invention of Tradition* (Cambridge: Cambridge University Press).
Hulton, W. (1995) *The State We Are In* (London: Jonathan Cape).
Jessop, B., K. Bonnett, S. Bromley and T. Ling (1988) *Thatcherism: A Tale of Two Nations* (Cambridge: Polity Press).
Mannheim, K. (1952) 'The problem of generations', in K. Mannheim, *Essays on the Sociology of Knowledge*, 276–320 (London: Routledge and Kegan Paul).
Martin, B. (1981) *A Sociology of Contemporary Cultural Change* (Oxford: Basil Blackwell).
Merleau-Ponty, M. (1964) *Sense and Non-Sense* (Evanston: Northwestern University Press).
Mills, C. W. (1970) *The Sociological Imagination* (Harmondsworth: Penguin).
Pearson, G. (1983) *Hooligan: A History of Respectable Fears* (London: Macmillan).
Rojek, C. (1993) *Ways of Escape* (London: Macmillan).

Rojek, C. (2000) *Leisure and Culture* (Basingstoke: Palgrave).
Rojek, C. (2005) *Leisure Theory: Principles and Practice* (Basingstoke: Palgrave Macmillan).
Sivanandan, A. (1990) 'All that melts into air is solid', *Race and Class*, **31**: 1–31.
Stedman-Jones, G. (1983) *Languages of Class: Studies in English Working Class History 1832–1982* (Cambridge: Cambridge University Press).
Thompson, E. P. (1963) *The Making of the English Working Class* (Harmondsworth: Pelican).
Tilly, C. (1981) *As Sociology Meets History* (New York: Academic Press).
Urry, J. (1990) *The Tourist Gaze* (London: Sage).
Williams, R. (1963) *Culture and Society 1780–1950* (Harmondsworth: Penguin).
Willis, P. (1990) *Common Culture* (Milton Keynes: Open University Press).

2

Double Measures: The Moral Regulation of Alcohol Consumption, Past and Present

Chas Critcher

***Chapter 2** is about the politics of the historically central leisure world of **drinking**. It is important in two further, specific respects: first, it shows the recurrent nature of the 'moral panic' in the history and social transaction of leisure and, second, it illustrates the tension – which is the principal purpose of this book to explore – between personal freedom (in this case to drink alcohol) and constraint. This tension is captured via the concept of 'moral regulation' and an analysis of the eighteenth century 'gin crazes' and contemporary political concern over 'binge drinking'.*

Introduction: Opening time

On September 10th 2008 the *Guardian* newspaper produced a special supplement titled 'Alcohol and Us'. It was essentially driven by an agenda about 'binge drinking'. The articles were heavily research based. One half-page article provided an historical perspective, comparing contemporary concerns about 'binge drinking' with the 'gin craze' of the early eighteenth century. Sociologist Fiona Measham identified 'a cycle of moral panics' about alcohol consumption with recurrent themes: working-class and female consumption as problematic, a politically powerful drinks industry catering to their needs and governments reluctant to act.

There is more than meets the eye in such historical parallels. Many years ago I argued that leisure studies required a more substantial grasp of history (Critcher, 1989). I thought then that its significance was what it revealed about two contradictory tendencies in leisure: its 'development' and its 'control'. I now prefer the term moral regulation.

This chapter develops the comparison between the 'gin craze' of the 1730s and 1740s and 'binge drinking' in the present. The aim is to evaluate the concept of moral regulation, its ability to account for the processes involved in reacting to a perceived problem of excessive alcohol consumption. The account is heavily dependent on three sources, for the gin craze Warner (2002), for binge drinking Plant and Plant (2006) and for moral regulation Hunt (1999). We start with the more recent and familiar case of binge drinking.

Binge drinking: One for the road

The binge drinking controversy emerged in Britain from 2004 onwards. It had two significant antecedents. The first in the 1980s was the police identifying problems in town centres and in rural areas at weekends. Young men would congregate, get drunk and then start fighting. The term 'lager louts' passed into common usage.

The second antecedent occurred in the mid-1990s. Brewers started to produce drinks which mixed soft beverages with alcoholic spirits, a concoction known as 'alcopops'. They were allegedly marketed at very young and possibly underage drinkers. After warnings from government, some marketing excesses were curbed and the furore died down as new energy drinks were introduced.

'Binge drinking' incorporated both of these previous problems. Young men were drinking lager, becoming intoxicated and perpetrating violence. Young women were drinking flavoured spirits in order to get drunk and then making spectacles of themselves. It was initially identified as a serious problem by the police and other emergency services who had to deal with drunken behaviour and its aftermath. In February 2005 Prime Minister Blair called binge drinking the 'new British disease' (Plant and Plant, 2006: 23).

A coherent policy on alcohol was an objective of New Labour when it came to power in 1997. Policies for Wales, Scotland and Northern Ireland were developed and implemented in the late 1990s; progress in England was slower. The Department of Health failed to produce a policy for alcohol so the task passed to the Prime Minister's Strategy Unit. This produced in 2003 an Interim Report of evidence with an Alcohol Harm Reduction Strategy for England following in 2004.

The twin objectives were to liberalise licensing hours and tackle binge drinking. The Prime Minister's Strategy Unit defined binge drinking as consuming more than twice the recommended daily intake: six units for females (equivalent to three pints of beer, cider or lager; six glasses of spirits or wine); and eight units for males (equivalent to four pints of beer, cider or lager; eight glasses of spirits or wine) (Plant and Plant, 2006: xii).

There had indeed been an overall increase in the amount of alcohol consumed in England and Wales, though per capita consumption remained a little lower than in the rest of Western Europe (Plant and Plant, 2006: 47). Adults were now drinking more at home and less in pubs and clubs. The opposite was true of young people. The Strategy Unit estimated that there were nearly six million binge drinkers in Britain, around 15% of all adults, including 75% of women and 84% of men aged 16–24 (Plant and Plant, 2006: 49). Real changes in drinking patterns in Britain had occurred, especially amongst the young including girls, ushering in a new 'culture of intoxication' (Measham and Brain, 2005). Older drinkers who binged at home were not visible on the streets.

The government was equally determined to 'modernise' England's licensing laws. The outcome was the Licensing Act 2003 which permitted pubs, clubs and other licensed premises to set their own opening hours. This was passed despite press hostility and scepticism from pressure groups and experts. Extended opening hours were expected to generate an extra £500 million sales per annum for an industry anticipating lost custom with smoking banned in pubs after 2005.

Action on binge drinking was, by contrast, feeble. In the 1990s the police had been granted additional powers to confiscate alcohol from underage drinkers in public places and immediately close disorderly clubs or pubs for 24 hours. Penalties for selling alcohol to underage children had been increased and a new offence introduced of the 'proxy purchase' of alcohol by adults for children. In the summer of 2004 such powers were enforced in a much publicised 'crackdown' on binge drinking. Otherwise, government ministers threatened to ban 'happy hours' or charge licensees for increased policing necessitated by binge drinking but the primary strategy remained voluntary self regulation by the drinks industry. Government would not contemplate those measures which experts argued would actually decrease the demand for alcohol: increasing prices, banning adverts and limiting the number of licensed premises. Such prevarication had not been evident in the parallel campaign against gin in the early eighteenth century.

Gin: Keeping spirits up

Until the beginning of the eighteenth century spirits in England were imported and expensive. Then several factors combined to make one particular spirit accessible and affordable. War with the Dutch provoked laws to restrict imports, including gin. A series of good harvests produced a surplus of grain. The real wages of the working poor rose. Parliament abolished the monopoly enjoyed by the London Company of Distillers. English brewers started to produce their version of gin, a name derived via 'geneva' from the Dutch term 'genever'.

English gin was so rough that it had to be made palatable by added fruit flavouring. In that form it became instantly popular. Unlike beer, its effects were instantaneous. Gin consumption became widespread amongst London's working poor, although 'gin did not constitute a serious threat outside London' (Warner, 2002: 40). Consumption was particularly high in Westminster, where it was inevitably noticed by politicians and other worthies as they passed through.

The ruling classes expressed disquiet. Gin consumption and drunkenness were visible on the streets. Laws governing the licensing of public houses were largely irrelevant because gin was sold on the streets. It seemed that the nation was being undermined by habitual drunkenness, especially amongst women expected to produce the next generation of labourers, soldiers and sailors. Small groups of reformers advocated legal regulation.

As Warner (2002) points out, parliament at this time did not much concern itself with social problems. This necessitated a concerted campaign against gin which became one of the first and most apparently successful efforts at moral reform. It used recognisably modern methods: tracts and sermons, horror stories in newspapers, appeals to the national interest and the sophisticated exploitation of statistical evidence.

There resulted eight acts of parliament between 1729 and 1751 designed to control the mass consumption of gin. Each act used a combination of increased excise tax, licensing requirements, prosecutions for illegal street trading and, notoriously in four successive acts, the use and reward of informers. However, legislation proved unenforceable without an organised police force. Street trading continued unabated. Requirements for licensing were ignored. Informers were often attacked when discovered. Occasionally protests developed into full scale riots. Eventually the final act of 1751 repealed previously ineffectual provisions and concentrated instead on raising excises on gin by more than 50%, banning sales of gin in prisons and other lockups, and barring distillers and street hawkers from selling gin. Gin consumption did subsequently fall but it is unclear how far this resulted from the act and how far from other factors such as a fall in real wages and shortages of grain (Clark, 1988).

Gin consumption peaked around 1743, after most of the acts had been passed, and only declined in 1757 when the government reacted to successive crop failures by banning the use of grain for distilling spirits. Hence there had been a consistent mismatch between the levels of consumption and the degree of concern:

> concerns over drunkenness bore very little correspondence to actual consumption, begging the question of whether a reforming elite was reacting to gin *per se* or rather to larger and more intractable threats to their society and way of life (Warner, 2002: 4).

Such overreaction might be termed a moral panic. But first a careful comparison of the two episodes is revealing about the processes of definition, reaction and regulation.

Similarities: The same again, all round

Similarities and differences between the two examples will be discussed under three headings: the issue as perceived at the time; claims-making, how it was done and by whom; and measures adopted, their form and effectiveness.

The issue

For both gin and binge drinking, there was great emphasis on the ease, speed and cheapness with which people could get drunk. The fundamental objection was to drunken disorder on the streets. Females seemed as culpable as males: 'it was the behaviour of women who drank gin that most offended polite society' (Warner, 2002: 63). Mothers and nursemaids neglecting children jeopardised the production of the next generation. Binge drinking produced shock that young women should imitate their male counterparts, making themselves vulnerable to sexual assault. Drunken women on the streets scandalised opinion both in the 1730s and in the 2000s.

Claims-making and makers

In both cases, newspaper coverage portrayed and indicted the behaviour for those otherwise unaware of it. Public opinion was constructed by enraged sectors of the middle class but popular feeling was indifferent or tacitly supportive.

The damage allegedly caused by excessive intake of alcohol was presented in statistical terms. Anti-gin campaigners balanced facts and feelings.

> This new type of advocacy was both irrational and seemingly rational at the same time: naturally it appealed to the emotions – as ultimately all forms of advocacy must – but it also relied on numbers, thus lending the imprimatur of scientific proof to what were in fact highly emotional claims (Warner, 2002: 103).

Two hundred and fifty years later the Prime Minister's strategy unit characterised the problem of binge drinking in essentially similar, if more sophisticated, statistical terms:

> The Prime Minister's Cabinet Office reported that alcohol misuse cost the National Health Service £1.7 billion per year. It was also noted that £95 million was being spent on specialist alcohol treatment. The report acknowledged that there were over 30,000 annual hospital admissions for alcohol dependence and up to 20,000 premature deaths associated with alcohol. It was calculated that the cost of alcohol-related crime was £7.3 billion and that 360,000 incidents of domestic violence were alcohol related (Plant and Plant, 2006: 67).

Measures adopted

Reaction to both problems extended powers of state officials to immediately apprehend offenders. In the gin craze the use of paid informers and later citizens' arrests designed to exert control instead provoked minor disturbances and occasional riots. Action against binge drinkers included on the spot fines for retailers serving underage drinkers; the declaration of alcohol disorder zones where trouble recurred; and proposals to charge bars and clubs for additional policing. Existing legal powers might have been more effective. In principle binge drinkers could all be arrested when drunk and disorderly, and fined on the spot, but no government could contemplate such a move. In both cases the power of the state could be increased in principle but proved difficult to realise in practice.

Differences: Shaken, not stirred

Between the gin craze and binge drinking the media system expanded, pressure groups proliferated and parties exacted discipline in parliament. Political priorities also changed. During the gin craze Britain was paradoxically a stable country with fragile institutions: 'even the slightest threat to the status quo was a source of enormous alarm to the nation's governing class' (Warner, 2002: 7). For Britain under 'New' Labour political disruption was less important than apparent moral disintegration for which the solution was the creation of a new moral order, based on rights and responsibilities, with great hostility towards those threatening it.

The issue

The geography and generation of each issue differ. The gin problem was confined to London. By contrast, binge drinking has been perceived as a problem in all towns and city centres. Binge drinking has been primarily an issue about young people and drink. In the gin campaign, age was not an issue, children not being excluded from buying drink or entering public houses until 1901 (Plant and Plant, 2006: 17).

Claims-making and makers

The deleterious effects of excessive alcoholic consumption were stressed for both gin and binge drinking, but with different emphases. Gin harmed the health of the nation and future generations. Binge drinkers damaged their own health. Claims makers about gin were moral and especially Christian reformers but for binge drinking more prominent critics were law-enforcement agencies with some support from anti-alcohol groups. Claims about gin were largely uncontested because brewers and distillers were unevenly organised in the face of opposition to gin. By the time of binge drinking the trade is highly organised, public relations conscious and allegedly very influential on government policy. It has 'enormous economic and political power and appears to have dominated much of the policy-making process' (Plant and Plant, 2006: 83).

Measures adopted

Attempts to suppress the gin trade were stymied by the absence of a police force. The police have been crucial to the definition of the binge drinking issue. The government of the day acted reluctantly over gin whereas government has been the prime mover on binge drinking.

Moral panic: Barring trouble

Another way of appreciating the continuities and changes between the two episodes is to utilise the concept of a moral panic. Two major accounts explicitly discuss its relevance. Plant and Plant pose the question: 'Is the campaign about "binge drinking" and the problems associated with heavy drinking just a "moral panic"'? While recognising a familiar pattern of media exaggeration, they suggest that there was 'ample justification for media interest and public concern' (2006: 28). Later they note how in 2005 'media interest built upon and probably amplified a moral panic about binge drinking that had been running for several months' (2006: 87).

Warner confronts the problem of what to call those who campaigned against gin. She rejects 'moral entrepreneur' as too modern, 'moral reformer' better reflecting 'the larger goal of the war on gin: to reform the morals of the working poor' (Warner, 2002: 11). She sees the gin problem as the very first drug scare, despite occurring in a society 'as yet lacking a police force and larger political and bureaucratic structures' (Warner, 2002: 7). Identified as common to past and present drug scares are six features: (1) a new, cheap and widely available drug emerges: (2) it is stronger and more dangerous than its predecessors; (3) it is taken by outsiders but with vulnerable people, especially women and children, as direct or indirect victims; (4) its distribution and consumption are located in places of notoriety within the city; (5) experts provide scientific evidence about the drug's prevalence and dangers and finally (6) extreme legislative measures must be taken to solve the problem.

This list bears an uncanny resemblance to that originally identified by Cohen as characteristic of a moral panic:

> Societies appear to be subject, every now and then, to periods of moral panic. (1) A condition, episode, person or group of persons emerges to become defined as a threat to societal values and interests; (2) its nature is presented in a stylised and stereotypical fashion by the mass media; (3) moral barricades are manned by editors, bishops, politicians and other 'right-thinking' people; (4) socially accredited experts pronounce their diagnoses and solutions; (5) ways of coping are evolved or (more often) resorted to; (6) the condition then disappears, submerges or deteriorates and becomes more visible (Cohen, 2002 [1973]: 9, numbers added).

By such criteria the gin campaign does seem to have approximated to a moral panic. Its media and experts were more restricted than in modern times but otherwise all the conditions are fulfilled. This is less the case with binge drinking. The media have helped to label and condemn binge drinking but few, outside government, law enforcement agencies and newspapers, have perceived it as a major issue. Established pressure groups have participated fitfully and no new ones have emerged. Experts have been thin on the ground. Medical authorities have appeared only occasionally. The distinctions are fine but gin seems to have been more of a moral panic than binge drinking.

As I have suggested elsewhere (Critcher, 2003), moral panic as an ideal type facilitates comparisons between different social problems but is less able to explain variations from the expected pattern. An additional problem is specifying in which wider field of activity moral panics should be situated conceptually. One answer is moral regulation.

Moral regulation: Drink rules

An essential guide to applying the concept of moral regulation to leisure is Alan Hunt. His main work (1999) examines movements for sexual reform in the eighteenth and nineteenth centuries in Britain and the USA. Hunt identifies four key deficiencies in the moral panic framework: (1) its negative connotations; (2) assumptions that reaction is irrational; (3) a tendency to subscribe to conspiracy theories, especially about the role of the state; and (4) the insistence that reform movements are invariably conservative. More promising is the idea of moral regulation. As originally developed by Corrigan and Sayer (1985), moral regulation referred to the way the evolution of the modern state constructed the identity of its citizens. It aimed to regulate their conduct according to the moral principles enshrined in the state. Because the process involved privileging some ideals over others, the regulation was always moral. This brief review (for a fuller version see Critcher, 2008) will explain and apply to the case studies the definition and focus of moral regulation and then its typical forms, sources and outcomes.

For Hunt moral regulation 'is a discrete mode of regulation existing alongside and interacting with political and economic modes of regulation' (Hunt, 1999: 17). It is defined as a distinctive type of regulation, involving 'practices whereby some social agents problematise some aspect of the conduct, values or culture of others on moral grounds and seek to impose moral regulations on them' (Hunt, 1999: ix).

Historically moral regulation has had a focus on three areas: alcohol, gambling and sex. The twentieth century discovered new topics: child abuse, welfare scroungers and psychoactive drugs. Traditional themes persisted, such as the perception of urban space as populated by 'not just unruly people, but unruly categories of people…that is, those who lived outside the structures of authority' (Hunt, 1999: 29).

Our case studies are two moments in a longstanding historical struggle over the moral regulation of alcohol consumption. The objective of campaigners is to present the activity (gin drinking/binge drinking) as intrinsically a moral issue. There is a recurrent preoccupation with the unacceptability of drunken behaviour on the streets.

Hunt sees the forms of moral regulation as having five elements: (1) agents who agitate for regulation; (2) targets of such regulation; (3) tactics and techniques to bring regulation about; (4) discourses which validate regulation; and (5) the wider political context with potential for resistance. Moral regulationists argue that the activity is harmful for those involved or the society as a whole but often their underlying assumption is that the activity is intrinsically immoral.

The gin issue agents comprised a loose alliance of moral reformers with the target of gin drinking. Their tactics involved extensive propaganda and intricate parliamentary manoeuvring. The discourse was one about the (military and labouring) health of the nation. The political context revealed a weak government and a disorganised trade. For binge drinking the crucial agent was government itself with the target of binge drinkers. The tactics were to threaten legal action against perpetrators and suppliers, supported by discourses about disorder and (personal) health. The drinks industry exerted considerable influence in the political context.

Hunt regards the sources of moral regulation as complex, involving any possible combination of state agencies, the 'respectable classes' and occasionally the populace at large. The middle class, especially women, remain the most active moral reformers, sometimes critics of their own class, more often of the working class.

The gin drinking issue was defined largely by men from inside the establishment, upper not middle class. Women were targets rather than agents of the agitation. The class dimension was transparent. Binge drinking is defined by the state, aided by some pressure groups. Some prominent politicians happen to be women but it is not a female-defined issue. Binge drinking is not a class issue, since all classes participate, but a generational one: it pits the middle aged against young adults.

For Hunt the outcomes of moral regulation are predictably changes in the law or its enforcement but there are always 'tensions between regulationism and anti-regulationism' which prove to be 'a central dilemma of moral regulation' (1999: 12). Hunt evokes Foucault's concept of governmentality to explain how moral regulation aimed at controlling the behaviour of others simultaneously involves control of the self.

The anti-gin laws addressed anti-social behaviour, in itself unusual for the time. There was no obligation on reformers or their class to govern their own conduct. Binge drinkers were exhorted to govern their own behaviour but sanctions were lacking. State and industry would only encourage consumers to 'drink responsibly'.

Moral regulation exists independently of economic and political regulation but sometimes the three converge. Political and economic considerations constantly intruded into efforts to enforce moral standards on both gin and binge drinking.

The political economy of moral regulation

In the early eighteenth century a prime political objective of government was to maintain public order, frequently jeopardised by popular protest. An important objection to anti-gin legislation, and a key factor in the repeal of much of it, was that it would and did provoke riotous protest. Moral regulation could not be permitted to jeopardise political regulation. An important economic consideration, with political side effects, was to increase funding of the 'Civil List', effectively the budget for the royal family. Excise taxes, including those on alcohol, were an effective means of achieving this. Any new law supposed to raise revenues by increasing excise taxes or licence fees would soon become discredited if it was disregarded or evaded: 'the traditional dictates of economic and fiscal policy always carried more weight than even the most pressing social policy objectives' (Davison, 1992: 26). Parliament was at this time dominated by the 'landed interest' who stood to benefit directly from any new market for low-grade grain, such as that offered by gin distilling. Brewers' and distillers' influence waxed and waned throughout. At other times, military considerations intruded. War provoked the original protectionist measures which banned Dutch gin. The lack of suitable recruits for the armed forces was an alleged consequence of excessive gin drinking by potential mothers. Raising money, armies and navies for war periodically focused parliamentary attention away from matters of morality. The absolute ban on the use of grain in distilling was eventually if temporarily brought about in 1757, not by moralistic agitation but by a disastrous harvest. The success and failure of moral regulation did depend in part on the tactics and strategies of campaigners but the outcome was equally affected by the quality of the opposition, the variety of economic interests and the preoccupations of legislators (Davison, 1992).

By the early twenty first century we encounter a 'New Labour' administration with a clear set of objectives: to use the apparatus of political regulation to deregulate economic activity and remoralise civil society. There was thus no contradiction between campaigning against binge drinking and promoting the deregulation of licensing hours. However, faced with substantial and well-organised economic interests, the moral urge faltered. At every turn government policy has been subservient to the economic interests of the drink industry. Regulatory measures have been directed against the consumer, not the producer nor the retailer. New Labour was hemmed in by its commitment to economic self-regulation, by the tax revenues derived from alcohol and by the status of the drinks industry as a major employer. During the whole campaign no measures were adopted which required brewers or retailers to change their behaviour or bear the cost of the excessive drinking which they encouraged.

An even sharper contradiction was evident between moralistic attacks on binge drinking and policies fostering night-time economies in city centres which depended wholly on consumer hedonism, a transformation of 'our town centres into liminal spaces in which individuals are encouraged to play with the parameters of excitement and excess' (Hayward and Hobbs, 2007: 438). Reacting to changes in consumer behaviour and economic regulation, brewery chains designed new kinds of city centre drinking hall quite unlike the traditional pub. They were bigger, had fewer seats and stocked different kinds of drinks, implying more raucous behaviour. They and their pricing polices were designed for young people to consume lots of alcohol, very fast. Alongside the drinking halls went night clubs and fast food establishments. 'Going up town' offered new meanings and experiences, built upon excess. This leisure environment consciously created by the industry and some local authorities invited the behaviour which the government was now decrying.

Conclusion: Last orders

This chapter has compared the construction of and reaction to the 'gin craze' of the early to mid-eighteenth century with the construction of and reaction to 'binge drinking' in the early twenty first century, both in England. Such a comparison is limited, geographically and historically. What happens in England does not necessarily happen elsewhere, though some comparative work suggests more similarities than differences (Gerritsen, 2000). In the period between our two case studies discourses about alcohol were dominated by temperance, absent from the controversies surrounding both gin and binge drinking (Harrison, 1971; Greenaway, 2003).

Generalisations must thus be tentative. The case studies do nevertheless seem to validate reconceptualising moral panics as extreme forms of moral regulation. The 'gin craze' did provoke a moral panic justifying highly repressive, albeit unenforceable, legislation. Binge drinkers, too numerous and heterogeneous to be designated successfully as folk devils, were vilified yet specific sanctions against them were limited, not least because of the political clout of the drinks industry. Falling well short of a moral panic, the campaign against binge drinking aspired to moral regulation but achieved only moral exhortation, to 'drink responsibly'. Political and economic interests overrode moral ones.

This contradiction goes to the heart of moral regulation in late modern societies. Consumer freedoms to pursue pleasure become problematic when they create disorder on the streets and costs to the state. The consequent need to define limits on the pursuit of pleasure denies the essence of the activity. Exhortations to 'drink responsibly' suppress the idea that one strong motivation for drinking alcohol is indeed to become intoxicated:

> problematic activities are managed and discussed in ways that deny or silence the voluntary and reasonable seeking of enjoyment as warrantable motives (O'Malley and Valverde, 2004: 26).

Such a Foucauldlian analysis of the governance of drugs and alcohol stresses the 'felicity calculus' where happiness as pleasure is allied with 'freedom, good order and pleasure' in opposition to 'disorder, compulsion and pain' (O'Malley and Valverde, 2004: 27). The definition of legitimate pleasure has to be kept under constant review because

> pleasure is a problem where its pursuit – as in the imagery of 'hedonism' – conflicts with the other key requirements made of liberal subjects, notably 'responsibility', 'rationality', 'reasonableness', 'independence' and so on (O'Malley and Valverde, 2004: 27).

Panics about drink are integral to wider processes of moral regulation, centring on the question of what is and is not legitimate pleasure. In late modern societies a combination of individual freedom and consumer choice can prove to be an unpredictable brew. The pursuit of pleasure in public may create disorder on the streets. The logic of moral regulation should lead to the suppression of such disruptive behaviour but the political and economic costs of this are too high to bear.

In March 2009 the Chief Medical Officer of Health for England, Sir Liam Donaldson, used his annual report to draw attention to the costs of excessive alcohol consumption. He called for a minimum price of 50p per unit of alcohol. Prime Minister Gordon Brown responded. 'We do not want the responsible, sensible majority of moderate drinkers to have to pay more or suffer as a result of the excesses of a minority.' (BBC website 16/03/09) Through such phrases is re-enacted the complex history of the moral regulation of alcohol. In her major study Marianna Valverde argued that this history was too complex to approximate to any simple model of changes in disciplinary modes or risk consciousness. It revealed instead

> that disciplinary control over minute detail coexists with hedonistic consumption, that regulation is not necessarily the opposite of prohibition, and that governing *through* an object such as alcohol may often involve governing all sorts of activities, spaces and identities without doing much to govern drinking itself (1998: 170, original emphasis).

These two case studies bear out these conclusions. We might also add that a precipitating factor in both cases was the transgression of a behavioural boundary. Both gin and binge drinking seemed to lead to a state of abandonment where all the conventions of daily public life were breached. Even in a society based on excessive pleasurable consumption, there are still some limits. When they are reached the apparatus of moral regulation, at once intransigent and cumbersome, is called into action. Order, or the semblance of it, will be restored.

References

Clark, P. (1988) 'The "Mother Gin" controversy in the early eighteenth century', *Transactions of the Royal Historical Society*, 38: 63–84.

Cohen, S. (2002) [1973] *Folk Devils and Moral Panics* 3rd edn (London: Routledge).
Corrigan, P. and D. Sayer (1985) *The Great Arch: English State Formation as Cultural Revolution* (Oxford: Basil Blackwell).
Critcher, C. (1989) 'The politics of leisure: Social control and social development', in F. Coalter (ed.) *Freedom and Constraint* (London: Comedia/Routledge).
Critcher, C. (2003) *Moral Panics and the Media* (Milton Keynes: Open University Press).
Critcher, C. (2008) 'Widening the focus: Moral panics as moral regulation', *British Journal of Criminology*, 49(1): 17–35.
Davison, L. (1992) 'Experiments in the social regulation of industry: Gin legislation 1729–1751', in L. Davison et al. (eds) *Stilling the Grumbling Bee Hive: The Response to Social and Economic Problems in England 1689–1750* (New York: St. Martin's Press).
Gerritsen, J-W. (2000) *The Control of Fuddle and Flash: A Sociological History of the Regulation of Alcohol and Opiate* (Leiden: Brill).
Greenaway, J. (2003) *Drink and British Politics since 1830* (Basingstoke: Palgrave Macmillan).
Guardian newspaper (2008) 'Alcohol and Us', special supplement, September 10th.
Harrison, B. (1971) *Drink and the Victorians* (London: Faber and Faber).
Hayward, K. and D. Hobbs (2007) 'Beyond the binge in "booze Britain": Market-led liminalization and the spectacle of binge drinking', *British Journal of Sociology*, 58(3): 437–456.
Hunt, A. (1999) *Governing Morals: A Social History of Moral Regulation* (Cambridge: Cambridge University Press).
Measham, F. and K. Brain (2005) '"Binge" drinking, British alcohol policy and the new culture of intoxication', *Crime, Media, Culture*, 1: 262–283.
O'Malley, P. and M. Valverde (2004) 'Pleasure, freedom and drugs: The uses of "pleasure" in liberal governance of drug and alcohol consumption', *Sociology*, 38(1): 25–42.
Plant, M. and M. Plant (2006) *Binge Britain: Alcohol and the National Response* (Oxford: Oxford University Press).
Valverde, M. (1998) *Diseases of the Will: Alcohol and the Dilemmas of Freedom* (Cambridge: Cambridge University Press).
Warner, J. K. (2002) *Craze: Gin and Debauchery in an Age of Reason* (New York: Four Walls Eight Windows).
http://.bbc.co.uk/1/health, date accessed 17 March 2008.

3

Outdoor Recreation and the Environment

Neil Ravenscroft and Paul Gilchrist

***Chapter 3** is about the politics of the key leisure activity of **walking**. It details the legislation governing **access to the countryside**, known in the history and politics of leisure as **the right to roam**. It assesses the political power of the owners of the land over which comparatively landless members of the public seek to walk and judges to what extent recent Acts of Parliament which promised greater access to rural Britain have actually done so.*

Introduction

In the context of outdoor recreation and the environment, the 'forbidden fruit' has long been equality of access to all rural environments: landscapes have been there for the public to see (from a distance), to read about, and to be preserved, but (largely) not to be touched, far less used for anything as ephemeral as recreation and leisure. While leisure in capitalist Britain may have brought limited rewards for the 'good citizen' (Ravenscroft, 1993), there was never – certainly when *The Devil Makes Work* was written – a question of 'unforbidding' the fruits of rural property for the good of ordinary people (Shoard, 1987; Stephenson, 1989; Ravenscroft, 1996, 1998a; Parker and Ravenscroft, 1999, 2001). Indeed, the rhetoric of the day was largely that rural property required a level of 'stewardship' that made recreational access and use inappropriate in all but the most robust locations (Ravenscroft, 1995). This was widely contrasted with the position elsewhere – especially 'Europe' – where, it was claimed, people could exercise 'citizen rights' of access over private land (see, in particular, Shoard, 1987). However, as Curry (2002) noted in his work on recreational access in New Zealand, intercountry comparisons are notoriously hard to make, even when the countries share similar legal foundations.

Despite the exclusive claims for stewardship, the period between 1997 (when Labour came to power) and 2000 (the enactment of the Countryside and Rights of Way Act 2000) witnessed the elitist superstructure of rural exclusivity seemingly being torn down in favour of a legal 'right to roam' on the uplands, moors, commons and downs of England and Wales. The Land Reform (Scotland) Act of 2003 swept in more fundamental and wide-ranging rights in Scotland. The upshot of these two Acts, we were told, was that all people would now enjoy the long-denied, unfettered recreational access to the countryside (DETR, 2000). Thus, it would seem, this chapter could very much be about the now 'unforbidden fruit' of postindustrial capitalism, with a new emphasis on environmentalism, health and economic growth replacing tired Marxist debates about rights and duties (see, for example, Land Use Consultants, 2008; Parker, 2007, 2008; Sempik, 2008). Yet, as we argue in this chapter, the construct of the forbidden fruit was, and remains, rather more complex than was initially understood, meaning that – while some fruit may have been rendered 'unforbidden' – the delineation between forbidden and allowed/deviant and good remains as vital as it was when *The Devil Makes Work* was published.

The campaign for access to the 'forbidden' lands

Most histories of the campaign for access to the countryside of England and Wales commence with the original enclosure of land (see Shoard, 1987, for example). As Clarke and Critcher's (1985) reading suggests, such enclosure – and the resulting diminution of opportunities for outdoor recreation – has been constructed as a function of the structural dominance of capitalism. This is particularly in the ways in which it 'captures' space, appropriates it and invests in it, but also restricts unwanted and unregulated public movement over it (Darby, 2000). This interpretation, informed by Locke's labour theory of private property (1690/1988), gives credence to landowning elites and other rural business interests in determining the uses of land and ensuring the interests of the rural economy (however narrowly defined) are not diminished (Haddad, 2003; Woods, 2005). Restrictions to public access for recreational activities have been justified on the basis of the trickle-down economic benefits that their management brings to the economy. The public are called upon to observe property boundaries and periods of exclusion for the conservation and management of land or risk seeing economic impacts and local infrastructure decline (Hillyard, 2007; Wightman, et al. 2003; Land Use Consultants, 2008).

The success of this regulatory arrangement is often dependent upon a sustained sense of an organic community. In the context of the countryside depictions of the organic community as part of a 'rural idyll', a space of harmony, social cohesion and continuity (see Halfacree, 1995; Lowenthal, 1991; Little and Austin, 1996) have been used to justify the exclusion of 'outsiders', be they urban working class, youth subcultures, or ethnic minorities (Chakraborti and Garland, 2004; Cloke and Little, 1997; Garland and Chakraborti, 2006; Sibley, 1997), or to warrant resistance to social and political change (Wallwork and Dixon, 2004). It is a vision of a community that is captured by Tonnies' (1957) concept of *gemeinschaft* – a small, stable society sustained through close interactions which are based on kinship ties, local proximity and traditional values.

As studies into transgressive leisure and the carnivalesque have noted, the relatively unchanging form and function of local community relationships is still central to understanding forms of countryside recreation (Presdee, 2000; Ravenscroft and Matteucci, 2003; Gilchrist and Ravenscroft, 2008; Ravenscroft and Gilchrist, 2009). Activities like night time (or boy) racing, raves and dogging, the temporary licensing of 'deviant' leisure, do little to challenge local systems of territorial governance (Ravenscroft, 1995; Ravenscroft and Matteucci, 2003). Further work has also shown how landowners have exerted subtle forms of discipline and governance over unwanted recreational use through extended notions of social responsibility (e.g. through codes of conduct, for example, see Parker, 2007); granting public access on terms of their own choosing which limit wider moral rights of freedom to roam (Parker and Ravenscroft, 2001; Parker, 2006; Ravenscroft, 1995, 1998a).

Nevertheless, as Newby et al. (1978) noted, rural patronage and the 'benevolent stewardship' of the landowner disguises antagonistic social relations and conflicts of interest between the dominant and subordinate (see Gilchrist, 2007). It would be wrong to suggest these relations are immoveable. In Britain, the history of outdoor recreation reveals processes of contest and conflict that have challenged the rights of landowners to constrain access to rural space, subsequently shaping the nature of public legislation (Darby, 2000). Strict adherence to legal doctrine rather than the exercise of occasional benevolence led to mass trespasses of moorlands in northern England in the 1930s, provoking a change in the law through the National Parks and Access to the Countryside Act 1949 which established some access to the countryside for recreational purposes, although not quite the rhetorical and forward-looking 'right to roam' claimed by the architects of the Act (see Hill, 1980).

Half a century on and organisations such as the Ramblers' Association realised that sustained campaigns still needed to be waged to guard against encroachment by landowner interests. In pressuring for legislative change in the access situation through the 'right to roam' promised but not delivered in 1949, the Ramblers escalated their campaigns in the 1990s by conducting mass trespasses of privatised wilderness and recruiting a voluntary corps of members to keep footpaths clear and report signage that discourages walking (Anderson, 2007a: 404; Shoard, 1989). In so doing, the Ramblers' Association recognised that small numbers of agitators had no more than a marginal effect and could not secure the political pressure required for reform. Instead, they required a closer identification with sought-after access land and routes, and so they now encouraged forms of settlement through local volunteers constantly passing and repassing over ancient footpaths and rights of way, thus challenging the legal and moral constraints imposed by landownership. This was part of the tactical innovations witnessed in environment-based protest movements in the 1990s designed to disrupt landowner hegemony through occupation (Doherty et al., 2000).

The pressure generated by such tactics led Labour to pass a new piece of legislation that, rhetorically at least, expunged the failure of the 1949 Act by extending rights to enjoy the countryside for recreational purposes (DETR, 2000). Through creating a statutory right of access on foot to 'open country', the Countryside and Rights of Way (CROW) Act 2000 gave 'greater freedom for people to explore open countryside' (DETR, 2000: 1), opening up all private land classified as 'mountain, moor, heath or down', for open-air recreation with no compensation for the landowner. When the Land Reform (Scotland) Act 2003 followed suit some three years later, it seemed very much that the 'forbidden fruit' of access to the countryside in the UK was no more and that, in its place, there was a new 'postcapitalist' equality of access to, and use of, the countryside. This was supported by new policy initiatives, such as the Countryside Agency (2005) diversity review, which was aimed at encouraging more people to visit the countryside for recreation, and by the Marine and Coastal Access Act 2009 which will create a new legal right of access on foot to the coast of England (see Defra, 2009).

Forbidding the unforbidden: The maintenance of exclusivity

As Parker and Ravenscroft (2001) highlighted, soon after the enactment of CROW, the rhetoric of unfettered access, of making available what was once forbidden, has remained highly problematic in a number of respects. The principal issue is that the delineation of 'access land' relies on a fixed spatial settlement, through the use of maps, to explain and disseminate where access can be enjoyed. Not only does this continue to deny access to the majority of the land for the population, since they cannot read maps with the expertise required to use – with confidence – 'access land', but it also does not specify adequately the forms of recreation that can be enjoyed, nor the nature of the 'exploration' that people can undertake. In addition, the Act retains several discretionary rights to exclude the public in the interests of safety and to protect economic (predominantly agricultural) activity (see Parker, 2008). This is a significant issue, and one that separates CROW from other apparently similar forms of access in other countries. Unlike access rights in much of Scandinavia, for example, where there is an assumption that all land is open to public access unless there is good reason to think otherwise (see Mortazavi, 1997), CROW rests on the opposite assumption: that land remains exclusive unless mapped to the contrary. The difference in the gestures of the access codes is unmistakable; in terms of CROW it is very much about keeping out of property unless it is absolutely clear that it is access land. For those who lack confidence in making such judgements (most of us), this construct underlines not only the continuing hegemony of private property rights, but also the subject/other dichotomy between those who 'belong' and those (at whom the Act was rhetorically aimed) who do not.

It is hard, therefore, to claim that the CROW Act signals the postcapitalist reversal of the enclosure of the commons and the (re)introduction of a universal right to roam. Rather, and in common with the 1949 Act, it retains hegemonic interests through a balance of public and private rights to enjoyment of rural space (Parker and Ravenscroft, 2001; Ravenscroft, 1998b). Such hegemony has been exercised through subtle forms of control that frame who can and cannot enter rural space. This in effect marks a shift from bureaucratic determinations of access through rights, to forms of moral governance which designate permitted forms of 'countryside citizenship', allied to moral guidance in the form of the Country Code (Parker, 2007).

Yet this fails to capture both the subtlety of the construct of 'forbidden' and the delicacy with which the CROW Act has helped sculpt a new exclusionary landscape. Other than the Rambler's Association claims about the lack of access to the countryside, there was very little evidence, prior to the Act, that more access was actually required, certainly to open countryside in the sparsely populated uplands of England and Wales (in this respect also being redolent of the 1949 Act). For example, the House of Commons Environment Committee (1995) concluded that there was no robust evidence that the demand for outdoor recreation had increased beyond the capacity of existing access opportunities. Similarly, the Country Landowners' Association claimed that its members already provided access to much of the land to be covered by the Act, and that the 'quasi-market' in which much of this access was granted was better suited to determining access requirements than a statutory and bureaucratic process (see Curry, 1998). Moreover, there has actually been a decline in the number of people visiting the countryside in the years since the new access land became available, although Natural England, the government body set up in 2006 to conserve and enhance the country's natural environment, is unwilling to speculate why this should be (see Countryside Agency, 2004; Natural England, 2006, 2008).

The point here is that, quite simply, the deeper rural landscapes of England and Wales never were forbidden fruit for all people, despite claims to the contrary. Rather, access was highly differentiated, between those with the cultural capital to use open land in a way that did not contravene local (owner) sensitivities, and those who were constructed as 'other' and were thus made unwelcome (see Ravenscroft, 1998a). Prior to the CROW Act, the former group were able to gain access to specific areas of land over which there were no formal rights; they simply chose spaces that were unknown to the mass, or were hard to access, and which did not have any major agricultural or commercial uses. In contrast, those without sufficient cultural capital to adopt such a strategy were 'denied' the access that they sought (although, as Curry and Ravenscroft, 2001, have suggested, there was never any evidence that there was demand from this section of the population, let alone 'denial').

Of course, it suited all those who wished to maintain the exclusivity of the land (as opposed to the forbidden fruit of access to the land) to construct the problem as one of legal rights of access, such that legislation could be enacted that would, in effect, make unforbidden a class of access (open, unmarked and unguided) that none of those at whom it was rhetorically aimed wanted anyway. Thus, the CROW Act came about, heralded as a major breakthrough in securing citizen rights to formerly forbidden lands, but in actuality offering very little that was not already available, *de facto*, to those who wanted it – hence the failure to generate increased access to the countryside since 2005, when the new access maps became available. But the really clever manoeuvre was in constructing the CROW Act as a major concession to the Ramblers' Association and the citizens of England and Wales, won at great cost to the thousands of land owners and farmers who had given up some of their rights, privacy and ability to make a living on already marginal land, as set out by the President of the Country Landowners' Association, Mark Hudson, on its web home page (accessed 25th September 2008):

> 'The challenge is to ensure that those exercising their access rights are responsible too. We urge people to remember that this is not a general right to roam: it is access to mapped areas of land which may be temporarily closed for safety and land management reasons. Visitors can help care for the countryside by respecting the Countryside Code and by supporting local shops and businesses. But the most important thing that people can do is to check the Countryside Access website, check the new maps and check the local information points to see where they can go.' (http://www.cla.org.uk/Policy_Work/Access_-_CRoW/Access_to_the_Countryside//25.htm/)

As Mr Hudson makes clear, while the Government and the Ramblers' Association may construct the CROW Act as the harbinger of a new era of freedom to roam, and the removal of the forbidden fruit of access, the landowners see it in an entirely different light: as a responsibility to exercise access rights, where they are allowed, within the codes set down and within a framework of service to the local economy. How many people who have never experienced open access are really going to feel that they have a right to experience it now? Rather, the warning – backed by the Act itself – is stark: you dare to taste the forbidden fruit and we – those who belong here – will ensure that you abide by every element of the Act and the codes of conduct that it endorses. Thus, some of the fruit of access may have been unlocked by the CROW Act, but they are not the ripest or sweetest fruit; indeed, they are a bitter reminder that the exclusion and demarcation so eloquently observed in *The Devil Makes Work* are still alive and well in the English and Welsh countryside.

The new forbidden fruits

As if the central deception of the CROW Act were not enough, it is now emerging that, in legalising a small class of access that was already happening, the architects of the Act and subsequent legislation have demonised other forms of recreational use of the countryside that have at least an equal claim to legal protection. In particular: paddlers (a generic term for canoeists and kayakers) have found that their claims for access to all inland waters have been marginalised by the 'demands' of walkers; climbers have found that some of their favourite climbs are now 'land-locked' because there is no right of way or access land to enable them to reach the climb (and they are implored by the landowners to adhere to the CROW Act); and off-road drivers and trail riders are increasingly blamed for the poor state of the routes that they share with walkers (and local authorities are using highways legislation to close these routes because of their poor state). Furthermore, the Natural Environment and Rural Communities (NERC) Act of 2006, which created Natural England, has also ushered in legislation that has effectively closed, without appeal, many of the remaining non-metalled routes used by off-road vehicles and motorbikes.

Paddlers and forbidden water

If the distrust between landowners and walkers had run deep prior to the enactment of CROW, it was nothing to the enmity that was often experienced between paddlers and anglers – with many of the angling clubs owning property interests in the waters that they were fishing. While there have undoubtedly been many paddlers who have carried out their sport away from the eyes of the anglers and landowners, there have been numerous high-profile incidents in which there has been outright conflict between paddlers and anglers. Actions on the Rivers Dee and Seiont in North Wales, for example, are highlighted in much paddler folklore (see Gilchrist and Ravenscroft, 2008), and there was clear pressure prior to CROW to ensure that no new rights of navigation were created. This pressure remains, with the Country Landowners' Association continuing to press for a voluntary approach to canoe access (http://www.cla.org.uk/Policy_Work/River_Access_for_Canoeists/, accessed 25 September 2008) and the Angling Trust – the new representative body for all forms of angling – seeking to restrict the number of rivers on which voluntary agreements are appropriate (Angling Trust, 2009).

In an attempt to address the issue of how to treat water-related activities within the proposed CROW Act, the Government asked the Countryside Agency, the Environment Agency, the Forestry Commission and English Nature – prior to the publication of the CROW Bill – to consider whether, amongst other things, access to inland waters and watersides should be included with the CROW legislation (Roberts and Johnson, 1999). They concluded (Countryside Agency et al., 2000) that the evidence available about the legal position with regard to access to inland water was both complex and incomplete, and thus inappropriate for inclusion in CROW. As a result, the CROW legislation did not extend to the use of inland waters for water-based recreation (although it allowed access on foot to waters that were within access areas, for fording purposes). Interestingly, the Rivers Access Campaign, which is funded by Canoe England, claims that provisions extending the public right of access to inland waters were 'pulled at the eleventh hour' from the CROW Bill, prior to enactment (http://riveraccess.org, accessed 25th September 2008).

The 1999 report also concluded that there was no overwhelming case for including woodland in the CROW Act, despite the fact that a lot of woodland is already publicly owned – by the Forestry Commission – and the legal issues are no different to those relating to open land. Rather, the Government sought to include a provision (section 16) within the CROW Act that would allow owners (presumably the Forestry Commission) to 'dedicate' their woods in perpetuity as access land, if they so wished. Section 16 dedications were also made available to the owners of rivers and lakes, together with other provisions that allowed owners to relax the restrictions that limited CROW to access on foot. These provisions could thus allow owners to dedicate land and water that would otherwise not be access land, and to widen the use of that land or water to include, for example, horse riding, cycling and, in the case of water, canoeing, rowing or other water-based sports. Perhaps not surprisingly, there was no rush of private owners seeking to use the powers of dedication, particularly for water.

In the aftermath of the CROW legislation, therefore, paddling (and other uses of inland waters) had effectively been dropped from the political agenda in England, with Government making it clear that no new legislation would be forthcoming. In responding to a petition to the Prime Minister calling for legislation on a free right to navigate all rivers and canals in England and Wales, the Government responded:

> ... demand for access would more effectively be met by a targeted approach, which involves identifying where access is needed, and then creating access agreements with the landowner and other interested parties. Creating access via agreements will undoubtedly require goodwill and hard work on all sides and nothing will be achieved overnight. But we firmly believe that this is the right approach. Given the commitment of all interested parties, particularly water users and landowners, this managed and targeted approach should, over time, result in a significant increase in the amount of inland water accessible to all water users (http://www.number10.gov.uk/Page14435, accessed 2 October 2008).

Interestingly, the Secretary of State for the Environment (2009) subsequently wrote to the Chairs of all Local Access Forums, pointing out that their remit for improving access to the countryside did not extend to water-based recreation. Thus, while championing negotiated access improvements on the one hand, the Government has simultaneously denied the potential parties to those agreements the strategic overview and advice that is available to similar parties considering public access to land.

Whereas prior to the CROW Act the British Canoe Union had been able to claim political parity with the Ramblers' Association, at least insofar as both were seeking more access to resources over which they felt that they had a reasonable claim, after the Act it was isolated. And isolation meant moving out of the policy arena, into confrontation with Government, and – if access was to be achieved – into direct negotiation with landowners over specific voluntary, contractual agreements for access to water – just as they had done prior to the Act – with recognition that dedication, although an option made available to landowners, was not one that was going to be taken up (Church et al., 2007). Indeed, the only known dedications of water for recreation have been on the River Mersey, where a number of golf clubs dedicated their waters as a contribution to a wider scheme to negotiate a long distance canoe route, from Stockport to the river's confluence with the Manchester Ship Canal, at Carrington (see University of Brighton, 2005; Environment Agency, 2006). The canoe lobby has now been divided internally by the decision of the Petitions Committee of the National Assembly for Wales (2009) to consider a case made by the Welsh Canoeing Association for creating a public right of access along rivers in Wales, in the process further isolating paddlers in England and their association, Canoe England.

There is no public evidence to suggest that the decision not to include a public right of navigation within CROW was related to actions or 'deals' made by other parties. However, it is interesting to observe that, at the moment that the – largely irrelevant – withdrawal of the 'forbidden fruit' of public access on foot was overturned, the forbidden fruit of access to inland waters (which would have made a significant difference to a great many paddlers, wild swimmers and many others) was ever more tightly defined. Not only is this evidence of the asymmetry of power relations between landowners and paddlers described by Church et al. (2007), but it is also an indicator that the postindustrial 'leisure landscape' has yet to permeate property relations such as those involving the control of the recreational use of inland water.

Climbing and forbidden hills

Alongside the Ramblers' Association, the British Mountaineering Council (BMC) has been a long-time advocate of 'freedom to roam'. However, given the much more specific access needs of climbers (access on foot to the base of crags that are suitable for climbing), the BMC has also been active in negotiating access agreements that allow climbers to reach the crags from public rights of way. It has also been active in developing management agreements that address the environmental impacts of climbing, particularly with respect to nesting birds (Anderson, 2007a, 2007b). Thus, when Labour came to power in 1997, the BMC had to make a choice about whether or not to join the negotiations over CROW. It did so, and claims to have won important concessions on night-time access and occupier liability (CROW – the BMC viewpoint, http://www.thebmc.co.uk/Pages.aspx?page=112, accessed 25 September 2008). In supporting the introduction of CROW, the BMC Access & Conservation Committee recognised that there was an element of 'emptiness' about it, because climbers had traditionally had 'free access' to crags (in many cases negotiated by BMC), meaning that the Act added very little new access. However, it was felt that legal backing would improve the situation, while also clarifying other issues such as landowner liabilities for accidents and injuries incurred by climbers.

In the event, it appears that the enthusiasm that the BMC Access & Conservation Committee displayed for CROW may well have been misplaced. This is because, in replacing person-to-person negotiation with legislated access, some landowners have felt able to take advantage of the provisions in the CROW Act for closures to access land, as described by Rick Gibbon, the BMC's Peak Park Access Officer:

> Some aspects of CRoW, such as land dedication, have never approached their full potential for benefiting both the outdoor community and landowners alike. And other aspects need constant attention to make sure that our hard-won rights don't slip away from us. There are disturbing reports of agreed access points being stopped up, open access signs masked by Beware of the Dog notices, and fences popping up on open moorland, replete with barbed wire. All these barriers aim to deny by stealth the access we have won by statute. No doubt each one will have been justified by 'land management', but surely there should be some distinction between that and an apparent right to erect feudal barricades. Cumulatively their impact is draconian ... Article prepared by Henry Folkard (http://www.thebmc.co.uk/Feature.aspx?id=2591, accessed 25th September 2008)

A specific case in point is Vixen Tor, in the Dartmoor National Park, but not in an area designated as access land under CROW. Access to and use of the Tor – which is in private ownership – by walkers and climbers has been allowed, without charge, for the last 30 years. However, a new owner has seen fit to prevent access, and the combined weight of the BMC, Ramblers' Association and the National Park Authority cannot, at present (September 2008) find a solution. And this is despite the Tor being in a National Park and despite all the assurances that came with the CROW Act, about improving access to important sites such as the Tor. This is an example of the issue that concerns the BMC, that land outside CROW, to which climbers have traditionally had free access, will be closed to them, on the grounds that the public now have ample land and rights under CROW and should not expect to continue voluntary arrangements outside of CROW. In addition, the BMC claims that some hostility towards climbers is now experienced on CROW land, with landowners seeking to find reasons for closing CROW land, whether on nature conservation or other grounds (http://www.thebmc.co.uk/Pages.aspx?page=121, accessed 26 September 2008).

The case of climbers is, arguably, of more concern than that of paddlers, because it seems that, unlike paddlers, climbers are losing access that they had established legally – mainly through contract – and that the provisions of CROW are unsuitable for helping them. While evidence is scarce – there are relatively few climbing sites in England and Wales, compared to the areas of access land under CROW – it does seem quite possible that, in accepting the implementation of CROW, some landowners determined that they would no longer support voluntary access arrangements on land that is not designated as access land. In addition to crags, of course, this also includes many areas of land close to urban populations.

Thus, rather than use the dedication provisions included in the Act to create specific access areas according to demand, it would seem that the opposite is happening: that the Act itself has, in effect, created a new category of forbidden fruit: often small areas of land, or routes to crags and monuments, that have traditionally been available free of charge, but which are now closed, in 'recognition' of the much greater (and presumably more valuable) access rights that the public has gained through CROW. The fact that most of the public have no desire – nor ability – to use most CROW land is irrelevant; the principle is that the forbidden fruit of access on foot to open country has been made available to consume, in part return for which people should not expect to consume fruit from areas that are not designated for access, even if these are more highly valued by some members of society.

The position in which the climbers find themselves, therefore, is very much a part of an elaborate 'gift relationship' between landowners and government. As Mauss (1980) has suggested, non-mercantile barter is often based on reciprocation: a gift given for a gift received, with the giver under a moral/customary obligation to give more than they received. In this case, as the earlier quote from the President of the Country Landowners' Association suggests, the 'gift' of recreational access has been given by landowners to the state (let us here gloss over the legislative aspect, because it seems apparent that there was at least an element of voluntarism on the part of the landowners when negotiating the detail of the CROW Act). In return, not only has the state had to accept a limited and bureaucratic approach to public access, but it has also had to accept the delegitimation of access arrangements that have been negotiated by those wanting access and that, prior to CROW, were considered to be the core of access policy (see Feist, 1978, for example). This is very much a case of exchanging the forbidden fruit of access that few people want for previously unforbidden fruit that individuals and groups had actually arranged to suit their requirements. As Mauss (1980) candidly observes, the gift relationship suits only the party with the deepest pockets, with the other party (or parties) standing to lose all.

Off-road drivers and forbidden lanes

It was never the intention that the CROW Act should create immutable public rights of access to private land for 4x4s and trail bikes. Equally, it was an apparent cornerstone of CROW that no current activity, commercial or not, would necessarily be displaced by the Act (Defra, 2001). That did seem to be the initial stance, with off-roader groups working with local and National Park authorities to find acceptable ways of managing vehicular access. This included, for example, continued co-operation between off-roaders and park authorities in the Lake District, through the Lake District Hierarchy of Trail Routes initiative (LARA and LDNPA, 1997), as well as elsewhere. Within the Lake District it seems clear that there was an understanding that off-roaders had a legitimate right to use some of the unmetalled tracks in the park, but that there needed to be management over which ones, and when. The first report on the management system (LARA and LDNPA, 1997) confirms this approach.

Yet, just six years after CROW was enacted, off-road drivers and riders found that changes to the law regarding the use of certain types of highway (within the Natural Environment and Rural Communities (NERC) Act 2006), particularly the reclassification of all Roads Used as Public Paths (RUPPs) to Restricted Byways, had reduced significantly the number of unmetalled (i.e. unsurfaced) routes available to them. The Green Lane Association (2006) has claimed, in this respect, that 62% of unmetalled routes available to off-roaders are no longer available, and that there is pressure to close many of the remainder. What has compounded the inequity of this, in the eyes of off-roaders, is that all RUPPs should already have been reclassified by local highways authorities (and many of those that have been reclassified have retained rights of vehicular use), meaning that those authorities which had failed to carry out their duties were suddenly relieved of them, and not in favour of off-roaders. In addition to the reclassification of routes under the NERC Act, many local authorities have also imposed – or are in the process of imposing – Traffic Regulation Orders (TROs) to further restrict off-road use due to issues such as poor maintenance of the routes.

The result of this 'double whammy' is that off-road drivers and riders, having been assured throughout the passage of the CROW Act that their interests would be safeguarded (Defra, 2001), have now been demonised to the extent that the majority of the routes formerly open to them are no longer legally available. This is particularly acute in the Lake District, where off-roaders feel that they have worked hard to cooperate with the authorities, only to find themselves isolated by interest groups such as the Friends of the Lake District

(FLD) and the Ramblers' Association. In a programme for the BBC, Brian Jones of the FLD claimed that a single motorbike can 'destroy' the 'luxury' of a large number of people, and thus should be banned (http://www.bbc.co.uk/go/insideout/northwest/series4.html, accessed 26 September 2008). While accepting that motorised access can cause erosion, the argument is made by the off-roaders that walking similarly causes erosion (Helvellyn, in the Lake District, is widely cited in this respect), but this does not lead to a ban on access. For the Green Lane Association, the action taken against off-roaders is part of a broader attempt to define who can and cannot enjoy access land:

> Since the NERC bill was introduced in 2006 many of the lanes that were open to us are now closed. Slowly but surely the National Parks are closing some of the nicest lanes we have left. The powerful Ramblers Association lobby will not stop until all lanes are closed (what ever happened to live and let live?). We need to stand united with one voice, one forum, and one place to discuss our hobby and what we can do to stop the rot (http://glag.co.uk/information–downloads/green-lanes-access-group.html, accessed 26 September 2008).

For others, such as those with disabilities who rely on motor vehicles to get into the countryside, the action against off-road vehicles is another sign of the reductionist exclusivity embraced by CROW and it successor legislation and statutory instruments (see Manuel Bernardez in a response to the BBC Inside Out programme on the Lake District, http://www.bbc.co.uk/go/insideout/northwest/series4.html, accessed 26 September 2008).

In many ways, the actions of central and local government, with respect to off-roaders under NERC and the highways legislation, has been some of the most blatant repositioning of good and deviant access that has been witnessed in England and Wales. While the paddlers may complain that they have been frozen out of the policy community, they were never really in it. In contrast, the off-roaders were, and they seem to have been quite unceremoniously dumped out of it, and for no apparent reason. The idea that such a ban will prevent off-road access – or the equivalent for paddling – is naïve; those who wish to continue will do so, safe in the knowledge that the chances of getting caught are relatively low. Rather, in both cases it is the loyal citizen – the person who abides by the rules – who is demonised. And demonisation of off-roaders is a clear signal that, just as the once forbidden fruit of access on foot is made less forbidden, activities that were previously accepted have assumed the mantle of forbidden fruit.

Discussion and conclusions

As we have argued, the rallying cry of 'freedom to roam' as the cornerstone of a new social contract covering leisure uses of the countryside is not in the least new. The National Parks and Access to the Countryside Act 1949 was duplicitously labelled thus (see Hill, 1980) in an attempt to convince ordinary people that the private uplands of England were no longer so legally inhospitable. Forty years later, the Countryside and Rights of Way Act 2000 was heralded in a similar fashion, this time in recognition of the new democratising forces at the heart of 'New Labour'. But, while both Acts were passed by idealistic Labour regimes and both shared a similar strap-line, neither delivered – by design – what had rhetorically been claimed. The 1949 Act was certainly more blatant in the disparity between the claims made for it and the actuality, in that not even walkers achieved unfettered access. Yet, the more breathtaking sleight belongs to the CROW Act, with its central purpose apparently being to inscribe on rural land a new relationship between people and property in which even basic 'rights' have been reconstructed as responsibilities that few can aspire to fulfil.

Central to this deception has been the ownership and control of rights in property. Using revisionist homilies about guardianship and custodianship, rights owners have been able – successfully – to argue that none other than they are capable of managing a fragile resource of such importance to the state, even if there are public access 'rights' over it. And, of course, such homilies play well in postwar Britain, where everyone either is, or aspires to be, an owner of property rights. Thus, linking the rhetorical unforbidding of access to highly prescribed responsibilities that are really only suited to those who appreciate what it is to own property, completes a circle of deceit in which the public is all too ready to believe that they have been rewarded for their good citizenship. The fact that few of them actively want what they have been given is of no matter (neither, it seems, is the fact that they have paid massively – through taxes – for what they have been 'given'); it is the sign that is important: the giving of access clearly means that landowners recognise the gains that they – the public – have made in achieving equal, property-owning, status. And the fact that many fellow citizens (paddlers, climbers, drivers, riders...) have lost out in this rather sordid deal seems equally to be of little concern.

Clarke and Critcher concluded *The Devil Makes Work* with an intellectual and political agenda. Of the former, an analysis of leisure, they argued, should be historically attentive and driven by connecting individual experience to public issues with a critical eye cast toward the complex relationship between the individual and social structure. Further, the analysis needs to be attentive to the ways in which capital disguises (class-based) forms of disadvantage, but subtle enough to capture individual and collective meanings attributed to leisure as an aspect of cultural power. We have shown these aspects within the consumption of outdoor space as a continued historic working of this dynamic. The outdoors may figure as a cornerstone of the national psyche, reaching new markets through the diversification of leisure pursuits and tastes for the extreme, but the essential and emancipatory promise of unfettered outdoor recreation remains utopian; caught within still-functioning power asymmetries that have done further damage to the interests of the weak and marginalised. In pursuit of leisure experience, especially where market solutions cannot be found, the tactics for consumption of the outdoors appear more feudal than postcapitalist: evade the surveillance of the dominant, take by stealth, or seek enjoyment from beneficent landowners when the moral curtain is lifted (Ravenscroft and Gilchrist, 2009). Not quite the festive, participatory and solidaristic leisure Clarke and Critcher longed for.

Where a political agenda is concerned, there is still a need for theorisations of leisure and critical policy interventions that engage with the classic concerns of socialism (the nexus of state, market and civil society in leisure provision; economic injustice; property relations; social disadvantage; and, the nature of work and family life), even if it remains more uncertain as to how such concerns can be delivered. This may not be a socialism of sophisticated (perhaps tangential) theoretical abstraction, or with a powerful labour movement to pursue its agenda, but one that perhaps has succeeded in providing a series of intellectual and critical coordinates for an empirically-based cultural politics rooted in the realities of contemporary leisure. As this chapter has shown, and in the spirit of Clarke and Critcher, we must continue to be mindful of new legislation that on the surface extends consumer choice, promises freedom and suggest social progress. If countryside leisure in capitalist Britain was about exclusion from enjoying the fruits of access, leisure in postcapitalist Britain is precious little different, even if the rhetoric of CROW suggests otherwise.

References

Anderson, J. L. (2007a) 'Britain's right to roam: Redefining the landowner's bundle of sticks', *Georgetown Journal of International Environmental Law*, 19(3): 375–435.

Anderson, J. L. (2007b) 'Countryside access and environmental protection: An American's view of Britain's right to roam', *Environmental Law Review*, 9(1): 241–59.

Angling Trust (2009) *A Statement on Inland Navigation*. www.anglingtrust.net (accessed 27 July 2009).

Chakraborti, N. and J. Garland (eds) (2004) *Rural Racism* (Cullompton: Willan).

Church, A. P. Gilchrist and N. Ravenscroft (2007) 'Negotiating recreational access under asymmetrical power relations: The case of inland waterways in England', *Society and Natural Resources*, 20(3): 213–27.

Clarke, J. and C. Critcher (1985) *The Devil Makes Work: Leisure in Capitalist Britain* (London: Macmillan).

Cloke, P. J. and J. Little (eds) (1997) *Contested Countryside Cultures* (London: Routledge).

Countryside Agency (2004) *Great Britain Day Visits Survey 2002/2003* (Cheltenham: Countryside Agency).

Countryside Agency (2005) *The Countryside Agency Diversity Review and Recommended Outline Action Plan* (London: Defra).

Countryside Agency, Forestry Commission, Environment Agency & English Nature (2000) *Improving Access to Woods, Watersides and the Coast. A Joint Report to Government on the Options for Change*. Publication CA 33 (Cheltenham: Countryside Agency).
Curry, N. (1998) 'Permitted access in England and Wales, in Country Landowners' Association', *Access to the countryside*, Volume 2. Submission to the consultation on Access to Open Country. London: Country Landowners' Association.
Curry, N. (2002) 'Access rights for outdoor recreation in New Zealand: Some lessons for open country in England and Wales', *Journal of Environmental Management*, 64(4): 423–35.
Curry, N. and N. Ravenscroft (2001) 'Countryside recreation provision in England: Exploring a demand-led approach', *Land Use Policy*, 18(3): 281–91.
Darby, W. J. (2000) *Landscape and Identity: Geographies of Nation and Class in England* (Oxford: Berg).
Defra (Department of Environment, Food and Rural Affairs) (2001) *Consultation on Rights of Way Improvement Plans* (London: Defra).
Defra (2009) *Marine and Coastal Access Bill Policy Document* (London: Defra).
DETR (Department of the Environment, Transport and the Regions) (2000). *Countryside and Rights of Way Act 2000: Explanatory Notes*, Chapter 37 (London: TSO).
Doherty, B., M. Paterson and B. Seel (eds) (2000) *Direct Action in British Environmentalism* (London: Routledge).
Environment Agency (2006) *A Better Place to Play. Our Strategy for Water Related Sport and Recreation 2006–2011* (Bristol: Environment Agency).
Feist, M. J. (1978) *A Study of Management Agreements*. Publication CCP 114. (Cheltenham: Countryside Commission).
Garland, J. and N. Chakraborti (2006) '"Race", space and place: Examining identity and cultures of exclusion in rural England', *Ethnicities*, 6(2): 159–77.
Gilchrist, P. (2007) 'Sport under the shadow of Industry: Paternalism at Alfred Herbert Ltd', in A. Tomlinson and J. Woodham (eds) *Image, Power and Space: Studies in Consumption and Identity*, 3–26 (Aachen/Oxford: Meyer & Meyer).
Gilchrist, P. and N. Ravenscroft (2008) 'Power to the paddlers? The Internet, governance and discipline', *Leisure Studies*, 27(2): 129–48.
Green Lane Association (2006) *Media Resource: The NERC Act*. http://www.glass-uk.org/index.php?option=com_content&task=view&id=391&Itemid=1160, accessed 26 Sepember 2008.
Haddad, B. M. (2003) 'Property rights, ecosystem management, and John Locke's labor theory of ownership', *Ecological Economics*, 46(1): 19–31.
Halfacree, K. H. (1995) 'Talking about rurality: Social representations of the rural as expressed by residents of Six English Parishes', *Journal of Rural Studies*, 11(1): 1–20.
Hill, H. (1980) *Freedom to Roam: The Struggle for Access to Britain's Moors and Mountains* (Ashbourne: Moorland Publishing).
Hillyard, S. (2007) *The Sociology of Rural Life* (Oxford: Berg).
House of Commons Environment Committee (1995) *The Impact of Leisure Activities on the Environment*, HC 246–I (HMSO: London).
Land Use Consultants (2008) *The Environment, Economic Growth and Competitiveness – the Environment as an Economic Driver*. Policy Paper ENV002 (Peterborough: Natural England).
LARA (Land Access and Recreation Association) and LDNPA (Lake District National Park Authority) (1997) *Lake District Hierarchy of Trail Routes – First Report* (Market Drayton: LARA).

Little, J. and P. Austin (1996) 'Women and the rural idyll', *Journal of Rural Studies*, 12(2): 101–11.
Locke, J. (1690/1988) *Two Treatises of Government* (New York: Cambridge University Press).
Lowenthal, D. (1991) *British National Identity and the English Landscape* (Cambridge: Cambridge University Press).
Mortazavi, R. (1997) 'The right of public access in Sweden', *Annals of Tourism Research* 24(3): 609–23.
Mauss, M. (1980) *The Gift. Forms and Functions of Exchange in Archaic Societies* (London: Cohen & West).
Natural England (2006) *England Leisure Visits Survey 2005* (Peterborough: Natural England).
Natural England (2008) *State of the Natural Environment 2008* (Peterborough: Natural England).
Newby, H. C. Bell, D. Rose and P. Saunders (1978) *Property, Paternalism and Power: Class and Control in Rural England* (London: Hutchinson).
Parker, G. (2006) 'The country code and the ordering of countryside citizenship', *Journal of Rural Studies*, 22(1): 1–16.
Parker, G. (2007) 'Leisure citizenship, constraints and moral regulation', *Leisure Studies*, 26(1): 1–22.
Parker, G. (2008) 'The politics of access under New Labour: Nothing to CRoW about', in M. Woods (ed.) *Rural Politics in Britain since 1997* (Bristol: Policy Press).
Parker, G. and N. Ravenscroft (1999) 'Benevolence, nationalism and hegemony: Fifty years of the National Parks and Access to the Countryside Act 1949', *Leisure Studies*, 18(4): 297–313.
Parker, G. and N. Ravenscroft (2001) 'Land, rights and the gift: The Countryside and Rights of Way Act 2000 and the negotiation of citizenship', *Sociologia Ruralis*, 41(4): 381–98.
Petitions Committee of the National Assembly for Wales (2009) *Access Along Inland Water* (Cardiff: National Assembly for Wales).
Presdee, M. (2000) *Cultural Criminology and the Carnival of Crime* (London: Routledge).
Ravenscroft, N. (1993) 'Public leisure provision and the good citizen', *Leisure Studies*, 12: 33–44.
Ravenscroft, N. (1995) 'Recreational access to the countryside of England and Wales: Popular leisure as the legitimation of private property', *Journal of Property Research*, 12: 63–74.
Ravenscroft, N. (1996) 'Access to the countryside of England and Wales: Public/private partnership or the privatisation of public rights?', *Journal of Park and Recreation Administration*, 14(1): 31–44.
Ravenscroft, N. (1998a) 'Rights, citizenship and access to the countryside', *Space & Polity*, 2(1): 33–49.
Ravenscroft, N. (1998b) 'The changing regulation of public leisure provision', *Leisure Studies*, 17(2): 138–54.
Ravenscroft, N. and P. Gilchrist, (2009) 'Spaces of transgression: Governance, discipline and reworking the carnivalesque', *Leisure Studies*, 28(1): 35–49.
Ravenscroft, N. and Z. Matteucci, (2003) 'The festival as carnivalesque: Social governance and control at Pamplona's San Fermin Fiesta', *Tourism, Culture and Communication*, 4(1): 1–15.
Roberts, B. and P. Johnson (1999) *'Access to "Other" Open Country'*, Countryside Agency Board Paper 99/29 (Cheltenham: Countryside Agency). Available at www.country-

side.gov.uk/LAR/archive/board_meetings/boardPapers/CA_AP99_29.asp (accessed 6 October 2008).

Secretary of State for the Environment (2009) *Guidance on Local Access Forums in England, revised March 2009* (London: Defra Recreation and Access Team).

Sempik, J. (2008) 'Building the evidence base for green exercise – Challenges and opportunities', Paper to the Pavilion/Mind/University of Essex conference: *Ecotherapy and the Green Agenda for Mental Health.* ORT House Conference Centre, London, 2 June 2008.

Shoard, M. (1987) *The Land is Ours* (London: Grafton).

Sibley, D. (1997) 'Endangering the sacred: Nomads, youth cultures and the English countryside', in P. J. Cloke and J. Little (eds) *Contested Countryside Cultures*, 218–31 (London: Routledge).

Stephenson, T. (1989) *Forbidden Land. The Struggle for Access to Mountain and Moorland* (Manchester: Manchester University Press).

Tonnies, F. (1957) *Community and Society: Gemeinschaft und Gesellschaft*, translated and edited by C. P. Loomis (Chicago: The Michigan State University Press).

University of Brighton (2005) *Improving Access for Canoeing on Inland Waters: A Study of the Feasibility of Access Agreements*. Final Report to the Environment Agency. Chelsea School and School of Environment, University of Brighton.

Wallwork, J. and J. A. Dixon (2004) 'Foxes, green fields and national identity: The rhetorical location of Britishness', *British Journal of Social Psychology*, 43: 1–19.

Wightman, A., R. Callander and G. Boyd (2003) *Common Land in Scotland: A Brief Overview*. Securing the Commons No. 8 (London: International Institute for Environment and Development).

Woods, M. (2005) *Contesting Rurality: Politics in the British Countryside* (Aldershot: Ashgate).

4

Television, Deregulation and the Reshaping of Leisure

Philip Drake and Richard Haynes

Chapter 4 *is about the politics of* ***television*** *programming – an increasingly popular, almost defining, leisure activity in Western societies over the last 60 years and the subject of often vexed political-cultural debate throughout that time. When television ownership expanded in the 1950s, it was regarded by some as a threat to literacy and civilised values. Many looked to* ***public service broadcasting*** *to safeguard these values. But the deregulation of broadcasting in many countries in the 1980s and 1990s opened television up to the market and, in the process, transformed the business of making television programmes. Many now decry the proliferation of TV gardeners, cooks, property renovators and talent contests. Others praise the greater choice and more truly democratic nature of contemporary television. This chapter assesses the cultural consequences of the* ***deregulation of television****.*

When John Clarke and Chas Critcher published their critical analysis of leisure in the mid-1980s they noted that one of the more significant features of the British population's leisure was watching television. Along with drinking, smoking and sex, television was one of the main focal points of the nation's habitual leisure practices. We would like to revisit the place of television in British society and trace some of the contemporary ways in which television is produced, regulated and consumed at the start of the twenty-first century.

Back to the 1980s

Academic research in the 1980s saw television as structuring society in some common and familiar ways. It shaped national culture (Schlesinger, 1987), developed rhythms of reception (Modleski, 1983) that connected with everyday life and, as a media form, was embedded in the bureaucratic and centralised concept of broadcasting (Williams, 1974). Audiences remained, more or less, nationally monolithic, with the exceptions of specific moments of regional output. In Britain and across Europe, television remained dominated by a restricted choice of channels, many with a long legacy of public service broadcasting. In terms of textuality, television's mode of address was communal and 'cosy', built on institutional and technological structures that were organised to achieve televisual flow (Williams, 1974) so as to keep the viewer watching.

During this period academic research on television began to recognise the importance of the contextual place of television in leisure lifestyles (Morley, 1980). The notion of the 'active audience', the suggestion that viewers interpret television through the prism of their own social and psychological circumstances opened up new ways of understanding the medium. Nevertheless, the communicative flow of television remained strictly one-way, *broad*casting in its wider sense.

Fast forward three decades and television in the twenty-first century is characterised by rather different institutional, technological and televisual structures that produce far more complex relationships between television and the viewer. Although television remains firmly rooted in the visual culture of most societies it has been transformed economically, institutionally, culturally and socially by the Internet and associated digital technologies. This has led some analysts to suggest television is moving 'beyond broadcasting' (Meikle and Young, 2008) as it is characterised by convergence (Jenkins, 2006) as well as by increasing forms of intertextuality and mobility in its programming.

In 1985, when *The Devil Makes Work* was published, television in the UK consisted of only four terrestrial channels. BBC 1 and BBC 2 were the backbone of public service broadcasting, funded by the television licence fee and attracting up to 50% of the national audience. The BBC's main rival was ITV, still organised through regional franchises spread around the United Kingdom, producing programmes for national and local audiences alike. Finally, there was the relatively new broadcaster Channel 4, launched in 1982, a public service broadcaster funded by advertising and operating as a broadcast publisher. Channel 4 helped institute a new economy of production where programmes were made by independent producers rather than in-house by the broadcaster. The introduction of Channel 4 dispersed television production capacity and in many ways represented a new industrial policy for television in the UK, forging a commissioning model now common in the multi-channel landscape. The model proved successful in opening up creative talent in British television production and augured a new era of youth-oriented programming and experimental television formats. It represented the loosening of the hegemony of the BBC/ITV duopoly and offered a glimpse over the horizon towards a new broadcasting ecology.

The deregulatory pressures that led to Channel 4 were partly driven by dramatic technological possibilities in the delivery of television and these would ultimately free up the transmission spectrum. The Cable and Broadcasting Act of 1984 offered evidence of the deregulatory zeal of the then

Conservative government and enabled the introduction of direct satellite broadcasting into British homes. However, the costs of launching satellite and cable services, including the demands on consumers to invest in new equipment, initially severely restricted its development. Nevertheless, the potential for a new era of multi-channel television services was established. This was subsequently capitalised on by the fiercely competitive News Corporation who, as major shareholder of British Sky Broadcasting – formed from the merger of British Satellite Broadcasting and Sky TV in 1989 – transformed the landscape of British television throughout the 1990s and beyond.

It is important to rehearse this brief history of British television in the closing decades of the twentieth century as it is clearly the bridging point from what some commentators have noted as a 'golden era' of British television in the 1980s to one where television is being radically shaped by emergent and increasingly pervasive digital technologies, not least the Internet, and by audience segmentation. Television remains a powerful force in the household and in many people's lives but it is produced, delivered and consumed in some quite radically different ways from previous eras.

This chapter now maps out in more detail examples of where such paradigmatic shifts in television are taking place. First, we examine the interrelated technological, market and policy fields, where the promise of a 'networked' or 'information society' has led to transformations in the regulatory regimes of British broadcasting and has resulted in some specific technological and political problems for the industry, government, regulators and audiences alike. Secondly, we look at recent case studies of how the changing ecology of television in an increasingly deregulated environment has raised some broader questions in relation to quality, standards and audiences. The very nature of television as a medium is being transformed in new and quite unexpected ways, perhaps most clearly embodied in the rise of reality television programming in the 1990s and the popularisation of celebrity-driven television formats.

Digital television: Shifting markets and policies

Digital television was rolled out in the UK between 1998 and 1999 across three competing platforms: digital terrestrial (DTT); digital satellite (DST); and digital cable (DCT). Commercial terrestrial broadcasting plus the satellite and cable television industries in the UK had seen a certain amount of market consolidation through the 1990s and the development of digital television, operated under separate licensing terms with the UK government, meant that new battle lines were drawn in the race to attract new customers.

All three platforms were originally run as subscription-based enterprises. Both BSkyB and the cable operators Cabletel/NTL and Telewest worked hard to translate existing analogue customers into subscribers to their new digital operations. More intriguing was the development of a new digital offering and multi-channel operator OnDigital (later rebranded ITV Digital), jointly owned by ITV commercial broadcasters Granada and Carlton. When the licence for a terrestrial digital platform was first released the ITV companies had sought to go into partnership with BSkyB, who were then leading the way in converting households to the concept of pay-television. Initial plans for a marriage between the satellite broadcaster and Britain's long-standing commercial broadcaster were blocked by the European Commission on competition grounds. Therefore, the ITV companies found themselves in direct competition with BSkyB for both subscribers and television content. A range of logistical and technological problems stalled the development of ITV Digital and when BSkyB began to give its digital set-top-box away for free to new subscribers, ITV Digital were forced to follow their lead. This led to substantial debt eventually totalling more than £1 billion, underwritten by the parent companies Granada and Carlton.

Where ITV Digital might have stolen a march on their rivals was in securing premium content. Through the 1990s BSkyB had trailblazed a formula for success in attracting and holding subscribers. The dual carrots of sport (in particular live football) and movies had gradually enabled the satellite broadcaster to grow its business. Key to their success were exclusive rights deals with the governing bodies of leading sports and Hollywood distributors (of which News Corporation was a key player in the global market, owning Fox Entertainment Group and Twentieth Century Fox). Rupert Murdoch had himself famously been quoted as calling sport the 'battering ram' (Boyle and Haynes, 2009) to a new era of television: this echoed a broader held belief that 'content is king'. In spring 2001 an extraordinary portfolio of rights to televise English football was auctioned. Rights held by the Premier League, the Football Association (including England internationals and the FA Cup) and the Football League were all for sale during the same week. BSkyB picked up live rights for the first two packages and in a fit of pique ITV Digital, not wanting to be left out of the football rights frenzy, paid £315 million for the rights to Football League matches in the second, third and fourth tiers of the League system. Massively inflated, the cost of the rights was ultimately untenable and, amid a downturn in advertising revenue, faulty technology and the loss of subscribers, ITV Digital went into administration in April 2002 leaving a £178 million-sized hole in the Football League's rights fee (Boyle and Haynes, 2004).

The collapse of the DTT licensee had wider ramifications for the UK government's drive to digital more generally. In pulling the plug on ITV Digital Granada and Carlton not only made an embarrassing mess of a massive investment from shareholders but also derailed the government's plans for digital switchover that had in 2002 been planned for 2010 and had now to be moved back to 2012. The government had incentivised the drive to digital, providing Carlton and Granada with a rebate on the tax levy for their main Channel 3

(ITV) licence. This so-called 'digital dividend' was part of a wider government strategy to place the UK ahead of other European nations thereby boosting a new economy of digital channels, programme providers, technological companies and associated services.

The government's 'white knight' of digital terrestrial television came not from the commercial sector but the BBC. Under its then Director General Greg Dyke, whose commercial astuteness had been effectively applied to the public service broadcaster, the BBC remastered the technology of DTT to ensure wider coverage and in a massive marketing campaign launched the new DTT platform Freeview in October 2002. In helping launch the service the BBC gained free access to the platform and this enabled it to introduce a suite of new channels including BBC 3, BBC 4, CBBC and CBeebies.

However, in spite of its success in stabilising the digital terrestrial platform, the BBC's activities in the digital world have also been subject to widespread criticism. As the BBC's Royal Charter came up for review in 2007 there was increasing pressure from commercial competitors for the BBC's interactive and online services to be scaled back for fear of damaging creativity in the open market. In February 2008 the BBC Trust (the internal regulator of the BBC's public remit that replaced the Board of Governors in January 2007) made its first notable decision in the Corporation's digital content when it ended the £150 million digital education project BBC Jam. The scheme, conceived in 2000 but launched in 2006 had already cost £75 million in developing resources for a digital curriculum, but was seen to fall foul of European Commission policies on state aid and anti-competitive behaviour. Similarly plans to extend BBC Local were rejected by the BBC Trust in November 2008 as it was considered that the provision of greater local news might prove damaging to regional and local newspapers.

These rulings also represented the more cautious approach to the BBC's public affairs adopted after the 2003 Hutton Inquiry, in which claims by the BBC correspondent Andrew Gilligan that the Labour Government 'sexed up' an intelligence dossier on weapons of mass destruction in Iraq were strongly rebuffed. The Inquiry criticised editorial control inside the BBC and, despite aspects of Gilligan's report subsequently proving correct, the incident ultimately led to the resignation of senior management including Greg Dyke and the Chairman Gavyn Davies.

Finally, the profusion of digital channels and audio-visual content via the web have raised more profound questions about the role of the BBC. For example, issues such as the potential redistribution of income from the license fee, principally to Channel 4, and discussion of a 'public service publisher' model, where broadcasters other than the BBC might be funded to produce public service broadcasting, have become key to debates over the future of the BBC and public service media more broadly in the UK. The BBC has made several strategic moves to ensure it maintains its relevance in public life across the UK and has instigated plans to regionalise some of its core operations by moving the headquarters for sport and children's television to Salford Quay near Manchester, as well as developing new public initiatives such as BBC Democracy Live, a political webcasting platform designed to give people access to the inner workings of UK and EU political institutions. However there is no doubt that entering the second decade of the twenty-first century the BBC is under threat, attacked both by the commercial media who are struggling with the recessionary downturn in advertising expenditure, – and the political elite. Discovery of a number of voting and quiz show phone-in scandals that hit the television networks had the effect of weakening trust in the BBC and other networks. In 2008 a high profile scandal involving the making of obscene phone messages by celebrity chat show host (and the BBC's highest paid performer) Jonathan Ross and comic Russell Brand, broadcast on BBC Radio 2, its most popular network, led to an investigation and subsequent fine by the regulator, and further questions were then asked about the future course of the BBC.

As the landscape of British television has been transformed so the regulatory environment has also been adapted to suit an era of rapidly advancing digital technologies and varied creative enterprises. The Communications Act (2003) brought into line a range of dispersed statutory organisations concerned with radio, television, advertising and telecommunications under one super-regulator: the Office of Communications (Ofcom). The converged remit of Ofcom mirrors the convergent processes of digital media. The newly formed communications regulator was crafted with a 'soft touch' in mind. In the context of television, as digital channels proliferate so the ability and desire to regulate standards diminishes. This places some complex and contradictory pressures on Ofcom to act as both the conduit and guardian of Britain's digital future. This was reiterated in a Department for Culture, Media and Sport report *Digital Britain* (DCMS, 2009) published in January 2009 that mapped the UK's transition into a 'digital society'. The DCMS voiced noble intentions to invest in wired and wireless infrastructures, to provide near-universal participation in broadband and digital television and quality digital content – but all framed in the market vocabulary of increasing competitiveness in the global 'Knowledge Economy'. For some, such as the BBC's *Newsnight* presenter Jeremy Paxman, the digital age of television has been characterised by 'dumbing down': where entertainment values predominate and there is a general lack of investment in drama, documentaries and investigative journalism that has led to a general malaise of impoverished television content. However others have criticised this 'dumbing down' argument by suggesting it is either elitist or ignores the proliferation of choice in the current media environment (Temple, 2006). This particular discourse on the woes of contemporary television is firmly rooted in a notion of a previous 'golden' age of broadcasting. Like the technology, the economics of television have also been transformed from the heady days of extravagant expenditure on programme budgets in the 1970s and 1980s. As the independent sector has flourished in a multi-channel environment, so production budgets have significantly decreased in real terms. Throughout the 1990s the independent sector railed against the power of commissioning editors at major broadcasters as their ability to leverage income from the value of their programmes remained limited. After significant lobbying by the Producers Alliance for Cinema and Television (PACT), the UK film and television trade organisation, the Communications Act (2003) broke up the value-chain of broadcasting giving broadcasters ownership of the primary rights to television programmes but ceding secondary and tertiary rights (which allow licensing and exploitation across other windows such as global television markets and DVD) to the producer. This has allowed independents to exploit their ownership of rights in new and creative ways in partnership with commercial distributors (Steemers, 2004; Haynes, 2005).

One outcome of the change in the legal ownership of rights has been the consolidation of the independent production sector, with successful small-to-medium sized production companies bought up by 'super-indies' like Endemol and RDF. These large producer brands provide reassurance and confidence for commissioning editors because of their track records. Producers like Endemol, who own the rights to the *Big Brother* franchise (Channel 4, 2000–10), increasingly search for, and stockpile the rights to, winning formats that have international market appeal, with long-running series and potentially high returns from secondary rights agreements. Although some of the formats win respectable audiences for particular broadcasters – *Big Brother* consistently gained above-average audiences for Channel 4 – this has arguably caused a level of conservatism in programme making, leading to a glut of populist reality-TV formats that follow a successful formula.

Formatisation and the global market for television audiences

The global expansion of multi-channel and digital television structures has had some profound effects on the kind of programmes television now produces and also on the experience of viewing. In the multi-channel broadcasting environment there is an urgent need to generate attention, to avoid being lost in the schedules – something that was unnecessary 20 years ago. This has manifested itself in a number of ways. The first, as already noted, is a greater conservatism in commissioning. If a programme format is successful then it is highly likely that it will be reproduced, with slight variations, until it is exhausted. The serial production of reality television shows, quiz shows, property and lifestyle advice programmes, commonplace in British broadcasting by the late 1990s, reminds one of the old adage that nothing sells like a hit. Channel 4, which implemented its minority broadcasting remit in the 1980s with adventurous and often challenging programmes, had by the early 1990s refocused on the youth audience, one notoriously difficult for advertisers to reach. Notwithstanding the continuing *Channel 4 News* (1982–)

and long-running documentary series such as *Dispatches* (1987–) the channel gradually shifted its attention away from niche audiences towards more populist youth programming, with music shows such as *The Tube* (1982–1987) and *The Word* (1990–1995) and, most significantly, the reality television series *Big Brother* (2000–2010). The innovative programming that Channel 4 pioneered in the 1980s and 1990s began to face sharp economic realities in the mid-2000s due to a decline in advertising revenues. A crisis of funding forced a more commercial outlook and with the closure of its Film 4 studio offshoot in 2002 it reduced its previously important role in British film production. In order to regain audiences in a multi-channel market full of niche competitors (such as the digital channels BBC3 and Sky 1) the channel propagated a 'shock doc' approach in broadcasting provocative documentary style dramas: for example, showing the fictional murder in the 'near future' of former US president George W. Bush (*Death of a President,* Channel 4, 2007) and the hanging of 1970s glam rocker and convicted sex offender Gary Glitter (*The Execution of Gary Glitter,* Channel 4, 2009).

One of the main consequences of deregulation in the UK television industry has been that the rights to formats such as *Big Brother* and *The X Factor* are retained by their creators (Endemol and Simon Cowell's Syco TV respectively) rather than by the network that broadcasts the programmes. These formats are sold around the world, giving rise to numerous international versions: for instance the quiz show *Who Wants to Be a Millionaire?* (ITV, 1998–) was developed by Celador, and acquired by 2waytraffic (a Dutch company created by former employees of Endemol and subsequently sold on to Sony Pictures Entertainment in 2008). In 2005 it was reported that the format of *Who Wants to Be a Millionaire?* was the most replicated global television franchise with the format airing in 106 countries, followed by *The Weakest Link* (BBC, 2000–) which was sold to 98 countries (BBC, 2005). The importance of formats to television at the start of the century was such that an organisation dedicated to protecting format rights was formed, the Format Recognition and Protection Association (or FRAPA). In its report of 2009 FRAPA noted that from 2002 to 2004 there were 259 formats traded globally but by 2006 to 2008 this had increased to 445 formats, generating approximately €9.3 billion. Surprisingly, perhaps, given US dominance in film export markets, the UK has emerged as the largest exporter of television formats followed by the USA, the Netherlands and Argentina. In the period 2006–2008 the UK exported 275 formats and imported 66, compared to the USA which exported 159 and imported 116 formats (Lyle et al., 2009: 8).

This process of trading rights and formats contrasts with the production of most programmes in the 1980s where the rights to the programmes were retained by the broadcaster. This demonstrates both the increased globalisation of media markets and the commoditisation of television formats; it represents a shift from traditional UK 'cost-plus' models of programme funding towards a more US style 'deficit-financing' model. A good example of this is the huge commercial success of the *Idol* franchise. Airing on British television from 2001 to 2003 as *Pop Idol* it went on to gain even greater success overseas, especially in the US with the top-ranking US version, *American Idol* (Fox, 2002–) starring producer/panel judge Simon Cowell. The programmes were created by Simon Fuller (manager/agent for 1990s pop band The Spice Girls and footballer David Beckham) who, with Cowell, had become a leading figure behind the export of talent reality programme formats.

The arguments outlined above demonstrate that the national British television system of the mid-1980s has not only changed in terms of the multiplicity of programme choice but also shifted to a more global outlook. The trend towards formatisation and replication of successful programmes and the rise of reality television programming has led some critics to argue that television has become more 'homogenized' in spite of the diversity of viewing channels and media to view television output. Certainly popular reality television formats can be seen as an attempt to create what the industry terms 'appointment to view television', a difficult task in an age characterised by increased audience fragmentation and channel flipping. Another strategy to generate and retain audience loyalty is the widespread use of television celebrities. The rise of the media celebrity in the twentieth century has been startling and in the last quarter of a century this has only intensified. Celebrity has been at the very centre of the development of British television culture, a topic to which we shall now turn.

Television, celebrity and national culture

The narration of popular TV histories is very often conducted through accounts of its most recognisable performers – television celebrities – but these have not always been given much academic attention. Granted, a number of scholars in the 1980s analysed television celebrity by suggesting that television produced 'personalities' rather than stars and that the nature of television celebrity was of a different and lesser order to that seen on the cinema screen. Such an emphasis on the 'distance' of glamorous film stars and the enunciative power of the filmic apparatus, was in the 1980s, seen in contrast to the seeming accessibility of British television celebrity and the 'distracted' viewing of the television experience (Langer, 1981; Ellis, 1982). However, these distinctions now seem anachronistic and technologically-driven accounts of the television viewing experience appear inadequate. The days of most households owning only a single small screen television set have long departed and they have been replaced by the majority of households possessing many ways of viewing multi-channel television: multiple sets, including large high-definition liquid-crystal display and plasma screens with qualities closer to cinema than the televisions of old as well as a variety of computers and mobile devices that move television viewing away from the living-room sofa and into multiple environments. The differences between film 'stardom' and television 'celebrity' appear now to be less explainable in terms of the technology through which they appear, and more to do with the kinds of performances they enact and the cultural status that these different kinds of performances engender.

Contemporary television celebrity, like television itself, is now a cross-media and intertextual phenomenon. The tabloid news press and the phenomenal rise in the circulation and range of celebrity magazines have increased the prominence and visibility of journalism concerned with celebrity as well as allowing greater synergistic links between different media platforms, encouraged by increased cross-media ownership and higher market concentration that followed deregulation. Television celebrities have a commercial purpose in that they function to brand programmes for audiences in the multi-channel environment and reduce risk (or, for independent producers, increase the likelihood of a network commission). The intimacy that accompanies television viewing allows a particular, seemingly familiar relationship with celebrities. This has been termed a 'para-social' interaction (Horton and Wohl, 1956) in that it reproduces the effect of a relationship between celebrity and audience, despite being a predominantly one-way flow of communication (audiences, of course, rarely get to meet celebrities). Audience recognition, accumulated by viewing the performer across previous roles and appearances, increases the performer's value in television's celebrity economy (Drake, 2006). However, even this relationship is mediated in many more ways than it was in the 1980s. Now we are quite accustomed to following celebrities across a variety of media – magazines, websites, online fan groups, blogs – and some have large followings through recent social networking internet software. The television comedian and presenter Stephen Fry, for instance, had by 2009 nearly one million followers of his regular posts on Twitter.

The increase in media coverage of celebrity and the consequent expansion of celebrity culture in the digital media environment have not been universally welcomed. For some critics, the rise in the last 30 years of what we might term a highly 'celebritized' media culture is troubling, causing the wringing of hands over the alleged 'dumbing down' of British culture through celebrity-driven media such as reality television programmes, tabloid journalism and gossip magazines. Proponents of this view argue that the rise in audience demand for celebrity-driven entertainment formats has led to a consequential decrease in more high-brow and less populist forms of culture, and is symptomatic of a general cultural decline: as noted above, a good example was BBC TV presenter Jeremy Paxman's lecture at the Edinburgh International Television Festival in August of 2007. Claims often advanced in support of this include a loss of appropriate 'role-models' for children, the 'tabloidization' of the press and broadcast journalism and a general slide towards antipathy and disengagement from the social world (indicated, for instance, by historically low voting turn-out figures for political elections).

It is undoubtedly true that the media, including television, have followed a trend towards greater celebrity coverage. Television has long provided one of the most viable and visible sites of national celebrity. What is much less clear is whether the overall quality and range of television programming have changed for the worse as a result, or whether the broad range of television programming is simply more dispersed in the multi-channel environment. If we were to compare a typical television schedule from the mid-1980s with one 25 years later the most noticeable difference, alongside a far greater choice of channels, would probably be the ubiquity of reality television formats, ranging from property shows to talent contests. This has sometimes been seen as evidence of a greater 'democratisation' of fame, whereby people not usually given access to television can become celebrities, and as encouraging a discourse on the accessibility of media fame. Such a process can be seen most overtly in reality television programmes structured around the ritualised search for a star through elimination contests, such as the franchises *Big Brother* (Channel 4, 2000–) *The X Factor* (ITV 2004–) and *Britain's Got Talent* (ITV, 2007–) and the earlier *Stars in Their Eyes* (ITV, 1990–2006). The rise to global stardom of Scottish singer Susan Boyle in April 2009 via her appearance on *Britain's Got Talent* in April 2009 was quite remarkable in this regard. The finale of the programme attracted over 19 million viewers (nearly two-thirds of the national television audience) and nine days after it aired, online videos of her performances were reported as having attracted over 103 million hits: an online record, with pre-orders of her album breaking all previous records as well. Unexpectedly Boyle's ordinary appearance, working-class background and middle-age, strongly resonated with a viewing public, alongside her evident skill as a singer, and she was feted internationally, both in Britain and overseas, being interviewed by US chat-show superstar Oprah Winfrey and veteran newscaster Larry King, as well as featuring on *The Simpsons* (Fox, 2009).

Talent shows that feature ordinary people on television have of course long been popular with audiences. Programmes like *Britain's Got Talent* and *The X Factor* can trace a lineage to earlier programmes such as *Opportunity Knocks* (ITV, 1956–1990) and *New Faces* (ITV, 1973–1988), which used the concept of audience voting via post and then telephone. By the 1990s a number of format variations gradually emerged as celebrity spin-offs. In place of ordinary people were celebrities of varying magnitudes, given various challenges such as jungle 'bush-tucker trials' in *I'm a Celebrity...Get Me out of Here!* (ITV, 2002–) or competitive ballroom dancing in *Strictly Come Dancing* (BBC, 2004–).

Michael Billig (1995) has argued that national culture is reproduced through the banal use of everyday symbolism and categorisation: flags, maps, historical accounts of the nation, national anthems, and suchlike that instil a sense of 'us' (those belonging to the national culture) and 'them' (who do not). The sense of a national culture, and the popular discourse about its constitutive parts, is therefore also an important political question. In spite of the growing trend for international format replication as well as the online consumption of global television, a collective sense of 'our' national television may still remain important in understanding the contemporary British context. Benedict Anderson's (1983) analysis of the production of nation as an 'imagined community' has suggested how the nation may be partially imagined through its cultural output: this might well include television and television celebrities. Indeed, the collective sense of national television helps to reproduce interest in the discursive construct of the national – evident in national media sports events, soap operas or reality television and talent shows that galvanise a national audience.

In order to illustrate this point we will now turn to examine two British television celebrities. The first is television performer, Michael Barrymore, in the 1980s and 1990s a prominent national celebrity famous for variety show and quiz programmes before a scandal engulfed his career. The second is Jade Goody, a former reality contestant turned television celebrity who was also embroiled in career-threatening controversy before a tragic early death. The narratives of both celebrities help to reveal the interdependencies between contemporary celebrity and the media.

Narratives of rise, fall and redemption: Michael Barrymore and Jade Goody

In 2006 the British television star Michael Barrymore published an autobiography called *Awight Now: Setting the Record Straight,* a title derived from a phrase he popularised on British television and charting the rise and fall of his career. Barrymore was a talented and highly popular television entertainer and celebrity during the 1980s and 1990s, presenting light entertainment shows that showcased his lanky appearance, physical energy in front of the camera and his gifts for song and dance improvisation. His characteristic down-to-earth interaction with members of the audience was showcased in programmes such as *Strike it Lucky* (ITV, 1986–1999) *Michael Barrymore's My Kind Of People* (ITV, 1995), *Michael Barrymore's My Kind Of Music* (ITV, 1998–2002) and *Kids Say The Funniest Things* (ITV, 1998–2000) and this quality became an essential part of his perceived authenticity as a performer, emphasising his working-class origins and his comic relationship with his audience.

As we have observed, the intimacy that accompanies television viewing creates a para-social relationship with celebrities, a relationship that, although mostly one-way and mediated, may nonetheless be affective and intensely felt. As a rule, audiences rarely get to meet celebrities, and Barrymore was unusual in this regard in that his image as game-show host and presenter was built upon interaction with 'ordinary people' on camera (indeed Susan Boyle, mentioned earlier, was one such performer). In *Michael Barrymore's My Kind of People*, for example, he visited shopping centres around Britain inviting passing shoppers to sing or dance, unrehearsed, for the camera. His trademark cheekiness when interacting with the audience or contestants, and his improvised conversational style, recalled a long British tradition of music hall and variety shows, and attracted millions of family viewers during a prime-time television slot.

During the 1990s, at the height of his popularity, Barrymore revealed that he had alcohol and drug problems and came out publicly as gay, subsequently divorcing his wife (who was also his manager). This generated front-page newspaper headlines but was as nothing to the scandal that subsequently undermined his celebrity. In March 2001 a man was found dead at Barrymore's house. Barrymore claimed to have discovered him in his swimming pool during a party attended at his house by a large number of guests. In the police investigation Barrymore maintained his innocence and lack of knowledge over what had happened, and an open verdict was recorded by at the inquest. Barrymore faced a media barrage from journalists and paparazzi photographers eager to cover the story. The subsequent investigation caused severe damage to Barrymore's reputation, and led to his contract being terminated by ITV who wished to avoid any association with his now tarnished image. As a result Barrymore's television career, based as it was on family entertainment, collapsed and in order to escape the media barrage he moved to New Zealand where he toured a show to small theatre venues.

Barrymore's career as a British television celebrity did not end there, however. In 2006, he was invited to be a contestant on the fourth series of *Celebrity Big Brother* (Channel 4, 2006), a spin-off format that placed a diverse group of celebrities into the Big Brother house. This format is often used to revive the careers of 'Z list' celebrities whose profiles have declined since they became famous and was widely read by the media as an attempt to rehabilitate his career and stage a public comeback. His appearance demonstrates the media's complicity in shaping the redemptive possibilities available in the narrative of rise and fall that characterises media scandals. As the series progressed his mix of confessional vulnerability and comedic performance drew support from the viewing public and he remained a contestant in the house until the final day, ending as runner-up. His interview with presenter Davina McCall, upon leaving the house, was highly emotional, stating how he had gone in as a 'broken man' and how had finally 'found himself' during his 'journey' in the house. There are numerous observations that might be made about this; however what we wish to note is the 'redemption by media' discourse that has characterised recent reality television shows featuring an array of fallen or forgotten celebrities. Fame, it seems, is something that needs to be performed, a media ritual validated and confirmed by a mediated public.

The arc of Barrymore's 'fame narrative' is also taken up in his autobiography. Publicity for the book emphasised his emergence from humble beginnings, his rise to television fame, his coming out as gay, his addictions and his scandal, the loss of career, his descent into despair, and finally his redemption in the Big Brother house. The publisher Simon and Schuster trailed the book by stating on their promotional website that:

> AWIGHT NOW is Michael's story. He takes us through his rollercoaster life, from his poor London childhood, his work as a Butlin's Redcoat, through to TV and tabloid fame. At the height of his popularity, he suffered increasing alcohol problems and split from his wife of twenty years, Cheryl. In the fall-out, he admitted that he was homosexual and endured horrendous press intrusion into his private life. Then in 2001, came the terrible death of Stuart Lubbock. Michael tells of his despair following the tragedy and how his world fell apart. Despite tabloid rumours, Michael had no involvement with the death and has never been arrested or charged. Since 2001, Michael has endured some dark moments but through determination and the love of his friends, he is now picking up the pieces. His remarkable TV comeback on *Celebrity Big Brother* has catapulted him back into the spotlight and assured him that the British public want him back.

This description of the book reinforces a familiar fame narrative of rise, fall and redemption. In many ways, the journey through rituals of fame, scandal, despair, humility and return to the television fold have become an increasingly common part of the media landscape of the twenty-first century: the performance of a well-worn fame narrative. It is somewhat fitting, then, that the eventual winner of this series of *Celebrity Big Brother* was Chantelle, a fake celebrity tasked with convincing her celebrity housemates that she was genuinely famous by purporting to be a pop star from a fictitious girl band, turning the tables on the fame process – a conceit clearly enjoyed by a celebrity-savvy viewing audience.

Such celebrity controversies can have an impact upon the media and wider culture. One example of such a scandal is the case of Jade Goody, a celebrity made famous from appearing in (but not winning) the third series of *Big Brother* (Channel 4, 2002). Goody shot to fame through a negative press campaign orchestrated to vote her out of the Big Brother house, and subsequently became a regular feature of celebrity magazines. In 2007 she appeared in the fifth series of *Celebrity Big Brother* and caused a major controversy when she made a number of bullying and racist comments to another contestant, Shilpa Shetty, a Bollywood film star. The British television regulator Ofcom received over 44,500 complaints from viewers and its Content Sanctions Committee produced a 70-page report examining the issue. As a result the broadcaster was sanctioned for breaches of the Broadcasting Code and forced to broadcast the findings and issue a public apology, whilst Goody was vilified in the popular press. The Goody/Shetty controversy inadvertently opened up a broader public discussion about casual racism in British society. Goody's terminal cancer, announced to her live on the Indian television version of *Big Brother* in 2008 dramatically changed her media depiction to something far more sympathetic. Her widely covered illness and sad early death, aged just 27, was reported to have led to a large increase in women going for cervical screenings, potentially preventing other early deaths. These examples demonstrate how celebrity narratives marked by rise, fall and tragedy, are able to ignite wider debates in popular culture.

Audience fragmentation, the battle for trust and the rise of the 'prosumer'

As we have argued, the television environment of the 1990s and 2000s has been characterised by audience fragmentation across channels, celebritisation and formatisation. In the 1980s the viewing experience was characterised by Raymond Williams as that of 'flow': a seamless, ritualised evening of television viewing. Broadcasters agonised long and hard about scheduling: both in order to attract viewers away from their competitors but also to hold them once they were watching. One interjection into this temporal order of television was the innovation and eventual assimilation into popular viewing of the video recorder. This technology allowed households to 'time shift' their viewing in new ways. Most crucially, in the cultural context of television, VHS shifted some of the organisational control of viewing to the viewer. For the best part of 20 years televisual flow combined with moments of 'time shifting' characterised most people's experience of watching TV.

Now in the age of digital television and the Internet the balance of power between broadcasters and the viewer has started to change in quite radical ways. Viewers not only select their viewing material from the panoply of niche channels from which to choose their favourite programmes, but through time-shifting technologies and online catch-up television services, increasingly control when they watch it, where they watch it and on what technology. The industry itself has characterised this new relationship with its viewers as 'martini television': anytime, anyplace, anywhere. In a speech to his staff in 2006 called 'Creative Future' the BBC's Director General, Mark Thompson, relayed what this meant for those working in broadcasting: 'From now on, wherever possible, we need to think cross-platform, across TV, radio and web for audiences at home and on the move. We need to shift investment and creative focus to on-all-the-time, 24/7 services.' (Thompson, 2006). The speech was a rallying cry for television producers to adapt to an era of convergence. This shift in focus by the public service broadcaster has led to innovations like the BBC iPlayer – the on-demand seven-day archive of radio and television programmes available through the web and some television services. In programming, it has led to new ways of connecting television content from the screen, to interactive elements, to the web and to mobile devices. Increasingly, if new programme formats do not engage with what has been called a 360-degree approach to television, one which encompasses broadcast, web, mobile and interactive platforms, then commissioning editors are unlikely to take a second look at the idea.

The kind of rhetoric espoused by Thompson on the creative role of the BBC is prevalent at a time of dramatic change. It comes as no surprise when the leaders of the media industries, struggling to maintain their large institutional scale, grapple to find a vision of where these complex processes are taking us. As Henry Jenkins has noted, 'we are in an age of media transition, one marked by tactical decisions and unintended consequences, mixed signals and competing interests, and most of all, unclear directions and unpredictable outcomes' (2006: 11).

So exactly how is the experience of television changing and how can these changes be conceptually understood? In the digital age, at the intersection of established media like television and new media technologies, examples abound of where these unintentional consequences transform our communication systems in profound ways. For example, when short message services (SMS) were commercially introduced as a technology in the mid-1990s nobody predicted the massive explosion of its use in the everyday lives of millions of people around the world. Neither could television producers have envisaged the way in which text messaging would increasingly be integrated into programmes as a means of communication between presenters and the audience. The importance of text messaging to the aforementioned reality television formats like Channel 4's *Big Brother*, the BBC's *Strictly Come Dancing*, or ITV's *X-Factor* cannot be overstated. In May 2008 *American Idol* broke the world record for the number of text votes when 78 million messages were recorded by AT&T (Duryee, 2008).

As the volume and influence of audience voting have increased so new regulatory issues have also come into play. In 2008 the BBC programme *Strictly Come Dancing* was bombarded with accusations of vote-rigging and unfair practices through its phone-voting system. An anomaly in the format and problems with the lines enabled actor Tom Chambers to remain on the programme, and indeed win the show, even though many thought he had been evicted by the public vote. The programme also hit the headlines when the audience kept political correspondent John Sergeant in the competition. despite the fact that he had consistently secured bottom spot in scoring from the programmes panel of 'expert' judges. Sergeant, perhaps for many an unlikely hero of reality television, decided to leave the programme voluntarily as the story of his public triumph reached its zenith in the national press, and he has since capitalised on this new-found celebrity as a presenter on other entertainment shows.

This last issue points to the power of audiences in the digital age of television. The Sergeant case revealed that, where interactive programming formats allow the audience to shape the course of events, some unexpected consequences can occur. Dedicated audiences of such programming may not only communicate their desired influence on programmes through voting but may also coalesce in virtual spaces on the web to discuss key elements of the programme: its celebrities and narratives, heroes and villains. Programme makers now understand this process and so create dedicated blogs or social network sites for fans to converse on. Themes and conceptual ideas in such spaces can feed back into programme ideas, improving the format, encouraging further audience participation.

More crucially, the web has opened up wider less regulated spaces for fans to engage with each other. Such sites are potentially more critical and reflexive of the viewing experience and yet connected by the act of consumption to the original television texts. Again, Jenkins (2006) sees these converged spaces of media as a site of 'participatory culture', contrasting with the image of the television viewer as a passive 'couch potato'. Television producers and consumers no longer occupy distinctly separate roles, the television spectator is increasingly seen as a 'prosumer' of television content, a hybrid of consumer/turned/producer. This hybridity is even more visible in user-generated material in television news output or increasingly in children's television. There is also the extraordinary sharing of television programming on sites like YouTube, where corporate media are often illicitly and illegally reused, reshaped and edited for new, viral audiences (Jarret, 2008). Here the formal advertising-led economy of television sits in tension with the social economy of the web where creative, informal and customised versions of television are being produced.

Of course not all viewers experience television in this way and the arguments about the 'prosumer' bely the different ways in which different groups in society access television in their daily lives. In spite of the fact that in December 2008 the regulator Ofcom could announce that 88% of UK households had digital television of some kind, and the British Audience Research Bureau (BARB) had 297 UK registered channels, it also remains the case that levels of knowledge and ability to access the multiple ways in which television can be consumed are equally divergent and complex. It is an issue that contributes to the nation's 'digital deficit', and has prompted Ofcom to seriously consider its role in ensuring broadcasters and government engage with an educational programme of media literacy.

Conclusions: British television in transition

This chapter has mapped the contours of significant changes in the way television has been produced and consumed in British society over the last 25 years. Many other changes might also have been discussed, including how mainstream representations on television have shifted alongside changes in social attitudes. Television in Britain now offers a greater diversity of representations of gender, ethnicity and sexuality than in the 1980s, even if many minority groups remain substantially underrepresented. Altering attitudes towards language and violent or sexual images have also redefined what can be acceptably broadcast either side of the 9pm watershed, a cut-off increasingly anachronistic in an age of time-shifting and online viewing. Channel 4, in particular, was instrumental in testing these limits, as for instance in its broadcasting of a live human autopsy in 2002 (Miah, 2003).

The trends we have chosen to debate are 1) the deregulation of television, 2) the shift to a multi-channel broadcasting environment and the impact of digitalisation, 3) the fragmentation and segmentation of audiences, 4) issues relating to the quality and diversity of broadcasting, 5) the celebritisation and formatisation of television and 6) the rise of a televisual participatory culture. The rapid pace of both social and technological change makes predicting future trends difficult. The rise of the Internet and related technologies suggests that 'television everywhere' is imminent. Television content and audiences are likely to continue to be dispersed across different technologies, and the success of online, on-demand television delivery and webcast television suggests that the process of media convergence is likely to continue. As technologies develop and mature distinctions between consuming television via a television receiver, satellite or online, and on mobile or fixed technology, will become blurred. The launch of Freesat in 2008, a free-to-air digital satellite service and joint venture by the BBC and ITV offers an interesting test case of how technological collaboration across the commercial and public sector might operate.

Such changes create new challenges for television producers: for commercial television the challenge of how to address declining audiences and advertising revenues, for the BBC the challenge of how to maintain public funding and public service values against a backdrop of deregulation and marketisation, with increasing questions asked about the viability of the television licence fee. The globalisation of media industries also brings outside pressures to bear, raising pointed questions about media ownership and how it will be possible to maintain the quality and diversity of programming. What may be asserted, however, is that the next 25 years of television are likely to witness changes at least as seismic as those that have taken place over the last quarter of a century and examined here.

References

Anderson, B. (1983) *Imagined Communities: Reflections on the Origin and Spread of Nationalism* (London, New York: Verso).

BBC (2005) 'Millionaire dominates global TV', *BBC News*, 12 April 2005. http://news.bbc.co.uk/1/hi/entertainment/4436837.stm

Billig, M. (1995) *Banal Nationalism* (London: Sage).

Boyle, R. and R. Haynes (2004) *Football in the New Media Age* (London: Routledge).

Boyle, R. and R. Haynes (2009) *Power Play: Sport, the Media and Popular Culture*, 2nd edition (Edinburgh: Edinburgh University Press).

Clarke, J. and C. Critcher (1985) *The Devil Makes Work: Leisure in Capitalist Britain* (London: Macmillan).
DCMS (2009) *Digital Britain: The Interim Report* (London: HMSO). http://www.culture.gov.uk/what_we_do/broadcasting/5631.aspx
Drake, P. (2006) 'Celebrity, television and the tabloids', in D. Gomery and L. Hockley (eds) *Television Industries* (London: BFI).
Duryee, T. (2008) 'American Idol fans shatter previous years' text messaging records', *The Washington Post*, 7 May 2008. http://www.washingtonpost.com/wp-dyn/content/article/2008/05/23/AR2008052302402.html
Ellis, J. (1982) *Visible Fictions: Cinema, Television, Video* (London: Routledge).
Haynes, R. (2005) *Media Rights and Intellectual Property* (Edinburgh: Edinburgh University Press).
Horton, D. and R. Wohl (1956) 'Mass communication and para-social interaction: Observation on intimacy at a distance', *Psychiatry*, 19: 125–9.
Jarret, K. (2008) 'Beyond broadcast yourself: The future of YouTube', *Media International Australia*, 126: 132–44.
Jenkins, H. (2006) *Convergence Culture: Where Old and New Media Collide* (New York: New York University Press).
Langer, J. (1981) 'Television's personality system', *Media, Culture and Society*, 3(4): 351–65.
Lyle, D., E. Jäger and S. Behrens (2009) *The FRAPA Report 2009: TV Formats to the World* (Cologne: FRAPA).
Meikle, G. and S. Young (2008) 'Beyond broadcasting? TV for the twenty-first century', *Media International Australia*, 126: 67–70.
Miah, A. (2003) 'Dead bodies for the masses: The British public autopsy & the aftermath', *CTHEORY: International Journal of Theory, Technology & Culture*, Event-Scene, E119. http://www.ctheory.net/text_file.asp?pick=363.
Modleski, T. (1983) 'The rhythms of reception: Daytime television and women's work', in E. Ann Kaplan (ed.) *Regarding Television: Critical Approaches, an Anthology*, 67–75 (Frederick: University Publications of America).
Morley, D. (1980) *The Nationwide Audience: Structure and Decoding* (London: BFI).
Schlesinger, P. (1987) *Putting Reality Together: BBC News* (London: Sage).
Steemers, J. (2004) *Selling Television: British Television in the Global Marketplace* (London: BFI).
Temple, M. (2006) 'Dumbing down is good for you', *British Politics*, 1(2), July 2006.
Thompson, M. (2006) 'BBC creative future', in M. Magor (ed.) *Transforming Television: Strategies for Convergence* (Glasgow: The Research Centre).
Williams, R. (1974) *Television: Technology and Cultural Form* (London: Routledge).

5
Sex and the Citizens: Erotic Play and the New Leisure Culture

Feona Attwood

Chapter 5 is about the contemporary **politics of leisure and pleasure** *as they relate to* **sex**. *In the social, political and religious history of Western societies, sex and guilt have had a strong association. For previous generations sex was a dirty secret, a source of acute embarrassment and, in the teaching of some churches, to be engaged in solely for the purpose of procreation. Moreover, certain forms of sexual activity and expression were curtailed by law. As punitive philosophies have faded and legal restrictions either repealed or fallen into disuse, a new* **culture of sex** *as* **play** *and* **personal exploration** *has emerged. That culture, the consequence of a further, important area of* **deregulation**, *is analysed in this chapter.*

One sustained theme in this volume has been a decline in the moral judgement of private behaviour. In particular, as Jeffrey Weeks has also remarked, there has been an erosion of traditional authority over sexual behaviour (2007: 132) and in the West sexual practices have largely become matters of personal taste and lifestyle. The increasing individualisation, recreationalisation – and particularly, the commodification of sex – have attracted the attention of both academic and popular writers. What has come to be termed the sexualisation of mainstream culture – all the multifarious ways in which sex is now more visible in contemporary cultures – has become an object of discussion and some concern.

The new recreational sexuality provides a space for self-pleasure. Its characteristic features are 'adventure', 'experiment', 'choice', 'variety', and 'sensation' (Illouz, 1999: 176) and these are associated with particular kinds of sexual encounters – the affair, the one-night stand, forms of autoeroticism constructed around the use of pornography and sex toys, forms of commercial sex, and more recently, cybersex. Sex is also increasingly presented as an occasion for self-discovery and fulfilment, and articulated in terms of a 'therapeutic' culture (Plummer, 1995: 124–5).

In both instances, sex has been seen as assuming a disciplinary role, presided over by a range of cultural intermediaries, ready to provide us with 'training, instruction, counsel...recipes, drugs and gadgets' (Bauman, 1999: 24). More optimistically, it has been seen as a contemporary technology of the self, the term used by Foucault (1988) to indicate the ways in which individuals draw on cultural discourses in order to construct themselves. Both views depend on a notion of sex as a form of 'autosexuality' in which individual experience, identity, and what Anthony Giddens has called 'plastic sexuality' (1992: 58), are interlinked. While intimate relationships must be worked at and 'spiced up', our primary sexual relationship is with our self. This is a new late modern sexual sensibility.

The emergence of late modern recreational sexuality is linked to – and can be seen as emblematic of – a broad range of contemporary concerns with image, lifestyle and self-exposure, which have become means of self-care, self-pleasure and self-expression. As traditional values, sexual and otherwise, have lost their authoritative grip in many societies, individual choices about lifestyle are becoming more important in determining a sense of self, a way of life, and 'a sexual lifestyle' (Weeks, 2007: 110); all increasingly available through various forms of cultural consumption. In this sense, sex increasingly overlaps with other important spheres of contemporary life, and in particular that of leisure. In this chapter I trace some of the ways in which sex is becoming marked as leisure, focusing on its entanglements with commerce and technology, and on the growing significance of sex as a form of 'play'.

Intimacy, technology and commerce: A reconfiguration of erotic life

Of course, one person's leisure is another person's work. Forms of work that are concerned with sex often excite condemnation, and commercial sex and the commercialisation of sex in contemporary life are both frequently seen in negative terms. For example, Chris Rojek has described sexual services which are provided for the recreation of others as a *mephitic* form of leisure which depends on the objectification and commodification of workers – most commonly of female workers by male customers. Sex work, he writes, essentially 'involves treating the other as...an object in which trust and respect are not a precondition of interaction' (2000: 180–1). Commercial sex occupies a similar place in Anthony Giddens' account of evolving forms of intimacy (1992), where it is associated with episodic and uncommitted sexual behaviour and related to men's avoidance of intimacy and their violence against women.

But our understanding of commercial sex as labour has suffered from a tendency to treat it 'only as a moral issue' and to focus exclusively on 'a single commercial moment when two individuals exchange sex for money' (Agustin, 2005: 618). In contrast, contemporary research on sex work indicates a much more complex phenomenon which needs to be understood in relation to the late modern dynamics of work, leisure, identity, and relationship formation, and must be contextualised as part of a growing postindustrial service economy and a postmodernisation of sex.

As Laura Agustin notes (2007), commercial sex is part of a much broader range of intimate forms of labour which have been performed in Western societies for quite some time. These include childcare and domestic duties, as well as services such as hairdressing, massage and counselling. This pattern of intimate service provision has become intensified in contemporary First World societies where 'lifestyles...are made possible by a global transfer of the services associated with a wife's traditional role – child care, home-making and sex' (Ehrenreich and Hochschild, 2002: 4). In addition, a growing service economy 'serves to redirect an ever-expanding set of human needs from non-commodified, domestic space to the (newly privatized and domesticated) market sphere' (Bernstein, 2007: 175).

This, in turn, is part of a broader shift in paradigms of work and kinship and their accompanying sets of sexual ethics. New paradigms of work and kinship emerge and exist alongside earlier forms of organisation. An earlier modern-industrialist paradigm of wage labour, nuclear family and sexual double standard which privileged companionate love for heterosexual couples whilst permitting promiscuous sex for men, is now joined by a paradigm of service work, the pursuit of 'creative' and 'flexible' jobs, recombinant families, isolable individuals, and a sexual ethic of 'bounded authenticity' in which commerce and intimacy coexist (Bernstein, 2007: 173).

As a result some new characteristics of sex work have emerged. For example, in some large Western cities there has been a shift in its characteristic elements and its physical location. Sex work is no longer a specific and limited set of physical services delivered in strictly demarcated areas, but a 'diversified and specialized array of sexual products and services' which are 'dispersed throughout the city and surrounding suburbs' (Bernstein, 2007: 170). It is carried out in an increasingly wide range of material, media and virtual sites. A wide variety of people earn their livelihood from the sex industry in roles such as drivers, cooks, barmen, investors in property, entertainment and tourism, lawyers, telecommunication business, internet services, newspapers, banks, and vendors of a wide array of products such as costumes, food and sex appliances (Agustin, 2007: 65–6). Older forms of sexual performance such as burlesque have been revived, while others such as pole dancing are delivered to new audiences. What is particularly interesting in the case of pole dance is the way that a practice which was traditionally understood as a form of sexual service provided by female workers for a male audience is repackaged as pole exercise – a form of leisure pursuit for women's health and pleasure. Both burlesque and pole dance are also part of a broader gentrification of some aspects of the sex industry. All these changes complicate the dominant and moralistic view of what commercial sex is and who it is for.

Technologies play a particularly important role in developing new ways of articulating and experiencing sex, expanding the range and source of sexual materials available, making possible new forms of sexual interaction in text-based or multi-user virtual environments. Online, the rise of amateur porn and erotica, sexblogs, swinger sites, sex chatrooms and contacts pages indicates that traditional relations of production and consumption, professionalism and amateurism, commerce and leisure are increasingly elided. The kinds of episodic and uncommitted encounters associated with commercial sex are often now sought and offered for free. But technologies also make possible new locations for paid sex work, in professional online porn, camsites and sex worker advertising sites. The internet has served to make commercial sex more visible and accessible to potential clients and workers. It has contributed to industry professionalisation, providing sex workers with new opportunities for advertising their services and screening clients, as well as for building communities and activism. The sex industry has become simultaneously 'more private and more exposed', and at the same time, 'more of an identifiable culture' as a result (Ray, 2007: 177).

Other changes can be noted in the composition of the sex workforce and in the tenor of sex commerce transactions. One particular area of expansion has been the creation of 'a vast and visible middle class of sex workers who cater to middle-class men', sometimes accruing what Veronica Monet has called 'pseudocelebrity status' (in Ray, 2007: 184–6). And, as Elizabeth Bernstein describes, for some middle-class sex workers, sex work is seen in terms of 'an ethic of physical pleasure and a therapeutic language of personal growth' (2007: 97). Encounters are likely to include 'emotionally engaged conversation as well as a diversity of sexual activities', to require a larger investment of time with each client, and to take place within the home', drawing on notions of eroticism that were 'formerly relegated to the private sphere' (2007: 102). This is a form of the 'bounded authenticity' noted by Bernstein as a contemporary sexual ethic, in which the commercial aspect of the encounter provides an emotional boundary, as it does in other forms of service work such as therapy and massage (2007: 103).

Sex and culture

None of this is to say that many aspects of our work, sex and love lives do not carry on as before, that sex is not still related to kinship or to relationships, or that much sex work does not take the form that it did in earlier eras. However, the context in which we live is different now and it is clear that new forms and characteristics of sexual commerce are emerging alongside the older ones.

Some of the most striking changes are clearly discernible in the sphere of culture; in contemporary sex-cultural artefacts, in the ways these are produced and consumed and in the themes and figures they draw on for meaning. As with the transactions that mark new forms of sex work, these processes are marked by a middle class 'fun ethic' (Bourdieu, 1984) which incorporates a hedonistic and 'liberated' sexuality. This has become more and more prominent, not least because of the prevalence of the new petite bourgeoisie in occupations concerned with presentation and representation (Jancovich, 2001). Workers in these occupations represent a new service class (Lash and Urry, 1987); cultural intermediaries, 'involved in the provision of symbolic goods and services', such as journalists, designers, PR practitioners, advertisers, sex therapists, marriage counsellors, and dieticians (Nixon and du Gay, 2002: 496).

Although increasingly mainstream, this new sexual hedonism draws on some previously quite marginal sexual sensibilities. The first of these is derived from sex-positive and sex-radical writing and practice devoted to the reclaiming of sexual pleasure and to a revaluation of reviled practices such as masturbation, S/M, the use of pornography and sex work. The second is drawn from gay cultures, emphasiszing the celebration of diversity and the creation of communities based around sexuality. The third is a 'playboy' sensibility, embodied in the men's lifestyle magazine of the same name and in the development of clubs focused on leisure, straight men's entertainment and the availability of female 'playmates'.

Although quite dramatically different in many ways, all these share a view of sex as a good thing and a valid source of work and play. All three have also become more visible in contemporary popular culture. *Playboy* can be seen as the forerunner of much of the lifestyle media and the mainstream leisure venues that are currently aimed at men, most obviously in men's magazines and the 'gentlemen's clubs' which have become the high-street counterparts of backstreet strip joints. Following a period during which *Playboy* became deeply unfashionable, in the early twenty-first century, it has enjoyed a revival as part of 'a wider vogue for retro-cool' (Osgerby, 2001: 202). A sex positive/sex-radical stance is evident in a range of texts and practices; notably those associated with the sexy form of mainstream postfeminism embodied by performers such as Madonna, as well as more radical forms of feminism exemplified by 'girlie' and riot grrrl groups (see Munford, 2004; Monem, 2007). The notion of gay lifestyle has been mainstreamed as a form of cosmopolitan leisure and conspicuous consumption. Recreational sex has become part of what has been described as the 'ethical retooling' of consumer capitalism and its promotion of a 'morality of pleasure as a duty' (Bourdieu, 1984: 365–71).

Figures representing these new playful sexual sensibilities have become more visible as a result. They include hedonistic lads and ladettes, bi-curious girls and just-gay-enough boys, characterised by 'heteroflexibility' (Diamond, 2005) and 'metrosexuality', a term coined by Mark Simpson (1996, 2002, 2005) to describe the emergence of a figure for whom sexual identity is forged, not through sex acts or sexual orientation, but through mediation, consumerism and the development of lifestyle. Carrie Bradshaw, the central figure of the TV series *Sex and the City* (1998–2004) and films (2008 and 2010), is perhaps the most well-known fictional character who embodies many of the elements associated with a contemporary sexual sensibility; a postfeminism which combines financial independence and female friendship with a love of fashion, beauty and heterosexual romance; the pursuit of sex as sensation and adventure; and self-making through image, narcissism, consumption and lifestyle.

In this context, it is not surprising that new types of sex professional and erotic labour are emerging, or that sex and sex work have become the object of public fascination. Indeed, the 'rapid succession of...accounts of erotic labor' has not been restricted to activists and policy makers, but has also taken a 'cultural turn', drawing in 'film-makers, journalists, talk show hosts and fiction writers' (Bernstein, 2007: 12). While Carrie Bradshaw is widely acknowledged as a figure of the contemporary sexual consumer, what has been less often remarked on is that she also represents a new type of stylish sex professional for whom work, leisure and self are not easily separable. Carrie is a *sexpert* whose strapline boasts that she 'knows good sex'. Her characterisation draws on the older figure of the *flâneur* (Richardson, 2003), a type associated with the development of contemporary urban life and its mores, and the predecessor of contemporary journalists, social commentators and society columnists (2003: 149). As cultural intermediary, Carrie represents the cultivation of a cosmopolitan lifestyle and of 'sexpertise', an important new form of cultural capital. But sexpertise is also a new form of erotic labour, combining commentary and the giving of advice. Carrie's brand of sexpertise is aimed at a particular audience for whom sex and style are equally important. This is evident in publications like *Nerve*, an online magazine launched in the late 1990s as a form of 'smart sex' for 'young, urban, over-educated hipsters'. Similarly, the 2008 sex guide, *Sex: How to do Everything* is marketed on the basis of its 'explicit pics from Rankin, cult British fashion photographer, and words, wit and hip tips from US sexperts Em and Lo'.

New forms of sex-cultural production are partly aimed at meeting the 'extension of sexual consumerism' (McNair, 2002: 87) which can be traced not only in the expansion and diversification of the 'pornosphere', but more broadly in the emergence of a 'striptease culture'. Striptease culture embodies a widespread preoccupation with 'self-revelation' (2002: 81) and 'public intimacy' (2002: 98), evident in reality TV and other types of first person media, as well as in the development of new technologies for self-publishing and social networking. Sex also figures more visibly than ever before in forms associated with high culture. Erotica is sold in large book-chains. Sex is a recurring theme in contemporary art, and more recently in design, possibly because both 'offer a realm of sexual pleasure and hedonism...are treated as recreational activities...and...are viewed in openly self-preoccupied, consumerist terms' (Poynor, 2006: 7–8). We are, Poynor argues, 'in the process of designing a pornotopia' (2006: 9). Some kinds of sexual products have also become more visible as 'chic' objects, and glamorous sex shops for women have emerged, modelled on stylish boutiques and selling expensive lingerie and luxury sex toys.

Retro styles of sexual representation have become associated with hipster consumption where they signify 'authenticity, quality, creativity or individuality' (Epley, 2007: 53). The development of online 'smart smut' is also representative of this shift, offering communities where members subscribe, consume and participate, and working to locate sex as part of an apparatus of techniques for presenting the self and constructing links with others. As a magazine about 'sex and culture' for an audience which appreciates 'original, award-winning writing and photography as well as discerning coverage of the best films, television, books and music', *Nerve* sets sex alongside a range of other cultural interests including music, fashion and current affairs. By presenting the dating profiles of 'young, urban trendsetters' alongside professional work *Nerve* also presents its community as a collective of amateur and professional members, united by their sexiness and stylishness.

The growing significance of sex as 'play'

One of the most striking things in the development of recreational sex is its recuperation of the sphere of sex as something that operates like other forms of cultural appreciation or skill. This is evident in the way sex is represented in many sex advice books. In Alex Comfort's classic text, *The Joy of Sex*, learning about sex is likened to learning about cooking. It is a matter of culture and skill which, like 'chef-grade cooking', 'doesn't happen naturally', but requires effort and application; 'comparing notes, using some imagination, trying way-out or new experiences' (2003: 8). Alongside the cultivation of sexual taste and expertise, sex is also regularly promoted in sexual self-help literature for the purposes of self-care and self-development. Anne Hooper's *Great Sex Guide* gives advice on how to perform the 'Sets of Nine', 'a Taoist exercise designed to massage...genital reflexology zones evenly and thus benefit the rest of the body'. As Hooper remarks, this kind of practice is not 'for pleasure alone', but 'a preventative health measure' (1999: 123–4). What sex *is* also increasingly includes a very wide range of practices. *Sex Tips for Girls* by Flic Everett (2002) includes instruction on the use of dirty talk, phone sex, email, texting, taking photographs, making videos, using porn and erotica, kissing, massage, dressing up, stripping, dancing, using sex toys, sex games such as porn charades and strip poker, fantasy and role play. As in 'sex work', what sex is about or for is changing. It is a cultural pursuit, an exercise of taste, a set of skills, a communicative practice, a performance, a form of self-care, and a type of leisure in which media are produced and consumed.

Central to this view of sex is the concept of sex as play; a point made explicitly in *The Ultimate Guide to Strap-On Sex*.

> 'Once you've reached adulthood, there just aren't many places left to play. Swing sets won't hold you, slides mess up your work clothes, and you're more likely to be running Excel on your computer than Sim City. Though the kids may have ejected you from the sandbox, sexuality is a playground available to you for the remainder of your days. By being playful with sex, you can try out new personas, genders, and power dynamics. You can travel to any time or place you like and make real personae divergent from your work-a-day life. You can share love and affection with a partner, explore new kinds of stimulation, make discoveries about yourself, and receive affirmation for secret parts of yourself. In other words, *sex is a wonderful chance to have fun*.'
>
> (Karlyn Lotney, 2004: ix)

This emphasis on sex as play is also reflected in the new vocabulary that has grown up around sexual leisure pursuits, set out in the type of sexual glossary produced by sex writers, Em and Lo. An increasing interest in casual sexual encounters is indicated in new terms such as 'hooking up', 'fuck buddy', 'friend with benefits' and 'booty call'. Some self-help texts focus on the benefits of such encounters for women; a group not traditionally associated with casual sex (Dubberley, 2005), while Em and Lo's *RecSex* (2006) provides a guide to recreational sex, or 'fucking for the fun of it' (2006: 7).

Organised forms of public sex also draw on a view of sex as play. As Regina Lynn notes (2008), 'play' is fitting as a way of indicating the 'theatrical element' of group events, the 'joyful abandonment into sensation and imagination; and the distinction it marks between "party and friend sex" and "deep meaningful lovemaking with a committed partner"'. The rapid growth in social networking technologies for sex and romance has been influential in making sex-as-play more visible and accessible. Adult Friend Finder, a site devoted to swinging, claims over 20 million members worldwide (http://adultfriendfinder.com). Offline, casual sex pursuits such as 'dogging' – public sex outdoors – have attracted a great deal of media attention, especially in the UK. More glamorously, organisations like Fever offer 'upmarket swingers parties for slim and attractive young couples and single women under 40' at which there is 'nothing distracting, crude or seedy' (http://www.feverparties.com). Tim Fountain's search for casual sex around Britain, documented in his book, *Rude Britannia* (2008), describes his experiences of meeting people engaged in dogging, swinging, watersports, bestiality, furry sex, pony play, S/M and sex tourism,[1] concluding that the episodic forms of sex previously associated with gay men are being much more widely practised now.

'Play' is also central to sexual self-pleasure, to masturbation, and to mediated forms of sex such as pornography 'in which an autoerotic mode of consumption dominates' (Tuck, 2008). Where medical procedures related to sex were previously directed at the suppression or relief of sexual desires – for example in monitoring boys' nocturnal ejaculations or treating women's hysterical symptoms by stimulating them to orgasm, they are now aimed at encouraging arousal and sustaining performance, for example, through Viagra, or at enhancing appearance, for example, through genital surgery. Similarly, the technologies associated with masturbation have undergone a dramatic shift in the way they are conceptualised and employed, developing from types of medical implement to become recreational devices; the contemporary vibrator, once used in treatments for hysteria is now marketed as a toy (see Maines, 1999). The promotion of sex appliances as toys reinforces a view of sex as a form of 'play' and of 'individual liberation through bodily pleasure' (Juffer, 1998: 83). And as Clarissa Smith shows, in their most recent incarnation, like many other sex products, sex toys have become increasingly stylish, with the vibrator signifying not only as a toy, but as a fashion accessory. Sex toys for women are marketed on the basis of their combination of 'perfect industrial finishes', 'feminine' colours and curves, and 'surfaces and textures' that evoke sensuality, 'sexual individualism' and a 'sophisticated hedonism represented as understanding the slow burn of true pleasure' (2007: 180).

As with other aspects of commercial sex, developments in technology have been crucial to the improvement of sex toys. These now feature 'seamless, snag-free construction and dishwasher-safe cleaning' and there are 'remote controls, cordless models, more comfortable materials like silicon instead of hard plastic' (Lynn, 2005: 180–4). It has also become possible to combine sex toys with other existing recreational devices. For example, the OhMiBod, is 'a sleek, sophisticated new generation of vibrator' which syncs with an iPod to vibrate to whatever music is being played; a new type of 'acsexsory' (ohmibod.com), neatly combining sex with the expression of individual taste, commodities and recreation. Internet-enabled sex toys are also becoming available, and companies such as High Joy and Sinulate have developed online sites where toys can be integrated with instant messaging, phones and webcams. There is enormous potential for integrating playful technologies within both casual and committed forms of relationship through social networking.

Discussions of existing online sexual activities also draw on notions of play, and the internet is frequently seen as a 'sexual playground' (O'Brien and Shapiro, 2004: 116). Dennis Waskul and Philip Vannini (2008) argue that some cybersex encounters build a 'ludic' or game playing relationship which is 'casual, distant and noncommittal', perhaps as a response to the orderly kind of 'pure' relationship which Anthony Giddens has described as typifying late modernity (1992). Waskul's own study of cybersex practitioners found that they perceived their interactions as a kind of 'self-game', taken to 'an overtly playful level for the purposes of...sheer amusement or pleasure' (2003: 21). It represented an adventure, a free space for identity work (2003: 22) and 'a safe form of communication play' (2003: 23). First person accounts of cybersex also emphasise the element of play. In the book describing her cyber-encounters, 'Show-n-tell' writes, 'I have learned that I can connect to people in the same spontaneous way that I did when I was younger. It's as if I went bar hopping in this virtual playground and picked up a bunch of guys, swingers, cross-dressers, even gals, chatted, had sex and eventually came back home to my hubby' (2005: 142).

Like sex toys, cybersex is becoming more visible as it moves into animated platforms and the more familiar form of game playing (Lynn, 2006). Thus, 'play' provides a way of framing new recreational forms of sex as familiar and unthreatening. Online sex games, such as Sociolotron have appeared, combining more established game activities such as magic and 'monster bashing' with sexual themes.[2] Although not designed explicitly for sexual purposes – and rejecting the description of 'game', perhaps the most interesting new environment for online sex play is Second Life, a virtual environment used by people who see themselves as residents rather than players and who inhabit it in the form of avatars. Sex in Second Life allows participants to combine text or voice chat with avatar sex. In-world companies such as Xcite makes sex toys and genitals (XCite's online store offers male, female, hermaphrodite and animal genitalia)[3] while SexGen sells poseballs, animation sets and erotic furniture that allow avatars to be posed and animated for sexual action (the SexGen Birdseye Platinum unit enables sex for couples, threesomes and foursomes, bondage and D/s sex, fetish, girl/girl, cuddles, kisses, snuggles and 'you can even SMOKE AFTER SEX!'[4] There are opportunities for romantic sex, sexual performances and paid sex as well as 'orgies, BDSM, ageplay,[5] incest fantasy, rape fantasy, furry sex, bestiality, necro fantasy, forced feminisation, adult babies, and medical/nurse fantasies' (Carr and Ponden, 2007: 209).

The development of usable technologies for sex play has sparked a resurgence of interest in predicting the future of sex. Trudy Barber foresees the development of 'body augmentation', allowing for 'novel pleasure organs' (2004: 331), of cybersex encounters with 'sensorama celebrity' avatars (2004: 327), and new forms of sex tourism which will combine sex work, pornography and entertainment to create an 'immersive' experience. Regina Lynn suggests that new combinations of sex and technology will 'blur the lines of fundamental concepts like gender, fidelity, privacy and beauty', whilst providing 'variety and novelty...more ways to share pleasure...an outlet for personal expression and mutual connection unlike anything the world has ever seen' (2005: 210). David Levy's book, *Love and Sex with Robots* (2008) argues that there is reason to believe that in the not-too-distant future, robots will serve as companions, lovers and life partners; a logical extension of their existing roles in industry and service and their established success as virtual pets and caregivers. One outcome of this will be that 'what are perceived to be the natural levels of human sexual desire' will 'change to conform to what is newly available – great sex on tap for everyone 24/7' (2008: 310).

Rules of the game: Moral panics and sexual ethics

Like other forms of late modern leisure, sex in all its manifestations has become increasingly difficult to regulate, and perhaps as a result, its regulation has become more complex. The shift to a view of sex as a type of recreation has been accompanied by concerns and anxieties which take three distinct forms. One focuses on sexualisation, and particularly on the commodification of sex and of bodies, especially those of children and women. Another focuses on sex work, especially trafficking, for many of the same reasons. The third draws on long established fears of media technologies, often replacing material with representational concerns. For example, anxieties about child sex abuse frequently now manifest in concerns about 'sexualised images of children, including child pornography', and as Chuck Kleinhans notes, this is part of a broader shift in the ways that 'discourse around sexuality at many social levels has focused more and more on visual representations' (2004: 71). The recent 'dangerous pictures' legislation implemented in the UK in 2008 as part of the Criminal Justice and Immigration Act suggests both a heightened level of fear about media technologies and an uncertainty about the line between fantasy and reality. The legislation's focus on images that portray 'in an explicit and realistic way' supposedly 'extreme' forms of sexual behaviour which a 'reasonable person...would think...were real'[6] collapses the difference between reality and representation. There is no concept here of consensual sexual play which is central to much sexually explicit representation. And while the legislation is not explicitly concerned with sexual practices such as S/M, the threat to make their representation illegal clearly has implications for the way they are understood, and indeed for their performance.

Online, certain forms of sexual practice have also become a source of dispute. For example, in Second Life an announcement that apparently violent sex and sexual 'ageplay' in which users adopt child-like avatars for virtual sexual encounters would no longer be tolerated appeared in 2007. Director of community affairs, Daniel Linden, issued the following statement to Second Life residents:

> ...Real-life images, avatar portrayals, and other depiction of sexual or lewd acts involving or appearing to involve children or minors; real-life images, avatar portrayals, and other depictions of sexual violence including rape, real-life images, avatar portrayals, and other depictions of extreme or

> graphic violence, and other broadly offensive content are never allowed or tolerated within Second Life.
> ...Individuals and groups promoting or providing such content and activities will be swiftly met with a variety of sanctions, including termination of accounts, closure of groups, removal of content, and loss of land. It's up to all of us to make sure Second Life remains a safe and welcoming haven of creativity and social vision.

In many ways, these anxieties about dangerous pictures and virtual encounters rearticulate familiar concerns around the loss of childhood, commodification, technology and representation. Jeffrey Weeks has argued that their surfacing in many critiques of new developments around sexuality takes three typical forms. They are characterised by a defence of traditional values, by a 'sense of cultural despair', or by the burying of difficult questions in the face of moral panics which 'sweep over us, as they all too frequently still do, suggesting we are still desperately uncertain in confronting the complexity of contemporary mores' (2007: 124), especially as traditional authorities over sex lose their power (2007: 132) Yet, Weeks argues, we should not forget that we live in a world which is also 'a world of new freedoms, positive identities, genuine choices, a world we are striving to make for ourselves, a world of challenges and opportunities, dangers and pleasures' (ibid).

The recreationalisation of sex and its repositioning as a leisure pursuit and lifestyle, as 'a series of isolable techniques that provide personal meaning and pleasure' (Bernstein, 2007: 175), and as an opportunity to demonstrate taste, individuality and self-care is one amongst a number of indications that the significance of sex is changing. As intimate forms of sexuality are becoming commercialised, new forms of sex commerce are becoming more privatised and individualised. Both developments can be understood as forms of 'bounded authenticity' in which commerce and intimacy are combined. They are connected to much broader transformations in the economy and in culture in which the relations between production and consumption, work and leisure, public and private, emotion and economy, the commercial and the non-commercial are being redrawn. The need for a form of sexual ethics which can refuse despair, moralising and panic and take into account these new developments is the challenge that now faces us.

Notes

1 Watersports are sexual activities involving urination; furry sex is sexual activity involving dressing as an animal/human hybrid; pony play involves costuming and performing as ponies. A good overview of activities like these can be found in Gates, 2000.

2 http://www.sociolotron.com/. Accessed 28 October 2009.
3 http://www.getxcite.com/index.php. Accessed 28 October 2009.
4 Advertised at https://uncensored.xstreetsl.com/modules.php?name=Marketplace&file=item&ItemID=586329. Accessed 28 October 2009.
5 Erotic roleplay where a participant plays a character of a different age from their own.
6 http://www.opsi.gov.uk/acts/acts2008/ukpga_20080004_en_9#pt5-pb1-l1g63. Accessed 28 October 2009.

References

Agustin, L. (2005) 'Call for papers: The cultural study of commercial sex', *Sexualities*, 8(5): 618.

Agustin, L. (2007) *Sex at the Margins: Migration, Labour Markets and the Rescue Industry* (London and New York: Zed Books).

Barber, T. (2004) 'A pleasure prophecy: Predictions for the sex tourist of the future', in Dennis D. Waskul (ed.) *net.SeXXX: Readings on Sex, Pornography and the Internet*, 323–36 (New York: Peter Lang).

Bauman, Z. (1999) 'On postmodern uses of sex', in Mike Featherstone (ed.) *Love and Eroticism*, 9–33 (London: Sage).

Bernstein, E. (2001) 'The meaning of the purchase: Desire, demand and the commerce of sex', *Ethnography*, 2(3): 389–420.

Bernstein (2007) *Temporarily Yours: Sexual Commerce in Post-industrial Culture*, (Chicago: University of Chicago Press).

Bourdieu, P. (1984) *Distinction; A Social Critique of the Judgement of Taste* (London: Routledge).

Carr, P. and G. Ponden (2007) *The Unofficial Tourists' Guide to Second Life* (London: Boxtree).

Comfort, A. (2003) *The Joy of Sex* (London: Octopus).

Diamond, L. (2005) '"I'm straight, but I kissed a girl": The trouble with American media representations of female-female sexuality', *Feminism & Psychology*, 15(1): 104–10.

Dubberley, E. (2005) *Brief Encounters: The Women's Guide to Casual Sex* (London: Fusion).

Ehrenreich, B. and R. A. Hochschild (2002) *Global Woman: Nannies, Maids and Sex Workers in the New Economy* (London: Granta Books).

Em and Lo (2006) *RecSex: An A–Z Guide to Hooking Up* (San Francisco: Chronicle Books).

Em and Lo (2008) *Sex: How To Do Everything* (New York: DK Publishing).

Epley, N. S. (2007) 'Pin-ups, retro-chic and the consumption of irony', in S. Paasonen et al. (eds) *Pornification: Sex and Sexuality in Media Culture*, 45–57 (Oxford: Berg).

Everett, F. (2002) *Sex Tips For Girls* (London: Channel 4/Macmillan).

Foucault, M. (1988) 'Technologies of the self', in L. H. Martin et al. (eds) *Technologies of the Self: A Seminar with Michel Foucault*, 16–49 (Amherst: University of Massachussetts Press).

Fountain, T. (2008) *Rude Britannia: One Man's Journey Around the Highways and Bi-Ways of British Sex* (London: Weidenfeld and Nicolson).

Gates, K. (2000) *Deviant Desires: Incredibly Strange Sex* (New York: Juno Books).

Giddens, A. (1992) *The Transformation of Intimacy: Sexuality, Love and Eroticism in Modern Societies* (Cambridge: Polity Press).

Hooper, A. (1999) *Great Sex Guide* (London: Dorling Kindersley).
Illouz, E. (1999) 'The lost innocence of love: Romance as a postmodern condition', in Mike Featherstone (ed.) *Love and Eroticism*, 161–86 (London: Sage).
Jancovich, M. (2001) 'Naked ambitions: Pornography, taste and the problem of the middlebrow', *Scope*, http://www.nottingham.ac.uk/film/journal/articles/naked-ambition.html. Accessed 22 May 2001.
Juffer, J. (1998) *At Home With Pornography: Women, Sex and Everyday Life* (New York & London: New York University Press).
Kleinhans, C. (2004) 'Virtual child porn: The law and the semiotics of the image', in Pamela Church Gibson (ed.) *More Dirty Looks: Gender Pornography and Power* (London: BFI Publishing).
Lash, S. and J. Urry (1987) *The End of Organized Capitalism* (Cambridge: Polity).
Levy, D. (2008) *Love and Sex with Robots* (London: Duckworth Overlook).
Linden, D. (2007) 'Keeping second life safe, together', 31.05.07, http://blog.secondlife.com/2007/05/31/keeping-second-life-safe-together/. Accessed 10 November 2008.
Lotney, K. a.k.a Fairy Butch (2004) *The Ultimate Guide to Strap-On Sex: A Complete Resource for Women and Men* (San Francisco: Cleis Press).
Lynn, R. (2005) *The Sexual Revolution 2.0* (Berkeley, CA: Ulysses Press).
Lynn, R. (2006) 'Real sex, virtual worlds', Sex Drive column, *Wired*, 06.30.06, http://www.wired.com/culture/lifestyle/commentary/sexdrive/2006/06/71284? Accessed 10 March 2007.
Lynn, R. (2008) 'Random musing on sex and "play"', 22.07.08. http://www.reginalynn.com/archives/518. Accessed 10 November 2008.
Maines, R. (1999) *The Technology of Orgasm: 'Hysteria', the Vibrator and Women's Sexual Satisfaction* (Baltimore: John Hopkins University Press).
McNair, B. (2002) *Striptease Culture: Sex, Media and the Democratization of Desire* (London & New York: Routledge).
Monem, N. (ed.) (2007) *Riot Grrrl: Revolution Girl Style Now!* (London: Black Dog Publishing).
Munford, R. (2004) 'Wake up and smell the lipgloss': Gender, generation and the (a)politics of girl power', in S. Gillis, G. Howie and R. Munford (eds) *Third Wave Feminism: A Critical Exploration*, 142–53 (Basingstoke: Macmillan).
Nixon, S. and P. du Gay (2002) 'Who needs cultural intermediaries?', *Cultural Studies* 16(4): 495–500.
O'Brien, J. and E. Shapiro (2004) '"Doing it" on the web: Emerging discourses on internet sex', in D. Gauntlett and R. Horsley (eds) *Web Studies*, 114–26 (London: Arnold).
Osgerby, B. (2001) *Playboys in Paradise: Masculinity, Youth and Leisure-style in Modern America* (Oxford & New York: Berg).
Plummer, K. (1995) *Telling Sexual Stories: Power, Change and Social Worlds* (London & New York: Routledge).
Poynor, R. (2006) *Designing Pornotopia: Travels in Visual Culture* (London: Laurence King Publishing).
Ray, A. (2007) *Naked on the Internet: Hookups, Downloads and Cashing in on Internet Sexploration* (Emeryville: Seal Press).
Richardson, H. (2003) 'Sex and the city: A visible flaneuse for the postmodern era?', *Continuum: Journal of Media and Cultural Studies*, 17(2): 147–57.
Rojek, C. (2000) *Leisure and Culture* (Basingstoke: Palgrave Macmillan).
Show-n-tell (2005) *webaffairs* (Boston, Massachussetts: Eighteen publications).
Simpson, M. (1996) *It's a Queer World* (London: Vintage).
Simpson, M. (2002) 'Meet the metrosexual', http://www.marksimpson.com/pages/metrosexual.html. Accessed 20 January 2006.

Simpson, M. (2005) 'Metrodaddy V. Ubermummy', http://www.3ammagazine.com/litarchives/2005/dec/interview_mark_simpson.shtml. Accessed 20 January 2006.
Smith, C. (2007) 'Designed for pleasure: Style, indulgence and accessorized sex', *European Journal of Cultural Studies*, 10(2): 167–84.
Tuck, G. (2008) 'The mainstreaming of masturbation: Autoeroticism and consumer capitalism', in F. Attwood (ed.) *Mainstreaming Sex: The Sexualization of Western Culture* (London: I.B. Tauris).
Waskul, D. D. (2003) *Personhood in Online Chat and Cybersex* (New York: Peter Lang).
Waskul, D. D. and P. Vannini (2008) 'Ludic and ludic(rous) relationships: Sex, play, and the internet', in S. Holland (ed.) *Remote Relationships in a Small World*, 241–61 (New York: Peter Lang).
Weeks, J. (2007) *The World We Have Won* (London & New York: Routledge).

6

Rituals of Intoxication: Young People, Drugs, Risk and Leisure

Shane Blackman

***Chapter 6** is about **the politics of leisure and pleasure** in relation to the socially ambiguous activity of **drug-taking**. Here we return to the concept of **moral panic** and we turn, crucially, to public invocations of 'youth'. Historically, certainly since the late nineteenth century, governments and social commentators have been vocal in their concern about **young people** with time on their hands – **illicit** or **undesirable leisure** – and likely therefore to be up to no good. In the media discourse of postwar Western societies the spectre of feckless, drug-taking youth has often been raised. This chapter, once again, using an historical perspective, explores the tension between government **reproof** and **regulation** – here in the matter of drug-taking and the notions of **risk as leisure** frequently held by the young people that take them.*

Introduction

This chapter works within the theoretical framework developed by Stan Cohen (1972) to describe youth intoxication as a moral panic but my focus is not on negative representations. I shall examine how young people's intoxication makes for an effective strategy of regulation by government and the media through the promotion of representations of youth 'out of control'. This strategy captures a disturbing voyeurism that is both attractive and threatening, and one that supports prohibition policies (Royal Society of Arts, 2007). Government and youth are brought together through different positions focused on risk. Prohibition policies seek to define youth intoxication as pathological whereas young people understand their participation in risk as leisure.

One key argument I wish to challenge in this chapter is that too much free time leads to a 'heady mixture' of intoxication and deviance. The dominant explanation of young people's intoxication is that 'unstructured leisure' is a causal factor in the generation of social problems. The concept of 'unstructured leisure' defines youth as a passive subject, in contrast to self-directed leisure, which would allow them the agency necessary to interpret their social and cultural activities (McKay, 1998). This chapter combines historical context and contemporary news to generate theory and policy analysis. Initially I will look at the British historical context of intoxication and explore opposing understandings of youth intoxication. Then I shall critically examine government policy and media representations of youth intoxication; in the final section I will examine the concept of risk and argue that risk can be understood through Foucault's (1991) concept of governmentality as part of the new politics of science prevention.

Opposing positions

The experience of intoxication has been an integral part of human culture since before recorded history. Alcohol and drugs are sources of capital and power, intoxicants are used to back criminal organisations and act as a means of finance for secret state intelligence services (McCoy, 1991). Both now and in the past the British government has profited from the legal and illegal intoxicant trade (Blackman, 2004). Intoxicants are both an opportunity and a dilemma for people, society and nation states. The archaeological discovery of the remains of opium and cannabis demonstrate that these plants have an old association with humanity. There has been continuous cannabis cultivation for about 6,000 years and the use of alcohol in culture dates back to the fourth millennium. According to Richard Rudgley (1993: 31) the real diffusion of alcohol in Europe can be traced to the 'early Bronze Age cultures of the Aegean and Anatolia' and the first cultural use of cannabis drives from Neolithic sites in the steppe zone in central Asia. Intoxication including opium and wine was an integral pleasure of Classical culture, but in Greek and Roman society there is evidence of an awareness of the self-destructive nature of drug and alcohol consumption. For Homer in *The Odyssey*, hallucinogenic overindulgence leads to indifference and despair, while Pythagoras saw drunkenness as the route to madness. In England medieval monarchs sought to bring intoxication under increased control because it was identified as a potential source for political agitation (Griffiths, 1996). But with the *Gin Craze* of the 1700s government finally resorted to prohibition in the form of higher taxation, which ultimately failed. The paintings by William Hogarth of *Gin Lane* and *Beer Street* (1750–1751), and the engravings of Gustave Dore and Jerrold's (1872) *London* offer graphic visual evidence of the pains and pleasures of drunkenness and poverty amongst the populace.

These themes were further elaborated by Matthew Arnold (1869) in *Culture and Anarchy* and by Henry Mayhew (1851) in *London*. Different positions were apparent; Dore and Mayhew promoted a sympathetic view focusing on inequality, whereas Hogarth and Arnold saw the lewd behaviour of young adults as a threat to society and as evidence of the need of reform. In the

Victorian period, a lobby for prohibition emerged in the United Kingdom with the Temperance Alliance in 1853, and the Society for the Suppression of the Opium Trade in 1874. During this period political positions on intoxicants become polarised. J. S. Mill (1859: 158) observed the movement from comparative tolerance of the temperance movement to the political aspiration of total prohibition: now intoxication was identified as a pleasure not to be regulated but prevented. It was in the twentieth century that prohibition policies really flourished and the use of intoxicants served as a scapegoat for society's ills. It was easy to blame drugs and alcohol users for their wretchedness rather than tackle real social and material conditions (Inglis, 1975; Woodiwiss, 1998). Intoxicant leisure and pleasure for the elite class remained a private affair, whereas for the labouring classes this form of leisure was addressed as a public issue that required reform and prohibition (Berridge, 1978). The consumption of intoxicants by the aristocratic and bourgeois classes was primarily seen as a form of personal indulgence or an act of experimentation with little relevance to the public (Hayter, 1988). For example, certain key literary and intellectual figures including De Quincey, Ludlow, Baudelaire, Havelock Ellis and Huxley promoted intoxication as a visionary experience. Even today, this class-based public/private separation still continues within the tabloids. For example, see the *Daily Star* front-page headline 14.12.2007. WILLS & HARRY'S 15K BOOZE BENDER where the two princes are described as 'royal ravers' participating in a 'festive frolic'. The article notes that 'Prince Harry for his playful boozy antics has been named Party Animal of the Year by event organisers Confex'. The tabloids offer no such indulgent headlines for young working-class intoxication.

Constructing youth leisure as a social problem in the nineteenth century

The notion of 'unstructured leisure' as a cause of juvenile deviance emerged during the Victorian period and solutions to this problem were proposed by groups devoted to moral reform within the organised youth movement (Springhall, 1986). A common concern across these organisations was an apparent increase in the free time available to young people (Shore, 1999: 20). J. R. Gillis (1974: 97) argues there was a convergence between academics, policymakers and the media, which created a social stereotype of unorganised and maladjusted youth. Facts and fictions of youth leisure and crime became blurred. For example, Geoff Pearson (1983: 254) demonstrates how the daily papers and the music halls promoted social discontinuity and moral crisis focusing on the young hooligan who became the 'Other' for a fearful public. Lawless youth was the subject of Clarence Rook's *The Hooligan Nights* (1899), the story of young Alf first serialised in *The Daily Chronicle* earlier that year. The apparent problem of youth free time according to Bill Schwarz (1996: 105)

> became an issue addressed by a new profession of experts, often of Fabian or new liberal temperament. Educationalists, criminologists... all constructed their respective scientific discourses through the protean figure of the hooligan.

As for the original hooligan, Geoff Pearson (1983: 255) notes, 'no one other than Clarence Rook seemed to have heard of him'. From the late 1880s 'youth as trouble' became a journalistic representation based on fiction that has been elaborated by the popular and quality press and (the subsequently served as the foundation upon which stricter social policy and medical discourses were constructed (Humphries, 1981).

An alternative approach to understanding youth leisure is to focus on young people's self-directed leisure. For example in the 1800s considerable independence is found in young people's new leisure forms such as 'youth taverns' in northern towns and the Penny Theatres or 'gaffs' in London and other major cities which were 'distinguished by the presence of a distinct adolescent or apprentice culture'.[1] The Penny Gaffs (theatres) and subsequent Penny Dreadfuls (comics) were popular and positively chosen by young people in contrast to more formalised education. These shows and magazines focused on popular working-class anti-heroes including Dick Turpin, Jack Sheppard, Spring-Heeled Jack, Sweeney Todd and Elmira, the female pirate. Springhall (1998: 93) demonstrates that Penny Gaffs and Dreadfuls offered little real challenge to bourgeois society, but there were prosecutions and they became a scapegoat in the state creation of juvenile delinquency (Gillis, 1974). These two leisure spaces of the Penny Gaff and the public house 'Singing Saloons' demonstrate not only the burgeoning economic power of young people but also show how their leisure-based cultural activities became a target for middle class moral reformers. Frank Musgrove (1964: 67) states that from the 1830s onwards young employees were in demand and often paid better wages than their parents, 'what shocked middle class commentators on factory life in mid-Victorian England... was the independence of the young'. The spending power of youth promoted more self-directed recreational entertainment.

The rise of the singing saloons was crucial to courtship, for example at The Star Inn, Bolton, few parents were in attendance. Peter Bailey (1978: 31) states that in 1852 the 'estimated nightly attendance of 3,000 to 4,000 at Bolton's singing saloons' yielded handsome profits. The Factory Commissioners reported in 1842 that by the age of 14 young people: 'frequently pay for their own lodgings, board and clothing. They usually make their own contracts, and are in the proper sense of the word free agents.'[2]

Middle-class moral reformers were quick to react, Robert Storch (1976: 486) points out,

> No doubt they (magistrates) and the police knew that dancing saloons were important components of the working class marriage market and vital to courting, so that excluding those under eighteen was tantamount to a death sentence on these places.

The examples of penny gaffs, penny dreadfuls, the growth in singing saloons and increased economic independence for young people resulted in further regulation of young people's leisure activities (Stedman-Jones, 1971: 71). Richard Johnson (1976: 49) states that middle-class moral reformers saw youth leisure as a political danger and 'they did not express these observations in the language of cultural analysis: they wrote instead of idleness, drunkenness, pauperism, vice, improvidence and crime'. Henry Mayhew writing in the 1850s offers a corrective to the social stereotypes put forward. But in government, Kay-Shuttleworth was explicit about working-class youth who 'have unclear indefinite and undefinable ideas of all around them, they eat, drink, breed, work and die and the richer and more intelligent are obliged to guard them with the police'.[3]

The construction of youth leisure as a social problem in the nineteenth century depends on the argument whereby the issue of free time is seen as a generator of deviance, fuelled by intoxication. These arguments were integral for Cyril Burt (1925: 308) to produce his theory of the young delinquent. He states,

> Defective youths, too, of an unstable disposition show a perilous susceptibility to alcohol. Indeed, given appropriate opportunities, a deficient person of whatever type falls an easy prey to intemperance, and to all the vicious habits which intemperance gathers in its train.

Burt's ideas were influential in government policy and quickly became the orthodox position which was only superseded in the early 1950s by John Bowlby's (1953: 38) more advanced psychoanalytic theory of deviance which defined young people as in possession of an 'affectionless personality'. Later I suggest that traces of Burt and Bowlby's theories live on in contemporary risk theory of youth.

Leisure trouble: Contemporary representations of 'youth as a social problem'

According to Furlong and Cartmel (2007: 77) current leisure and life-styles of young people have drawn vocal disapproval from government, media and academy alike. These institutions often see young people and intoxication through a lens of adult indignation focusing on addiction and anxiety. The tabloid press generates images of youth 'out of control' and lurid accounts of drug and alcohol consumption, then present them as a major problem for society, community and the individual. Take, for example *The Sun* front-page headline 15.8.08, Out of our heads.[4] This pattern of sensational and stereotypical reporting of young people's involvement in crime, alcohol and drugs is described by Plant and Plant (2006: 28) 'as a chronic diet of moral outrage, often oddly combined with blatant titillation'. Plant and Plant confirm that this perspective established itself as the orthodox representation whereby young people are defined 'as trouble' (Cohen, 1972).

For a century both government and the tabloid media have been preoccupied with the notion that young people have 'too much free time' and that this freedom can lead to crime, anti-social behaviour and intoxication (Feinstein and Sabates, 2005; Sweeting and West, 2003; Feinstein et al., 2006). The *Make Space Youth Review*[5] 2007 report found that for many teenagers time outside school was unstructured and chaotic in circumstances that were often unsafe. At a local level, Coventry City Council's Whoberley Neighbourhood Plan 2005, put forward ideas for structured and unstructured leisure alternatives to divert young people away from engaging in dangerous behaviours. The Joseph Rowntree Foundation (February 2006)[6] work on parenting and children's resilience in disadvantaged communities put forward the idea that parents often went to considerable trouble to arrange organised activities for their young people; these were seen to be safer than unstructured leisure and promoted positive skills and relationships. At policy level, it is clear that self-directed leisure for young people is recognised as a potential social problem, especially where the media focus is on the dangers of 'hanging around' or 'doing nothing' (Corrigan, 1975).

The constant reporting of young people as a social problem within public leisure spaces such as parks, the high street, in shopping malls or on street corners has resulted in local and central policy initiatives such as Anti Social Behaviour Orders and curfews (MacDonald and Marsh, 2005: 70). A key reason why youth leisure is a problem for local communities, councils or government is because it occurs in spaces, which are subject to regulation and surveillance. Youth leisure is highly visible and can be easily depicted as irresponsible, but at the same time Cara Robinson (2009) suggests that councils and government can be seen as pushing young people into unsafe spaces, placing them at increased vulnerability. Public spaces are far from clean and the dirt and squalor are then associated with the young people who occupy such sites. With little alternative to the street, park or corner for consumption of alcohol, drugs, engagement in courtship rituals or physical acts of bravado, young people are presented as being different from others because their private leisure practice is performed in public places and has become subject to public consumption (Osgerby, 2004: 66). Due to the poor condition of free public spaces where young people gather they are often seen as having little respect for their social spaces. However, young people work hard to make such locations their own territory through physically marking it and metaphorically locating place in their narratives (Nayak and Kehily, 2008; Shildrick et al., 2010).

Youth appear threatening partly as they are massed in a group, representing the fear of the crowd, but the disturbing feature of youth leisure in public spaces is their apparent deportment. Their actions and activities do not follow the adult conventions of sitting and standing, thus they challenge the dominant expectations of how to 'do leisure'. Because young people appear in public they come to the attention of the police through the technology of surveillance cameras. Young people's deportment can be interpreted in relation to Marcel Mauss's (1934) theory of the 'techniques of the body' where a given group i.e. young people, have a ritualised series of bodily actions which express their culture or position in society. Due to the low regard in which society holds young people their appearance can be both aggressive and hostile or their bodily performance may embody degradation or humiliation (Blackman, 1997). Bodily techniques are learnt; they may be unspoken and adaptable to situation and setting. Youth have posture styles, skill in body techniques and knowledge about how they will be read by figures of authority.

The charity Barnardo's issued a press release titled: 'The shame of Britain's intolerance of children'. The report conducted by YouGov argued that there is an unjustified and disturbing intolerance towards young people in the UK which is extreme whereby 'society casually condemns all' youth. This critical and insightful report was turned on its head by the *Daily Telegraph* 17.11.2008 in the headline 'Public Scared of Children who "Behave like Animals"'. The newspaper speaks of Britain's streets as 'infested' by young yobs, who are described as 'animal', 'feral' and 'vermin'. The *Daily Telegraph* inverts Barnardo's report to heighten the social stereotype of 'youth as trouble' and thereby relinquish any sympathy towards young people who are collectively condemned and labelled as negative. Thus young people are defined as deserving no respect. On tabloid television and in newspapers we see normal behaviour by young people presented as dangerous or delinquent where they are employing their 'techniques of the body performing youthful communication and physical interaction' (Blackman, 1995: 37). Young people who engage in public leisure practices are defined as 'Other' for doing leisure, which is a majority practice, but adult leisure is not subject to the same public scrutiny.

The negative interpretation of young people's self-directed leisure is often linked to subcultural deviance and contrasted with desirable youth leisure. Within *Every Child Matters*[7] the government speaks of positive leisure[8] to demonstrate that leisure plays an important role in an individual's later life. Thus within certain government, academic and media accounts there remains an assumption that youth are constructed as a dangerous liability rather than an investment in the future. In sociology, from the work of the Chicago School to structural functionalism, youth leisure has been dominated by subcultural theory housed within deviancy theory (Roberts, 1983). But as Blackman (2004: 111) points out, in an attempt to break the association of subculture with crime and pathology, Phil Cohen (1972: 30) argued that it is 'important to make a distinction between subculture and delinquency'.

The subsequent Marxist theory of subculture developed by the Centre for Contemporary Cultural Studies (Hall and Jefferson, 1975) enabled youth leisure to be seen as a resistant form, but the separation from deviance failed to last significantly because youth leisure and intoxication came to the attention of the police and media (Barton, 2003). The consumption of intoxicants could be interpreted as recreational resistance within a Marxist, structuralist or postmodernist approach to youth, but the conservative model of drug and alcohol policy fuelled by media moral panic became the dominant explanation with a new focus defined as risk. As John Clarke and Chas Critcher (1985: 134) argued, 'The youth leisure problem has here become subsumed under this massive effort of social control'. Thus, opposing interpretations of youth cultural intoxication are apparent.

To summarise, I have sought to demonstrate that young people's unstructured leisure has agency and there exists a pattern of self-directed intoxicated leisure pursuits, which have been a stable part of British young people's culture. Thus, we saw the emergence of dancing and singing saloons in the nineteenth century, the growth of regional youth cultures in the south such as *hooligans* in the 1890s and the in the north around the 1900s of *scuttlers* (Roberts, 1971; Davies, 2008). Further, within sociology and cultural studies a range of ethnographic work has been done including Richard Hoggart's (1957) reflections on 1930s hard-working and hard-drinking cultures, 1950s everyday life for youth in northern towns binge drinking shown in social realist films, such as *Room at the Top* (1959), *Saturday Night, Sunday Morning* (1960), young mods in Soho London, 1963, pilled-up at *The Flamingo* or the *Allnighter* (Barnes, 1979) and ravers on ecstasy at *Spiral Tribe's Castlemorton Common Festival* May 1992 (McKay, 1998). From the 1890s to the twenty-first century we have seen the application of 'new' scientific experts focusing on young people's behaviour as a social problem for society and a pathological problem for the individual. In the examples above, young people have been defined as in need of rehabilitation and their use of intoxicants has been identified as a sign of their deviance within and disaffection for society. The expansion of the youth policy prevention industry not only appeared to have logic of its own; it required little justification because control of young people presented itself as a matter of common sense (Coles, 2000).

Parodies of encomium: Government policy and media reporting

During 2008 and 2009 there has been an emphasis on prohibition from both government and media. First, the Labour government set out its policy on young people and alcohol in the *Youth Alcohol Action Plan* (2008). Its aim is to regulate and to 'stop' excessive alcohol consumption by young people. The policy document makes reference to tabloid representations of the social problems caused by excessive alcohol consumption by young people. Secondly, the then Home Secretary, Jacqui Smith argued for cannabis to be reclassified as a class B drug.[9] She received support from the Prime Minister, Gordon Brown, who remains determined to tighten the law on cannabis.[10] Both politicians stood firm against the advice of the Advisory Council on the Misuse of Drugs and a Downing-Street spokesman for the Prime Minister maintained that 'cannabis use is illegal and unacceptable'.[11]

One common feature of British media reporting is to present intoxication through the inverted use of the metaphor of youth 'being the best' at getting 'wrecked'. A parody of encomium[12] is where British youth are presented as Henry Fielding (1743: 7) notes: as 'worthy of admiration or abhorrence'. With some irony the bad is celebrated and young people are cast as society's 'mock heroes'. The media's false praise of British youth as being at the top of the league for intoxication problems can be described as a parody of encomium[13] which upholds prohibition policy at the expense of young people who are understood as suffering from folly: young people are made into a joke. These media representations assert that youth have no respect, they are defined as reckless and a danger to themselves. These parodies of encomium strip young people of reason and intentionality, framing them as passive subjects who require punitive polices of prohibition and punishment.

For example, Plant and Plant (2006: 31) state that the recent research findings from ESPAD[14] confirm, 'The UK had high rates of periodic heavy (binge) drinking and illicit drug use'. BBC News, 14 September 2006 reported that British 15-year-olds are among Europe's heaviest users of alcohol (ACMD).[15] BBC News 16.11.2007 stated that nearly half of all ten to 15 year olds have tried alcohol, representing a figure of 48% of young people. In the UK young men between 16 and 25 are the heaviest drinking group in the population, The *Daily Telegraph* (19.6.01) ran the headline: 'British youth near top of under-age drinking league'. The *Mail on Sunday* (24.2.08) reported: 'Now it's 24-hour drinking for kids'. The BBC News (20.2.2001) had 'UK children top drugs league'. The Institute of Alcohol Studies (2008) states 'UK teenagers came at or near the top of the international league for binge drinking, drunkenness and experience of alcohol problems'. Government and media assert that a key challenge for society is to tackle the issue of first-time users becoming younger. Whilst there is some evidence to support this claim, the power of the argument rests at the level of rhetoric based on the construction of fear. During the 1990s the British tabloid press focused on young drug users and dealers,[16] and in 2008 the BBC News report focused on child-alcohol consumption, 10.11.08, 'Hospitals Treat Drunk Under-10s'. Focusing on illicit drugs the *Daily Telegraph* (11.1.2008) front-page states: 'Abuse of cannabis puts 500 a week in hospital', while the *Daily Mail* (15.8.2008) headline reads: SHOCKING TOLL OF DRUGS ON UNDER-16s. The BBC News reported on 7th November 2008, 'UK Top of European Cocaine League'. Annual figures from the European Monitoring Centre for Drugs and Drug Addiction, reveal that one in eight Britons under 35 had taken cocaine, giving Britain the highest number of cocaine users in the EU for the fifth year running. In 2008 the British preoccupation with intoxication became a concern for the United Nations. *The Times* front page on 5th March 2008 asserted that the 'United Nations condemns Britain's celebrity culture of intoxication'; 'offenders' such as Pete Doherty and Amy Winehouse were singled-out. The UN argued they 'can profoundly influence attitudes, values and behaviour, particularly among young people'. The *Evening Standard* (6.6.2008) headline announces: PUT COCAINE STARS ON TRIAL SAYS MET CHIEF, and although this is not new it is possible to see that during 2008 the tabloid newspapers wanted to dispense their own justice.[17] The London gossip newspapers e.g. Metro, London Lite, and The London Paper lead the way with printing pictures of celebrities in states of near 'unconsciousness' through their use of alcohol and drugs but the tabloids also had their own targets, for example the *Daily Mirror's* front-page headline KATE MESS:[18] tipsy star's 10hr drink marathon (20th September 2008).

It is with some irony that the *Daily Express* front page covered Dame Helen Mirren who played the monarch in the film, *The Queen*, 'Shock Confession: I LOVED COCAINE'. There appears to be a similarity between government policy and media representations on two broad patterns; first, a preoccupation with young people's new cultural practice of 'binge drinking' (Measham and Brain, 2005) and the consequent personal or social tragedy; secondly, an intense focus on media celebrities and their apparent bad behaviour. However, there have been counter hegemonic moments in the relationship between intoxicants and the media, for example, Carlton Television's fake drug documentary *The Connection* and also, Channel 4's satirical spoof drugs documentary, *Brass Eye*. In 1998 the Independent Television Commission fined Carlton Television £2 million for its documentary *The Connection*, which faked evidence of drug running where actors pretended to be drug traffickers. Michael Gillard and Laurie Flynn of the *Guardian* won Scoop of the Year at the 1999 Press Awards exposed TV's award-winning fraud. In 1997 the Channel 4, comedy Brass Eye, episode Drugs, exposed a dangerous fictional Eastern European drug called 'Cake' which was purported to be sweeping the nation. Prominent public figures were persuaded to pledge onscreen support for the campaign against the fake cake, these included Bernard Ingham, Bruno Brookes, Bernard Manning,

Rolf Harris, Noel Edmunds, David Amess MP, Claire Rayner, Jimmy Greaves and Jas Mann. Now in 2009 from the 8Ball Company you can buy designer CAKE T-shirts celebrating the humour of *Brass Eye*. These two examples demonstrate that intoxicants are fundamentally about power, privilege and the control of capital and communication. The exposed fantasised risk of youth and intoxication presented by *The Connection* and Brass *Eye* programmes does have real consequences when real young people and their families experience tragedies.[19] Here we have seen Carlton Television exposed for political corruption, while Brass Eye has penetrated the superficiality behind acts of popular drug prohibition, which are more concerned with marketing media celebrity status than preventing harm.

Risk

The politics of youth policy has been influenced by the take-up of the concept of risk across policy locations as part of a new science of prevention. Government has applied the notion of risk in the areas of Citizenship (Home Office, 2006), Education through *Aiming High* (HM Treasury, 2007), Health in the white paper (DoH, 2004) *Choosing Health: Making Healthier Choices Easier*, and drugs alcohol and crime (Seddon et al., 2008). In general, risk has formed the backdrop to government policy focused on wider social responsibility, for example in the New Deal employment programme and the Connexions Service, focused on crime, drug consumption, teenage pregnancy and anti-social behaviour. The Secretary of State for Crime Reduction, Alan Campbell (2009) states, 'young people are at risk; and for communities crime and anti-social behaviour causes fear and misery. These are compelling reasons to help people overcome their problems and stop the cycle of offending that prolific and drug-misusing offenders are in.' Recently, changes in government language have occurred. Where previously there was an over-concentration on exclusion, the new focus is on inclusion. The politics of this presentation is now confirmed by the application of the word *positive* in government youth policy for example *Youth Matters*. On 16th October 2009 the Minister for Children, Young People and Families, Dawn Primarolo, stated: 'The Government has invested £679m in *Aiming High* to increase positive activities and facilities for young people to go.'[20] The Home Office website *Tackling Drugs Changing Lives* (2009) states 'Positive Futures is a national social inclusion programme using sport and leisure activities to engage with disadvantaged and socially marginalised young people'.

According to France et al. (2010) these policy developments are based on a 'political deficit of the young' because it assumes that young people are making the wrong decisions. In Foucault's (1977) understanding young people now become subject to 'the correct means of training' to enable them to make the right decision through their own choice i.e. regulated from inside through auto-correction. Within youth policy risk has become a strategy of governance, making Foucault's concept of governmentality[21] highly relevant to a critical understanding of the application of risk. Through governmentality Foucault enables us to see how governments promote conditions of consensus through an individual's capacity for self-control on the basis of accepting responsibility. As Foucault (1993: 203) notes this is 'where the techniques of the self are integrated into structures of coercion and domination. The contact point, where the individuals are driven by others is tied to the way they conduct themselves.' Defining youth 'as trouble' the government and media organically create the motive for increased regulation over the whole population of young people and then target all as succumbing to the attractions and dangers of intoxication. The responsibility for increased use of intoxicants shifts to the individual where it becomes a matter of the young person's self-care. Thus, the new reasoned governmentality of the youth population is based on 'participatory' politics and 'positive activities', which promote the message for young people to discipline themselves. Here the politics of risk with its focus on Beck's (1992) 'invisible' or 'virtual' risks, ultimately leads to a position where young people are forever framed in a discourse of fear and insecurity. Thus risk theory advances that youth are always at risk, now and in the future and from invisible dangers. The politics of risk denies young people everyday normality. Government and media have been preoccupied with correcting young people's forms of intoxication from 'shots' to banning 'legal highs' (Measham, 2008). Negative images of underage drinkers have been common currency in the tabloid press and on TV news coverage. Here the category youth becomes inclusive of young adults who are also deemed to be at risk. As a result other populations which engage in youthful types of activities, are also open to increased surveillance and control. Foucault (1991: 100) states a new 'population is the subject... but it is also the object in the hands of government'. Thus, youth as a target population for government correctional policy is a strategy for wider regulation of population groups.

Nigel South (1999: 8) sees youth intoxication as being surrounded by policy and control discourses, including psychiatry, the psychology of dependence, public health concerns and international law. For Tim Rhodes (1997: 227) the control of drugs and alcohol has increasingly come under what he calls the 'dominant scientific construction of risk'. This newly emergent prevention science is based on psychology and criminology and presents risk theory as offering solutions to the problems of intoxication. Through the new politics of prevention, risk has become a key concept to advance positivistic solutions to the problems of youth intoxication (Muncie, 2000). Here risk is housed within powerful frameworks of psychiatry, criminology and medicine, which carry their own legitimacy and rationale to dissect young people and propose adjustment (Porter and Teich, 1995).

Risk theory posits the notion of the individual through two theoretical approaches, firstly, through theories of postmodernity and secondly, through developmental criminology and psychology.

The term risk emerged in conjunction with postmodern approaches towards understanding society, such as those of Lyotard (1984) and Baudrillard (1988). In particular, the work of Anthony Giddens and Ulrich Beck identifies social changes impacting on young people creating increased uncertainty, rendered as the *risk society*. The value of postmodern theory is that it has brought increased sensitivity towards interpreting young people's actions especially in terms of spatiality and location, providing individuals with more opportunities for freedom in creating identities. The idea of risk has entered the mainstream political discourse as a tool articulating the conditions of postmodernity characterised by increased levels of personal, social and cultural insecurity. The attraction of postmodern theory is that it suggests an individual acts with purpose and in response to perceived social changes, thus it commendably stresses the creativity of the individual agent. However, Beck's postmodernist conceptualisation of risk is primarily an individualisation thesis, which identifies society as in a process of transformation whereby individuals will be set free of constraining forms.

Rather than see youth intoxication in terms of a psychological understanding of risk, the work of Howard Parker and Fiona Measham explores an alternative understanding of youth culture and intoxication based on normalisation. They focus on new patterns of drug and alcohol consumption, which have become part of an everyday acceptance. Within their work there is an understanding of wider social and cultural structures, which play a significant part in the growth of normalisation. For them, increased forms of intoxication are not related to individuals but to collective youth behaviour. It is on this basis that their interpretation of risk is understood as grounded in relation to the participant's perspective. At a theoretical level Howard Parker and Fiona Measham are engaging with agency and they interpret drug and alcohol consumption as an intentional act even when the participants are aware that their intoxication could be dangerous or could have unknown consequences. They are investigating the contradictions of pleasure through excess, demonstrating that there has been a growth in the cultural acceptance of intoxication and recognition of the fact that young people like to get drunk or high (Wilson, 2006). Thus normalisation as an idea relies on 'sensible use', but both Parker and Measham also refer to youth who go beyond control. The theme of normalisation has been supported in the empirical work[22] of Hammersley et al. (2002), Jackson (2004) Moore and Miles (2004) Wilson (2006) and Saunders (2007) who are searching for meaning, decision making and rationality in young people's use of intoxicants based upon their everyday leisure. This could be understood as a key change in culture, which has promoted the political democratisation of intoxication in society based on a notion of freedom of choice.

Normalisation theory has been criticised by Shiner and Newburn (1999: 139) who specifically want to 'challenge the link... between drug use and post modernity'. Here it is possible to see a difference between the work of Parker and Measham focusing on post modernism. Parker (2003: 142) wants to rejuvenate the concept of risk by placing it in the 'post modernity-risk-consumption debate by rethinking what is normal and normative in today's new world'. Thus, Parker appears more open to postmodern ideas. Whereas Measham's analysis does not rely on postmodern theory, she (2006: 263) argues, 'A new culture of intoxication is emerging that features a determined drunkenness by young people as part of a broader cultural context of risk-taking'. Measham offers a more materialist understanding linked to the cultural theory of carnivalesque derived from the work of Makhail Bakhtin.

Although, Parker, Aldridge and Measham, and Shiner and Newburn do equally value the insights of qualitative research to the study of risk, there is an important difference between them at an epistemological and methodological level. In general, the normalisation theorists take on a more subjective symbolically grounded approach to understanding their research participants, whereas Shiner and Newburn pursue a more developmental and objectivist stance towards the presentation of data. They give priority to counting over interpretation. It is not the case that one approach is more or less scientific than the other; each approach is based on a different philosophical tradition. Work by Parker and Measham falls within a broad Weberian interpretative paradigm, where young people and their understanding of risk are at the centre of the analysis through an understanding of participation in intoxication as part of young people's everyday leisure pursuit. In contrast, from within criminology and development psychology, risk theory places more emphasis on the individual than on the social factors of youth intoxication. This political understanding of risk supports the construction of social pathology in youth policy (Farrington, 2002).

Recently, a new development within drug education has focused on the concept of desistance in the 'Blueprint' drug policy (Baker, 2006; UK Drug Policy Commission, 2007; Bennett and Holloway, 2008). The concept of drug desistance, defined as movement towards drug abstinence, comes from the discipline of criminology and its explanatory power is based upon psychological theories. Where desistance is directly used as a model of primary prevention, it is seen in conjunction with medical interventions to bring about the end to drug use. This means that desistance is often linked to addiction or drug dependency. According to Martin Frisher and Helen Beckett (2006: 133) the psychological studies, which apply the term desistance, support the 'dominant view of drug use as an illness which removes or reduces the capacity for voluntary behaviour'. The concept of desistance in this framework offers functional answers; it is used in attribution theory, matching theory, the study of twins and Pavlovian classical conditioning theory. In many of these psychological experiments on drug users there is a comparison of their behaviour with that of animals such as dogs, spiders[23] or in the case of Green et al. (1981) with pigeons. The aim is to explain situations where people act as a result of immediate contingency between behaviour and reward. The studies on animals are generalised to humans to explain the failure of will power or to show how individuals yield to temptation.

The political basis of the above prohibition policies is based on developmental criminology, which draws on scientific methods to see young people's problems of intoxication as 'curable'. Whilst these approaches systematically assert their scientific rationality and principles of investigation, they continue to apply stereotypical labels, for example 'intellectual disabilities'. Risk factor analysis has emerged out of a developmental psychology that starts from a theoretical position and a given set of assumptions dominated by the empiricist psychometric approach. Alan France (2008: 9) argues, 'It does not critically engage with debates about how the problem of youth is being constructed and used within a political context.' Certain contemporary risk theories of youth are preoccupied with the notion of 'intellectual disabilities', as the cause of deviant behaviour, for example Farrington, 2006; Holland et al., 2002; Lindsay and Taylor, 2005. According to France and Homel (2007) these new ideas are not only based on the positivist diagnostic testing derived from the work of Cyril Burt, but also the new ideological framework of values within which developmental criminology impacts on government policy through their creation of risk and protection models. As with the earlier work by Burt and later Bowlby, the attraction of these new risk theorists is the power of their stigmatisation of young people, which apportions blame and fails to offer choice or agency.

In everyday life the notion of risk taking is a normal and ordinary factor of young people's lives. However, the concept of risk has become a new theoretical tool, located within wider discourses of political control; here the concept has become a key strategy to define young people's leisure as 'Other'. As a result, what is normal is now defined as deviant for youth, but more than this, the concept of risk carries further consequence because it asserts that young people are defined as vulnerable and therefore require political intervention. For David Matza (1964, 1969) young people in mainstream culture engage in risk taking, in activities ranging from leisure and business to sport, through the use of profane and symbolic language wherein risk becomes the pursuit of excitement and pleasure. Within cultures of intoxication the competitive understanding of risk taking reflects the same sets of values as within normalised areas of society and culture. Furlong and Cartmel (2007: 12) state that risk theory has brought forward new ideas, but it shows a 'tendency to exaggerate changes and to understate many significant sources of continuity'. The idea of a risk society has become a normative explanation in political discourses, which not only lack substantial evidence but also suffer from excessive generalisation (France, 2008). Blackman (1998: 54) argues that 'risk is neither open to all people to the same extent, nor with the same consequential outcome: risk is an ideological construction of the deviant "Other"'. Further, I would argue that not only has there been an expansion of studies that employ the concept of risk and risk factor analysis, but also the term is used widely as shorthand for vulnerability or implicit use of Bowlby's theory of 'inadequate socialisation' (Downes, 1966).

At policy level risk theory has sought an individualistic understanding of youth intoxication and therefore the solutions put forward tend not only to assume an objective, atheoretical approach but also to assert the legitimacy for programmes of rehabilitation (Farrington, 2002; Wikstrom, 2006). In a similar vein to Cesar Lombroso's ideas on the individual and medical understandings of social activity, the new prevention science follows positivist solutions claiming both objectivity and neutrality. For the majority of young people, their leisure-based pleasure of intoxication is directed to promote solidarity. Young people's ritualised forms of intoxication conform to Emile Durkheim's notion of consensus rather than anomie. Youth leisure intoxication rituals are focused on strengthening internal group relations and identity, in an ironic sense it is intoxication, which is the solution to feelings of insecurity, which promotes self-identity for young people (Valentine et al., 2007). Furthermore, Durkheim[24] is against psychological theories or individualistic accounts of group actions, in this sense, postmodern theories of risk would be anathema to him. The new politics of prevention science constructs pathological models of social behaviour, which put forward common sense answers through the attraction of new Social Darwinism (Williams, 1980: 86). Durkheim's (1895: 70) focus is on the normal and he sees deviance as 'bound up with the fundamental condition of all social life'. Durkheim argues that group consensus is heightened through differences between people in groups; this bonds them together therefore through intoxicated leisure young people gain increased solidarity. Unstructured youth leisure is a way of engaging with social and cultural complexities, through occupying and marking space through telling stories young people can create the narrative of their social bonds (Blackman, 2007). Thus the inability to conceive of young people's leisure on a collective basis results in an atomistic and misleading assessment of young people's relation to intoxication, rather than looking at the support and consensus developed by youth themselves they are defined as a problem at a pathological level (Measham, 2008). In this section I have tried to show that risk linked to intoxication has become a normal and celebrated feature of mainstream contemporary leisure, but in youth and young adult culture, the thrill of intoxicated leisure is understood differently in terms of the 'Other'.

Conclusion

The chapter has sought to challenge the notion of youth leisure as a social problem using historical and contemporary analysis. Building on Clarke and Critcher's (1985: 239) argument that leisure has 'political significance' this chapter has examined how young people's 'unstructured leisure' is a challenge to the dominant political understanding of how leisure is undertaken.

Youth leisure should be theorised from an alternative position, which interprets young people's cultural activities of intoxication as a sign of agency and solidarity. The work of Parker on drugs, and Measham on alcohol was used to support an interpretation, which sees young people as conscious agents in their leisure practices. A weakness in this understanding of drug and alcohol normalisation is their uneasy combination of referring to both 'sensible' and 'stupefied' forms of intoxication, which results in different forms of behaviour, expectation and outcome.

In contrast, government, media, and academic discourses show a tendency to emphasise a negative view of young people's self directed leisure. It was argued that the concept of risk within government policy is part of a new political strategy defined as prevention science to bring about control of specific populations such as youth and young adults. The popular portrayal and reporting of young people at the 'top of the league' for alcohol or drug consumption was theorised as a *parody of encomium*, where young people are captured as being 'mock heroes'. This notion of false praise within tabloid news coverage has a link to recent ideas in developmental criminology and government policy through the application of the concept of risk. For young people, being defined as 'mock heroes' and then subject to risk theory results in a passive and objectified assessment of their understanding of leisure, which is in opposition to young people's experience of intoxication rituals as a form of solidarity and bonding.

Acknowledgements

In writing this paper I should like to thank the editors, Peter Merchant, Fiona Measham, Alan France and in particular Debbie Cox for her comments.

Notes

1 Springhall (1998: 16).
2 Musgrove (1964: 68).
3 Storch (1976: 141).
4 *Daily Mail* headline 15.8.2008. SHOCKING TOLL OF DRUGS ON UNDER 16s.
5 Make Space Youth Review is run by Charity 4 Children.
6 JRF February 2006 – Ref 0096.
7 Aiming High for Young People.
8 www.everychildmatters.gov.uk/youthmatters/thingstodo/

9 BBC News: 7 May 2008.
10 BBC News: 3 April 2008, PM 'stands by stance on cannabis'.
11 BBC News 3 April 2008.
12 Oxford English Dictionary, the definition of encomium, is formal praise or high-flown praise.
13 Henry Fielding's novel Jonathan Wild, 1743 an example of a mock hero.
14 European School Survey Project on Alcohol and Other Drugs.
15 Advisory Council on the Misuse of Drugs.
16 Below selected newspaper headlines:
The Mail on Sunday 11 February 2007. Children as young as 12 'on heroin'.
The Daily Mail 13 March 2007. Schools are forced to expel 230 pupils a day.
The News of the World 28 January 2007 Kids aged 7 arrested with guns.
Brighton and Hove, Argus 9 August 2006. Child Drug Dealers.
ETON BOYS BUST *The Sun* 17 June 1998. Front page.
BBC 19 May 1998. News. The Heroin Epidemic.
Evening Standard 16 May 1998. SCHOOL WAR ON PUSHERS.
Daily Mail 1 February 1998. Front cover Ecstasy Boy is hit by train.
Daily Express 17 July 1995. Just 14, but I was hooked on drugs.
Daily Record headline 21 February 1995. DRUG DEALER AGED NINE
Daily Express 20 February 1995. Children Hooked on Drugs and Crime.
Daily Telegraph 10 October 1994. Drugs a danger to primary schools.
17 *The Sun* 17 June 1998. Front page STOP POP AND FASHION ICONS PREACHING DRUGS TO OUR KIDS.
18 Kate Moss is the model's real name.
19 For example, the murder of husband and father Garry Newlove, *Daily Mail* 17 January 2008 by a group of youth who used intoxicants.
20 www.dcsf.gov.uk/pns/DisplayPN.cgi?pn_id=2009_0189, also see DCSF report calls for improved communication about positive activities to increase young people's participation, *Positive Activities: Qualitative Research with Young People*, Research Report DCSF-RR141, 2009.
21 Governmentality works on the basis of governing populations.
22 Jackson and Wilson take a more postmodernist approach.
23 Blackman (2004: 160).

Bibliography

Arnold, M. (1869) *Culture and Anarchy* (Cambridge: Cambridge University Press).

Bailey, P. (1978) *Leisure and Class in Victorian England-Rational Recreation and the Contest for Control, 1830–1885* (London: Routledge).

Baker, P. (2006) 'Developing a blueprint for evidence-based drug prevention in England', *Drugs: Education, Prevention and Policy*, 13(1): 17–32.

Barnes, R. (1979) *Mods!* (London: Eel Pie).

Barton, A. (2003) *Illicit Drugs* (London: Routledge).

Baudrillard, J. (1988) *Selected Writing* (Cambridge: Polity).

Beck, U. (1992) *Risk Society* (London: Sage).

Bennett, T. and K. Holloway (2008) 'Identifying and preventing health problems among young drug-misusing offenders', *Journal of Health Education*, 108(3): 247–61.

Berridge, V. (1978) 'Opium eating and the working class in the nineteenth century: The public and official reaction', *British Journal of Addiction*, 7(3): 107–12.

Blackman, S. J. (1995) *Youth: Positions and Oppositions – Style, Sexuality and Schooling* (Aldershot: Avebury Press).
Blackman, S. J. (1997) 'Destructing a giro: A critical and ethnographic study of the youth "underclass"', in R. MacDonald (ed.) *Youth, the Underclass and Social Exclusion*, 113–29 (London: Routledge).
Blackman, S. J. (1998) '"Disposable generation?" An ethnographic study of youth homelessness in Kent', *Journal of Youth and Policy*, Issue 59: 38–56.
Blackman, S. J. (2004) *Chilling Out: The Cultural Politics of Substance Consumption, Youth and Drug Policy* (Maidenhead/New York: Open University Press/McGraw-Hill).
Blackman, S. J. (2007) '"Hidden ethnography": Crossing emotional borders in qualitative accounts of young people's lives', *Sociology*, 41, 4: 699–716.
Bowlby, J. (1953) *Childcare and the Growth of Love* (London: Penguin).
Burt, C. (1925) *The Young Delinquent* (London: University of London Press).
Campbell, A. (2009) 'Reducing crime through offender-based interventions', National Conference, 18 & 19th June 2009. Birmingham ICC (Draft speech).
Clarke, J. and C. Critcher (1985) *The Devil Makes Work: Leisure in Capitalist Britain* (London: Macmillan).
Cohen, S. (1972/1980) *Moral Panics and Folk Devils* (Oxford: Martin Robertson).
Cohen, P. (1972) 'Subcultural conflict and working class community', in *Working Papers in Cultural Studies*, CCCS, University of Birmingham, Spring: 5–51.
Coles, B. (2000) *Joined-Up Youth Research, Policy and Practice* (Leicester: Youth Work Press).
Corrigan, P. (1975) 'Doing nothing', in Hall, S. and T. Jefferson (eds) *Resistance Through Rituals*, 103–5 (London: Hutchinson).
Davies, A. (2008) *The Gangs of Manchester* (Preston: Milo Books).
DoH (2004) *Choosing Health: Making Healthier Choices Easier* (London: DoH).
Dore, G. and B. Jerrold (1872) *London: A Pilgrimage* (London: Grant and Company).
Downes, D. (1966) *The Delinquent Solution* (London: Routledge and Kegan Paul).
Durkheim, E. (1895/1964) *The Rules of Sociological Method* (New York: Free Press).
Farrington, D. (2002) 'Developmental criminology and risk focused prevention', in Maguire, M., R. Morgan and R. Reiner, *The Oxford Handbook of Criminology* (Oxford: Oxford University Press).
Farrington, D. P. (2006) *Saving Children from a Life of Crime* (Oxford: OUP).
Feinstein, L. and R. Sabates (2005) 'Education and youth crime: Effects of introducing the education maintenance allowance programme', *Wider Benefits of Learning Research Report*, No. 14.
Feinstein, L., J. Bynner and K. Duckworth (2006) 'Young people's leisure contexts and their relation to adult outcomes', *Journal of Youth Studies*, 9: 305–28.
Fielding, H. (1743/2003) *Jonathan Wild* (Oxford: Oxford University Press).
Foucault, M. (1977) *Discipline and Punish* (London: Penguin).
Foucault, M. (1991) 'Governmentality', in Burchell, G., C. Gordon, and P. Miller (eds) *The Foucault Effect*, 87–105 (London: Harvester/Wheatsheaf).
Foucault, M. (1993) 'About the beginning of the hermeneutics of the self' (transcription of two lectures in Darthmouth on November 17th and 24th 1980, in M. Blasius, (ed.) *Political Theory*, 21(2) May: 198–227.
France, A. (2008) 'Risk factor analysis and the youth question', *Journal of Youth Studies*, 11(1): 1–16.
France, A. and R. Homel (eds) (2007) *Pathways and Crime Prevention: Theory Policy and Practice* (Abingdon: Willan Publishers).

France, A., L. Sutton, and A. Waring (2010) 'Youth, citizenship and risk in UK social policy', in Leaman, J. and M. Wörsching (eds) *Youth in Contemporary Europe* (forthcoming) (London: Routledge).
Frisher, M. and H. Beckett (2006) 'Drug use desistance', *Criminology and Criminal Justice*, 6(1): 127–45.
Furlong, A. and F. Cartmel (2007) *Young People and Social Change* (Buckingham: Open University Press).
Gillis, J. R. (1974) *Youth and History: Tradition and Changes in European Age Relations, 1770–Present* (London: Academic Press).
Green, L., E. B. Fisher Jr, S. Perlow and L. Sherman (1981) 'Preference reversal and self-control: Choice as a function of reward amount and delay', *Behaviour Analysis Letters*, 1: 43–51.
Griffiths, P. (1996) *Youth and Authority* (Oxford: Clarendon Press).
Hall, S. and T. Jefferson (eds) (1975) *Resistance Through Rituals* (London: Hutchinson).
Hammersley, R., F. Khan and J. Ditton (2002) *Ecstasy* (London: Routledge).
Hayter, A. (1988) *Opium and the Romantic Imagination* (Wellingborough: Crucible).
HM Treasury (2007) *Aiming High for Young People: A Ten Year Strategy for Positive Activities* (London: HMSO).
Hoggart, R. (1957) *The Uses of Literacy* (London: Chatto and Windus).
Holland, T., I. C. H. Clare and T. Mukhopadhyay (2002) 'Prevalence of "criminal offending" by men and women with intellectual disability and the characteristics of "offenders": Implications for research and service development', *Journal of Intellectual Disability Research*, 46(1): 6–20.
Home Office (2006) *The Respect Agenda* (London: Home Office).
Home Office (2009) *Taking Drugs Changing Lives* http://webarchive.nationalarchives.gov.uk/20100418065544 / http://www.homeoffice.gov.uk/drugs/
Humphries, S. (1981) *Hooligans or Rebels? An Oral History of Working Class Childhood and Youth 1889–1939* (Oxford: Blackwell).
Inglis, F. (1975) *The Forbidden Game: A Social History of Drugs* (London: Hodder and Stoughton).
Jackson, P. (2004) *Inside Clubbing* (Oxford: Ber).
Johnson, R. (1976) 'Notes on the schooling of the English working class 1781–1850', in Dale, R., Esland, G. and MacDonald, M. (eds) *Schooling and Capitalism* (Milton Keynes: Open University Press).
Lindsay, W. R. and J. L. Taylor (2005) 'A selective review of research on offenders with developmental disabilities: Assessment and treatment', *Clinical Psychology & Psychotherapy*, 12: 201–14.
Lyotard, J-F. (1984) *The Postmodern Condition* (Manchester: Manchester University Press).
MacDonald, R. and J. Marsh (2005) *Disconnected Youth* (London: Palgrave).
McCoy, A. (1991) *The Politics of Heroin: CIA Complicity in the Global Drug Trade* (New York: Lawrence Hill).
Matza, D. (1964) *Delinquency and Drift* (New York: John Wiley and Sons).
Matza, D. (1969) *Becoming Deviant* (Englewood Cliffs: Prentice-Hall).
Mauss, M. (1934) 'Les Techniques du corps', *Journal de Psychologie*, 32: 3–4, 271–93.
Mayhew, H. (1851) *London Labour and the London Poor* (London: Spring Books).
McKay, G. (ed.) (1998) *DiY Culture: Party and Protest in Nineties Britain* (London: Verso).
Measham, F. and K. Brain (2005) 'Binge drinking, British alcohol policy and the new culture of intoxication', *Crime, Media and Culture: An International Journal*, 1: 263–84.
Measham, F. (2006) 'The new policy mix: Alcohol, harm minimisation and determined drunkenness in contemporary society', *International Journal of Drug Policy*, 17: 258–68.

Measham, F. (2008) 'A history of intoxication changing attitudes to drunkenness and excess in the United Kingdom', in Martinic, M. and F. Measham (eds) *Swimming with Crocodiles: The Culture of Extreme Drinking*, 13–36 (London: Routledge).

Mill, J. S. (1859/1974) *On Liberty* (London: Penguin).

Moore, K. and S. Miles (2004) 'Young people, dance and the subcultural consumption of drugs', *Addiction Research and Theory*, 12(6): 507–23.

Muncie, J. (2000) 'Pragmatic realism? Searching for criminology in the new youth justice', in B. Goldson *The New Youth Justice* (Lyme Regis: Russell House Publishing).

Musgrove, F. (1964) *Youth and the Social Order* (London: Routledge and Kegan Paul).

Nayak, A. and M. J. Kehily (2008) *Gender, Youth and Culture* (London: Palgrave).

Osgerby, B. (2004) *Youth Media* (London: Routledge).

Parker, H. (2003) 'Pathology or modernity? Rethinking risk factor analyses of young drug users', *Addiction Research Theory*, 11: 141–4.

Parker, H., F. Measham and J. Aldridge (1998) *Illegal Leisure: The Normalization of Adolescent Drug Use* (London: Routledge).

Parker, H., L. Williams and J. Aldridge (2002) 'The normalisation of sensible drug use', *Sociology*, 36(4): 941–64.

Pearson, G. (1983) *Hooligan: A History of Respectable Fears* (London: Macmillan Press).

Plant, M. and M. Plant (2006) *Binge Britain* (Oxford: Oxford University Press).

Porter, R. and M. Teich (eds) (1995) *Drugs and Narcotics in History* (Cambridge: Cambridge University Press).

Rhodes, T. (1997) 'Risk theory in epidemic times', *Sociology of Health and Illness*, 19(2): 208–27.

Robinson, C. (2009) *Illegal Drug Use, Deviancy and Social Exclusion Amongst Youth*, Canterbury: Unpublished PhD Canterbury Christ Church University.

Roberts, K. (1983) *Youth and Leisure* (London: Allen & Unwin).

Roberts, R. (1971) *The Classic Slum* (London: Penguin).

Rook, C. (1899) *The Hooligan Nights* (Oxford: Oxford University Press).

Royal Society of Arts (2007) *Drugs – Facts* (London: Royal Society of Arts).

Rudgley, R. (1993) *The Alchemy of Culture* (London: British Museum Press).

Saunders, B. (2007) (ed.) *Drugs, Clubs and Young People* (Aldershot: Ashgate).

Schwarz, B. (1996) 'Night battles: Hooligan and citizen', in M. Nava and A. O' Shea (eds) *Modern Times* (London: Routledge).

Seddon, T., R. Ralphs and L. Williams (2008) 'Risk, security and the "criminalization" of British drug policy', *British Journal of Criminology*, 48(6): 818–34.

Shildrick, T., S. Blackman and R. MacDonald (eds) (2010) *Young People, Class and Place* (London: Routledge).

Shiner, M. and T. Newburn (1997) 'Definitely, may be not? The normalization of recreational drug use amongst young people', *Sociology*, 31(3): 511–29.

Shiner, M. and T. Newburn (1999) 'Taking tea with Noel: The place and meaning of drug use in everyday life', in N. South (ed.) *Drugs: Cultures, Controls and Everyday Life* (London: Sage).

Shore, H. (1999) *Artful Dodgers: Youth and Crime in the Early 19th Century* (London, Woodbridge: Boydell Press).

South, N. (1999) 'Debating drugs and everyday life: Normalization, prohibition and "otherness"', in N. South (ed.) *Drugs: Cultures, Controls and Everyday Life* (London: Sage).

Springhall, J. (1986) *Coming of Age* (London: Gill and Macmillan).

Springhall, J. (1998) *Youth, Popular Culture and Moral Panics: Penny Gaffs to Gangata Rap 1830–1996* (London: Macmillan).

Stedman-Jones, G. (1971) *Outcast London* (Oxford: Clarendon Press).
Storch, R. D. (1976) 'The policeman as domestic missionary: Urban discipline and popular culture in Northern England, 1850–1880', *Journal of Social History*, 9(4): 481–509.
Sweeting, H. and P. West (2003) 'Young people's leisure and risk-taking behaviours: Changes in gender patterning in the West of Scotland during the 1990s', *Journal of Youth Studies*, 6(4): 391–412.
UK Drug Policy Commission (2007) *A Response to Drugs: Our Community, Your Say Consultation Paper* (London: UKDPC).
Valentine, G., S. Holloway, M. Jane and C. Knell (2007) *Drinking Places: Where People Drink and Why* (York: JRF).
Wikstrom, P. O. (2006) *The Explanation of Crime Context, Mechanisms and Development* (Cambridge: Cambridge University Press).
Williams, R. (1980) *Problems in Materialism and Culture* (London: Verso).
Wilson, B. (2006) *Fight, Flight or Chill: Subcultures, Youth and Rave into the 21st Century* (Montreal: McGill-Queen's University Press).
Woodiwiss, M. (1998) 'Reform, racism and rackets: Alcohol and drug prohibition in the United States', in R. Coomber (ed.) *The Control of Drugs and Drug Users* (Amsterdam: Harwood).

7

States, Markets and New Media: The Contemporary Politics of Gambling

Jackie West and Terry Austrin

Chapter 7** is about the politics of another historic and often proscribed leisure activity –* ***gambling*. Once perceived, like drinking, as a curse of the poor, gambling, as the writers point out, is, like sex, an activity which people in Western societies are now given cultural permission to enjoy. Indeed it has become an integral part of the hospitality and online industries. The chapter, in keeping with the central theme and purpose of this book, explores the* ***tension*** *between the new* ***freedoms and opportunities*** *afforded to gamblers and the* ***regulatory regimes*** *that, via licensing and surveillance, now govern an erstwhile illicit activity.*

The legalisation of commercial gambling, the normalisation of a once pariah sector, has moved it from the periphery to a much more central position within a number of societies across Europe, North America, Asia and Australasia. Gambling, increasingly accepted as popular leisure, has been promoted as entertainment linked both to the global hospitality industry and to forms of new media and technology. Casinos and online gambling in particular testify to the erosion of class-based segmentation in gaming and to the integration of gambling into the mainstream economy. Casinos have become palaces of popular pleasure. The firms which own them include some of the best known international hotel and media chains, among them the Hilton group and MGM in Las Vegas. Online gambling is available at home, as well as in casinos and betting shops, and the internet company PartyGaming reached the FTSE 100 when it was first floated on the London stock exchange in 2005. In the case of the UK, turnover from all forms of gambling exceeded £84 billion in 2006/7. According to the Gambling Commission (2009: 3), '68 per cent of the population (about 32 million adults) had participated in some form of gambling' in 2007 and as many as 48% (that is 23 million adults) in forms other than the national lottery.

Accounts of legal gambling frame their arguments in different ways. In the UK, those in sociology have drawn on class analysis, linking the growth of gambling to opportunities for private capital to exploit working-class hopes and dreams in the face of economic, social and cultural disadvantage (Downes et al., 1976; Critcher and Clarke, 1985; Reith, 1999). More recently, the growing participation of women, particularly in the national lottery, has prompted more nuanced feminist accounts. For example, Casey shows how gambling is, for working class women, an integral part of 'respectable money management' (2003: 262). Some authors outside the UK have also emphasised the role of gambling legitimation in promoting wider neo-liberal economic agendas in contemporary capitalism (Cosgrave and Klassen, 2001) while others have highlighted the diverse economic – and political – interests to which states respond in promoting gambling of different types (Sallaz, 2009). By contrast we will follow gaming tables, electronic gaming machines and their reconfiguration on the internet to describe how state regulation is both complex and highly variable across all gambling sectors and across both national and local regions.

The growth of gambling as mass consumption has been fostered by two main developments. Following initial legalisation, when licensing replaced criminalisation to secure state control and raise revenue, governments have become increasingly interested in gambling as a tool of economic regeneration. Secondly, new gaming technologies have expanded markets while also facilitating accountability. Gambling continues to be regulated as exceptional rather than normal business, for the legitimacy of gambling as popular leisure is premised on forms of private and public control which 'sanitise' this erstwhile deviant activity (Austrin and West, 2005). So while older repressive models and prohibition have been cast aside, licensing allows not only taxation but regulatory regimes governing the number, scale and location of operations. Surveillance is also exercised – over workers and punters – by the very gaming technologies that increasingly underpin industry profitability. At the same time, political interests play a crucial role in legalisation, so regulatory regimes are always local and in tension with global tendencies in industry expansion. There is also scope for the continued sway of moral discourses and other interests opposed to gambling or to some of its forms.

The interplay of these diverse economic, socio-political influences and technologies on the normalisation of gambling is highlighted in this chapter using two case studies. The first, casino development in the UK, illustrates the shift from elitist to popular leisure, recently limited – to some extent – by political resistance to New Labour's modernisation agenda in gambling. The second, online gambling and its regulation, has been another principal government concern, particularly in the USA where political opposition has constrained growth in this sector. At the same time, new markets in casinos and internet gambling are proliferating. But the chapter begins by discussing the features that characterise gambling as popular leisure and its initial development in the USA and Australasia.[1] It is here, especially through the medium of the hotel casino, that legal gambling was first routinised as an everyday activity subject to state control.

Casinos as sites of consumption and control

The new model casino pioneered in 1930s Nevada legitimated gambling for a mass market in contrast to the more exclusive forms of gaming characteristic of Europe. Commercialisation banished both social exclusivity and connotations of addiction in the name of entertainment. This was to be dramatically extended from the 1970s and again in the 1990s through hotel resort casinos with 24-hour access to games, alcohol and credit. The casino became a space of desire and licence, epitomising new forms of experiential consumption (Reith, 1999), a world of play, illusion and 'spectacular simulation' (Ritzer, 1999: 121) played out among crowds of strangers. These new sites of pleasure entail the flattening of cultural hierarchies, a key process emphasised by postmodern theorists (Kingma, 1997). The rituals of play at gaming tables and uniformed dealers (croupiers) symbolise tradition and order alongside increasingly dominant gaming machines.

These twin sites of the action, tables and machines, are also means through which state as well as corporate interests are secured. Dealers are 'exquisitely' skilled in handling chips and cards (Lafferty and McMillen, 1989) but their scripted moves embody hierarchical control designed to rule out the deviance of operators as well as gamblers (Skolnick, 1978). All those present on the gaming floor are perceived as potential criminals, in legalised casinos no less than in illegal forms, for there is always scope for deceit in gaming (Goffman, 1967). Aside from various forms of state surveillance, dealers and punters are scrutinised by the intense and detailed observation of managers and inspectors, by the spatial arrangements in casinos and by the technologies deployed in both table games and gaming machines (Austrin and West, 2005). Specialised software provides management with the means to trace rule non-compliance as well as data on spend and turnover, while state inspectors audit daily printout and memory integrity (Crevelt et al., 1989).

Gaming machines had outnumbered tables in Las Vegas by the late 1990s, absorbing around 70% of floor space (Earley, 2001) and accounting for two thirds of gaming revenue (Eadington, 1999). Embodying new forms of control, both for operators and the state, as well as profitability, machines represent the archetypal repetitive production line for mass consumption, enabling casinos to become, in Goodman's words (1995: 124), 'theme-decorated warehouses...the new McGambling...populated by what the gambling industry calls "grind players", a clientele who sit with plastic cups of coins, pulling levers and pushing buttons'. Hence the identification of these machines by health professionals as the crack cocaine of gambling.

Electronic gaming machines transform games of skill, for both dealer and gambler, into disembodied games of chance, converting punters into self-servicing customers as in retail and finance generally (Ritzer, 1993; Lash and Urry, 1994; Knights and Tinker, 1997). Gaming machines are also essentially lotteries, blurring the distinction between table games and wagering (Austrin, 2002). Small-time consumers become potential jackpot winners as the outcomes of discrete games are linked by networks that extend beyond individual casinos, and beyond casinos into clubs and bars, and potentially the home. Products are also reconfigured, through constant innovation, expanding consumer choice and variety. Roulette, for example, can be played in one of three ways: first, as a game orchestrated by a dealer/croupier, on a single roulette table; secondly as a discrete machine game in which the dealer is simulated (displaced) by an automatic spinning device and punters place bets through a computerised 'pad'; thirdly, as a distributed video game in which punters place bets on a computerised, simulated table that is linked to a real-time game conducted in the same casino or potentially in another casino. This third form allows for the establishment of networked roulette games across terrestrial sites within a single firm and, by extension, to cyber casinos (Kelly, 2008). Since companies frequently own premises of different types (for example bookmakers/betting shops and casinos), new technologies help them overcome the segmentation of different gambling forms even without a relaxation of regulation.

Global capital, local regimes of regulation

Gambling as popular leisure is epitomised by Las Vegas, now a world resort entertainment centre (Parker, 1999; Rothman and Davis, 2002). Gambling was legalised in Nevada in 1931 by an impoverished state interested in raising fiscal revenues from the underground economy. Las Vegas underwent its first major re-invention in the post-war period, partly supported by local politicians and national mob (i.e. organised criminal) interests. Casinos were family-owned, racially discriminatory (blacks were barred as customers and from working as dealers and bar tenders) and anti-union (workers depended heavily on tips and privileges – 'juice').[2] But the full transformation of gambling as an engine of economic development was effected by corporate capital from the 1970s onward. The Circus Circus casino targeted a new market of truck drivers and factory workers in place of the high rollers, achieving this radical departure in five ways – discounted hotel charges, a cheap buffet, the hiring of women dealers, openness to inexperienced players and a substantial increase in gaming machines with their bigger jackpots (Earley, 2001).

Within three decades Las Vegas was dominated by 12 industrialised, mass-market hotel-casino resorts run on Fordist lines, each employing 6,000–8,000 shift workers, predominantly low-skill waiters (the largest category), maids, cleaners and casino dealers – an increasingly female labour force. Many of these workers (dealers aside) became well organised (Rothman and Davis, 2002), with the union a partner in the Las Vegas 'entertainment revolution' which enabled its marketing as a convention centre and international tourist destination. Links between casinos and mainstream hospitality run by multi-national firms have been especially evident since Holiday Inns bought into the Nevada-based Harrah's casino in the late 1970s, and by the 1990s, the Las Vegas strip hosted 17 of the 20 largest world hotels (Earley, 2001).

The volume and concentration of casinos in Las Vegas are untypical, in the US and elsewhere, with only Mississippi in the US adopting the Nevada model directly. But the number of casinos outside of Nevada increased from 53 to 300 between 1989 and 1995 (Thompson, 2002: 360), as other states and cities such as Detroit and Atlantic City adopted gambling tourism, albeit that tighter controls were imposed through zoning on numbers and location. And the use of gambling to promote economic (re)generation through labour-intensive employment in the service sector has also underpinned developments in US Native American reservations (Carmichael, 1998). Legalisation in Las Vegas entailed the licensing of unlimited casinos, which, in response to competition, have increasingly incorporated within them other forms of gambling such as horse and other sports betting, further enhancing the casino as a site of mass culture.

In other jurisdictions, government regulation has produced very different markets and the boundaries between distinct forms of gambling are eroded in other ways (Sallaz, 2009). In Australasia, for example, privately-owned casinos have been granted effective monopolies by state legislatures, in direct competition with state-owned or private monopolies in other types of gambling such as betting or lotteries. In Australia, a single hotel-casino was licensed in Tasmania's capital in 1973 and another centre in 1982. The model was followed in the Northern Territories, also a peripheral state, but subsequently extended to Melbourne (Victoria) and Sydney (New South Wales) with temporary casinos in 1994 and 1995. These were both made permanent in 1997, with licences of six and 12 years respectively. In New Zealand, in 1996, Auckland was granted one casino with a monopoly for two years over the entire North Island, and for five years within an area of 100 sq km (Austrin, 1998).

The launch of the Auckland casino and another in Christchurch, two years earlier, were televised live; both were quasi-state events attended by TV and sporting celebrities. Casino development in Australasia has been promoted as a key to regional employment and tourism, as elsewhere, and its clientele is a local manifestation of shifting class and cultural boundaries; indeed, casinos in New Zealand are one of the only settings where Asians, Maori and Europeans (Pakeha) come together. Casino operators are transnational players taking advantage of these new markets but on a smaller scale than in Las Vegas. The Melbourne casino, with 16 million visitors a year, employs up to 3,500 and that in Auckland up to 2,500. The Australasian regulatory regime has also forced state franchises in distinct gambling sectors into competition, leading casino operators to buy into betting firms and links between gambling and television companies. Rupert Murdoch's company, News Datacom, was among the first to develop dedicated gambling channels and interactive satellite gaming systems.

US and Australian casino interests have recently extended to China. The return of Macau (former Portugese colony) in 1999 made this off-shore location the only domain in China where gambling is legal. From 2002 the government offered concessions to six hotel casino operators including Las Vegas Sands, MGM Mirage, Wynn Resorts and an Australian-Macau partnership (Melco & PBL). Within four years turnover on gambling transactions in Macau was greater than that of Las Vegas. Distinctive features of what is now the world's largest gambling market are local to its Chinese setting (Garnaut, 2009). They currently include its state-guaranteed oligopoly, the persistence of table gaming (in this case baccarat) as the dominant form, dependence on visitors (three in five gamblers are from the mainland requiring visas) and heavy reliance on high rollers whose debts are collected by 'junket networks' which have long characterised gambling in Macau.

Casinos UK

The rhetoric of casinos as development tools recently emerged in the UK with the Budd Review (DCMS, 2001) and Labour government proposals (DCMS, 2002) to 'modernise' legislation, justified as necessary to recognise the gambling industry's contribution to the economy and to develop greater protection for vulnerable groups. This began to undermine the tradition of the small British urban 'club' table gaming casino and the principle of protecting players from the 'seductions of gambling' through paternalistic policy alone (McMillen, 1996). The government has rescinded its initial support for any major regional resort-casinos and has reduced the scale of newly legal operations established by the Gambling Act of 2005, but most of the 'modernisation' agenda has remained intact in the face of political opposition. This includes substantially more and larger casinos than were previously eligible for licensing, along with a considerable extension of machine gaming, albeit on a smaller scale than Budd had proposed.

The recent history of gambling (de)regulation in the UK is, then, a good deal more complex than it might appear from high-profile shifts in government policy on casinos (the same is true of online gambling which is the focus of our second case study). We detail these policy shifts and their implications below, but the background is a familiar one.

Decriminalisation of gambling in 1960 led to the unlimited growth of small table gaming casinos throughout Britain, but many were forced to close following the establishment in 1968 of a national gaming board which endorsed the principle of only licensing sufficient casinos to satisfy unstimulated demand. During the 1970s, some 125 clubs were operating, but one-sixth (25) were in London generating 75% of the 'total drop' (money exchanged for chips) until 1979/80 (Miers, 1983: 25). The London casinos, dominated by Ladbrokes and Coral, both with bookmaking interests, attracted high stakes international players, particularly from the Middle East. The 'club' casino was seen as a culturally and politically acceptable way to confine casino gambling to a small, affluent elite. Open only to members, casinos were limited almost exclusively to table gaming, with each allowed only ten slot machines. Additionally, they were not permitted to advertise or offer credit.

By 2000, gambling was recast as a popular and 'modern legitimate leisure activity', with the government envisaging large gaming machine-based hotel casinos as the catalysts to revitalise run-down seaside resorts. It was these aspects of policy that captured the attention of the mass media as well as the towns that stood to gain. Blackpool, for example, began planning for hotel-casinos as an integral component of new leisure and retail complexes to arrest local economic decline (BCP, 2000). It had experienced a one third drop in hotel bookings during the 1990s and its GDP was the lowest in the north-west, 12th lowest in the UK. Casinos were seen as a 'bold and imaginative' vehicle for community prosperity, directly through employment and also through local taxes on companies such as levied in many US cities.

In welcoming the Budd report, the government argued that regulation needed to be 'modernised' to make gambling safe (by tightening crime control and protecting the vulnerable, especially the young and those at risk of addiction) but also to enable the industry 'to respond rapidly and effectively to technological and customer-led developments in both the global and domestic marketplace...[so] increasing its already important contribution to the UK economy' (DCMS, 2002: 1). This initial government response stressed the financial benefits for UK operators including a share of the global market and the need to remove unnecessary barriers to both customers and new products.

These departures from traditional UK government policy attracted critical commentary from politicians and the mass media. While in keeping with New Labour's liberal economic programme, the proposals were seen by many as likely to increase problem gambling if not as openly inviting corporations to encourage this. However, as Miers (2006: 4) has argued, Budd's proposals were 'indistinguishable' from the legislation of the 1960s in aiming to 'guarantee probity of the market and to guard against inappropriate consumption', that is a crime-free industry, with fair and open conduct and protection of the vulnerable. Critical social researchers note that an expansion of counselling and other initiatives to deal with problem gambling testifies to its further medicalisation via an industry of professional experts in place of recognising its relationship to class and ethnic inequalities (Volberg and Wray, 2007). And it remains to be seen, as with the drinks industry, how far gambling operators meet their new obligations to encourage 'social responsibility'.[3] But the government was committed to balancing greater freedom for adults with tighter regulation, not just a simplified regime. This amounts, Miers (2006: 6) emphasises, to 'a much greater degree of regulatory discipline' available to the Gambling Commission than to the Gaming Board that it replaces. This includes operating and personal licences for managers and workers (as required under previous legislation), plus a battery of quality controls over operations, from the nature of transactions to the speed of gaming machines, and the power to revoke and amend licences, impose unlimited financial penalties and prosecute offences.

Government controls over operators were thus to be increased while those over premises and the forms of gambling allowed within each would be relaxed, recognising that new technologies were already promoting such developments in any case, with attendant risks. Yet parliamentary concern focused on the changing landscape of gambling, leading the government to adopt a more cautious approach, with the Bill presented giving more emphasis to 'social protection' and 'social responsibility' along with more restrictions (DCMS, 2004). Eight regional resort casinos were still proposed along with an additional 16 (eight 'large' and eight new 'small' casinos), but the number of gaming machines (and corresponding jackpots) in different locations was scaled back. Budd had recommended a machine: table ratio of 8:1 in small casinos but no caps on either the number in larger ones or the size of jackpots, hence unlimited winnings. The 2004 Bill proposed a ratio of only 2:1 in small casinos (with a maximum of 80) and 5:1 in large casinos (with a maximum of 150) albeit 25:1 in the largest regional casinos up to a limit of 1,250 machines in total. 'Community influence' on licensing was also emphasised and regional and local planning controls over the largest casinos would, with tough 'impact assessments', ensure their contribution to economic development as a condition of licensing. Nevertheless, it was still argued that social control and economic benefit could best be secured through the licensing of a few relatively large casinos in place of the proliferation of smaller ones (ibid: 50). In the event, the legislation enacted and the permissive, if contested, climate created during its passage through parliament, has seen expansion of both old and new casinos, and important changes in the former, despite lack of support for the biggest Las Vegas-style casino resorts.

The retreat on casino growth began in April 2005 when, on the eve of a general election, 15 legislative proposals were dropped including the parts of the Gambling Bill covering the regional casinos. By July it was clear that only one, not eight, would be allowed. The successful local authority, Manchester, was announced in January 2007, but its fortunes were dashed in July when Gordon Brown scrapped plans altogether for any regional casino shortly after taking over as Prime Minister from Blair.

However, this dramatic *volte face*, claimed as a moral victory by many MPs of all political persuasions and by leading tabloid and broadsheet newspapers especially *The Daily Mail* and *The Guardian*, obscures a more complex picture in which a good deal of the initially proposed legislation was in fact implemented.[4] The 2005 Gambling Act included measures covering 16 new casinos, all larger than those previously allowed.[5] Although the secondary legislative orders necessary to authorise these were rejected by the House of Lords in March 2007, they returned to parliament in February 2008 and were passed in May (Miers, 2008), albeit with a requirement for casinos to close for at least six hours a day and a ban on free drinks and the use of credit cards. Miers (2006: 41) has suggested that all new generation casinos are very much more likely to be 'large' than 'small' since fewer tables can be matched by more machines on account of the different ratios – tables with their associated labour costs (dealers and inspectors) are much more costly investments. At the same time, before the new law came into effect, firms used the existing legal framework to significantly expand their operations.

In the year to April 2006, the deadline for licences under the 1968 Act, the Gambling Commission had received 111 new applications and by the end of that year the number of new casinos approved or under consideration had taken the total in the UK to more than 200, double the number that had existed a decade earlier (when Labour assumed power). Yet by March 2008, there were only 144 in business, a good deal less than the expansion predicted (Miers, 2008). However, many of these new 'older generation' casinos were also larger than their predecessors and were designed to appeal to a mass entertainment market by being combined with cafes, bars and restaurants. At least one local authority, Nottingham, indicated its refusal, as specifically provided for under the 2005 proposals, to grant any licences for new style casinos but did grant one to London Clubs for an additional older style casino. This, together with a further five new licences in other cities, was set to more than double the firm's traditional casino floor space (Bowers, 2006a). In addition, advantage was taken of a loophole in the 2005 Act which, while restricting existing older generation casinos to 20 slot machines, placed no limit on the number of electronic roulette terminals, which were not classed as automated gaming machines since they use live data from tables in the casino even though they are played 'virtually' by the punter. Following the tendency to equalise licensing conditions for different sectors, this echoes a decision to allow gaming machines known as FOBTs (fixed odds betting terminals) in betting shops, a move that is credited with doubling the share price of William Hill (Bowers, 2006a). In 2005, Stanley Casinos, which owned 46 sites, was reported (*The Guardian*, 25.10.06, 21.01.07) to have doubled its provision of touch screen roulette with earnings confidently predicted to exceed those from traditional table games.

Online gambling

Writing in 1985, Critcher and Clarke emphasised how the class-based practices of gambling were intimately linked to location. For working-class culture, this was the street, the race meeting and the bingo club, with football pools played in the home. These forms of physical as well as social segmentation have been eroded by casino developments and the wider deregulation of gambling markets promoted by new media. In the UK, horse and greyhound racing, football and bingo are now to be found on bookmakers' websites and sports betting can be played from a computer, or some mobile phones, any day, any time. The same is true for gaming machines and table based games such as blackjack and poker. Gambling, once set apart from the routines of daily life as an exception, has become routine and distributed across diverse media and settings including the home. For a number of global companies, online gambling has underpinned new heights of profitability and stock market flotation. But they have also met with political and economic resistance which has stimulated further innovations.

The extraordinary proliferation of online poker is perhaps the most significant of these developments. Its scale is evident from the fact that at the time of completing this chapter (18.00 hrs UK time, 30.11.09), 211,352 players were taking part in 6,864 tournaments on the website PokerStars.com alone. Echoing the innovations of Circus Circus in Las Vegas, online poker provides an opening for inexperienced gamblers. But a further aspect of its popularity is the route it additionally provides to a place in face-to-face regional, national and global tournaments, epitomised by Chris Moneymaker (his real name), an online poker player and winner of the 2003 World Series of Poker. As Wilson (2005) observes, 'a low rent game for card sharks and hustlers' has been transformed into one whose 'competitors are as likely to be resource managers, accountants and students' while its 'top professionals are TV celebrities on a par with basketball players and racing drivers, similarly sporting sponsors' logos, usually for gaming websites'. Victoria Coren (2007), a professional journalist and winner of the European Poker Tournament in 2006, also testifies to the many transformational features of this new market.

Online poker makes popular television. Coren herself has acted as commentator and presenter of the television show *Late Night Poker* and *The Poker Nations' Cup*, both for the UK Channel 4. She also writes columns for *The Guardian* and *The Observer* and is the author of a book (Coren, 2009) chronicling her own career. In Europe, *The Poker Channel*, its largest gaming television network, broadcasts online the most popular tournaments including the World Series of Poker and the World Poker Tour. The channel attracts around one million TV viewers a month, with 65% playing poker online everyday (http://www.pokerchanneleurope.com). As Farnsworth and Austrin (2010) comment:

> [these developments] allow viewers not only to follow the game onscreen, or through websites or through wireless mobile servers, but also to participate directly. The game onscreen is provided with cameras embedded in the card table and in miniaturised chips embedded in each card. Like cricket, viewers can follow the game from multiple viewpoints and receive detailed, blow-by-blow statistics as the face-to-face play unfolds. Moreover, viewers can gamble themselves on the fate of the gamblers at the table, using deal histories and updated player form.

In turn the success of online poker has led to the establishment of pub poker leagues in the UK. These were made possible by the 2005 Gambling Act. The first of these, the Nuts Poker League, founded in Wrexham, North Wales, now has over 900 venues and runs more than 120 regional events. The largest, Redtooth Poker, with 50,000 players registered in over 1,000 pubs is sponsored by a brewery (Carlsberg) and bookmaker (Bet Fred Poker) (http://www.redtoothpoker.com/national.asp).

Shifting fortunes

Online developments have more generally contributed to the participation of gaming firms in mainstream financial markets. 1st June 2005 saw the flotation of PartyGaming on the London Stock Exchange, with a value of £5.16 billion at close of trading that day, equivalent to more than BA and EMI combined, and its subsequent membership of the FTSE 100. This was by no means the first gambling company to be listed and thus secure investment from major City financial institutions. For example, the shareholders of Sportingbet, floated in 1999, have included Merrill Lynch, Goldman Sachs and Fidelity (Richtel, 2005).

However, the industry's fortunes were curtailed by an intensification of legal action in the US during 2006. The US Wire Act of 1960 prohibits sports betting ('wagers on sporting events or contests') although a court judgement in 2002 ruled that it did not apply to non-sports internet gambling (that is games of chance rather than skill; the category to which poker belongs is disputed). The US accounted, until 2006, for around half of the industry's profit (and for some firms the great majority of sales), with an estimated 12.5 million Americans betting online, but attempts to outlaw this were eventually successful. By July 2006 a bill to this effect was passed in the House of Representatives by a 3:1 majority, followed by the arrest of the Chief Executive of another firm, BetonSports, on charges of racketeering using legislation passed in the 1960s against the mafia. In September, the Chairman of Sportingbet was arrested, this time using a Louisiana statute on 'offences affecting general morality', specifically updated in 2003 to cover internet gambling (Bowers and Clark, 2006). By October, a last-minute amendment to a bill on port security secured the Unlawful Internet Gambling Enforcement Act, passed by 409 votes to two (Pratley, 2006a). This banned banks and credit card companies from processing payments to online companies. Earlier falls in share prices (sometimes substantial, as firms sought to deal with what they called 'regulatory uncertainty' in the wake of the above arrests) turned into an overnight industry crash, analogous to the dotcom collapse (Wray, 2006), with most online gambling firms closing their US-facing websites if they had not already done so.

Political action was taken in the name of morality but inspired by more mundane interests. One US Congressman, Jim Leach (Republican), called gambling 'injecting drugs without needle marks: you just click on the mouse and lose your house' (quoted in Pratley, 2006a). But Leach had also cited concerns with money laundering, terrorist fundraising – and tax evasion. More significantly, the 2006 Act, like its predecessors, includes crucial exemptions, in this case horse racing, state lotteries and fantasy sports. The Safe Port Act, to which the gambling ban amendment was attached, confines ownership of ports to interests perceived as friendly to the US. The World Trade Organisation ruled in 2005 that US laws contravened its principles of competition and in March 2007 repeated this, claiming that the US ban amounted to discrimination against foreign companies. Similar issues have been evident in France. Arrests of gambling firm executives in September 2006 followed complaints from French lottery and horse betting operators, notwithstanding a European Court of Justice ruling in 2004 which was designed to limit such protectionism (Bowers, 2006b). Claims by the gambling industry that fiscal protectionism is masquerading as social welfare (Yeager and Blitz, 2006) would appear to be justified.

The impact of specific economic and political restrictions on internet gambling is, however, complex. Some firms have gone bankrupt, others have been bought out, and the reduction in the volume of players reduces the size of jackpots with corresponding knock-on effects on punter interest. The loss of the US market has prompted diversification of products (PartyGaming, for example, has launched a backgammon site and has encouraged poker players to play blackjack and *vice versa* through shared accounts), along with aggressive recruitment of customers in new markets in Europe and Australasia, and consolidation. Sportingbet in September 2009 had 3.9 million customers and over 40 websites in 21 countries. Share prices of the firms remaining have recovered, albeit not to 2005 levels. PartyGaming was reported to have lost four-fifths of its revenue in the wake of the US ban but this did not prevent it announcing rises in profits – for example, in net profits in June 2006 and in pre-tax profits in March 2008. And Pratley (2006b) has astutely observed that listing on the stock exchange was always less about securing legitimacy for an erstwhile pariah industry than a calculated move on the part of many firms' founders to take large winnings by selling part of their holdings – a classic gambler's strategy. As he emphasises, the regulatory risks were made abundantly clear in float prospectuses (over 30 pages in the case of PartyGaming on the likely violations of US law and the prospect of a ban even on businesses without a physical presence in the US). And yet mainstream investors (like naïve punters, Pratley notes) were willing to 'stay at the table'. The winnings accumulated were indeed good – two of PartyGaming's four founders were among the top ten in the Asian Rich list by April 2006 (at numbers three and seven).

The US ban was first challenged in April 2007 by Democrats who anticipated that a licensing regime would raise taxation of $26 billion over five years, and the change of administration in 2009 might make such a regime very much more likely. In the UK, the government saw its 2005 Gambling Act as a key instrument in raising revenue but also in tightening regulation. A precedent was set when high-street betting firms, including Ladbrokes, negotiated a major tax concession in 2001 – when Gordon Brown was Chancellor of the Exchequer – in exchange for moving phone and online sports betting back to the UK. Advertising by gambling firms is now legal (though on TV only after the 9pm watershed)[6] providing that they are based in approved locations, that is those where the regulatory regime is considered sufficiently robust. However, these locations include Gibraltar where low taxation remains a strong incentive for firms to remain offshore. The industry regards the UK's 'remote gaming duty' of 15% as prohibitive and low taxation regimes make it easier for firms to offer highly competitive bets on sporting events. So the UK market is now very open, if not to all, but firms are not required to pay UK taxes or comply with specific regulatory conditions including those on problem gambling.

Political interests do shape the form taken by corporate developments, as UK casino regulation shows, but they are always under challenge. Indicative of the adaptability of gambling to its political 'masters' is the announcement in December 2006 by Las Vegas Sands, the then largest US casino by market value, of its link with Cantor Gaming to launch an internet casino website. This would not accept US punters but would be aimed at the UK market.

Conclusion

Gambling as global entertainment is an integral part of what Hobbs et al. (2000: 702) refer to as 'an increasingly complex mass of night-time leisure options through which flow new economic and employment opportunities'. But this 'urban frontier' and the 'marketing of liminal licence' (ibid.), the replacement of illegal 'carousal zones' by 'diversion districts' (Judd, 1995), is not governed by commercial imperatives alone. National and/or local governments have become inspectors of leisure opportunities rather than policemen. But they retain the power to license and franchise operators, employees and machines. This means that the right to organise and sell gambling remains a privilege and that national and/or local, rather than purely global, influences are critical. The UK government's attempt to limit gaming machine-based casino expansion, its interest in regulating online gambling, as well as attempts to address problem gambling are indicative of this, as is the US government's ban on internet gaming, notwithstanding the distinct economic as well as moral interest behind such measures.

Interests in the segmentation of gambling markets thus remain, but they are continually challenged, in particular by the possibilities for transformation created by new technologies. It is these that have especially fuelled the UK government's interest in tougher regulation over operators, if not premises, and this interest should not be dismissed as purely rhetorical. It legitimises resistance to untrammelled casino growth by other constituencies such as local authorities and the outcome of the planned future review of new developments cannot be predicted. Nevertheless, the innovatory capacities of new media technologies cannot be underestimated. The gaming machine-based hotel-casino is constituted as a fixed city centre or beach-front setting to which state development strategy attempts to draw the mobile tourist. Large, if not regional, casinos have similar goals. But gambling games such as poker and roulette have themselves become mobile and now circulate through different distribution networks, including television, video and the internet. These media are transforming gambling and the entertainment sector more generally as casinos of all types promote networked table gaming (such as poker and roulette) as well as machines. Along with online gaming in the home in addition to designated leisure settings, gambling is increasingly normalised as a component of the entertainment industry. These new technologies are simultaneously central in guaranteeing 'secure' public and private means of surveillance for corporations and for states, although the balance of economic and political interests is frequently contested.

Notes

1 The chapter draws on two previous publications, Austrin and West, 2004 and 2005, together with documentary sources on the most recent developments. These include daily international financial news reports from the UK, US and Australia from June 2005 through 2009. Our 2005 paper on casino work and surveillance is based on fieldwork observation and interviews in Australasia, the USA and the UK between 1994 and 2003. Sites included casinos in Nevada, New Jersey, Melbourne, Sydney, Brisbane and New Zealand.

2 Tipping remains in US casinos but has not been imported outside. In Australasia and the UK the practice is viewed as directly encouraging the involvement of dealers in the financial arrangements of table games and is therefore prohibited.

3 The Responsibility in Gambling Trust (RIGT) was established in 2004, replacing an earlier industry charity. It is one of several routes through which the government is funding research on gambling behaviour and prevalence, treatments for addiction,

the social contexts of gambling and other questions. Miers (2008) notes that the industry's annual voluntary contribution to such activities stood at £4.5 million in March 2008, way short of the estimated £9 million or so needed, but that the industry's major players are among the 80% who subscribe. While there is the possibility of activating a statutory levy (provided for in the 2005 Act), the government prefers self-regulation. At the same time, as he also emphasises, a social responsibility code of practice is a condition of licensing and requires detailed policies and procedures for the identification of problem gambling behaviour and measures to address it. Breaches of this code are a criminal offence. A crucial High Court judgement has balanced an operator's 'duty of care' with an individual's responsibilities for their own actions (i.e. losses) even in the case of a problem gambler. And there have been difficulties enforcing the removal of certain gambling machines from places like fast-food outlets to which young people have access. But Miers (ibid.: 598) also suggests that the industry will accept real changes in the exercise of social responsibility as 'the price [it] pays for a regulatory environment that facilitates its commercial interests'.

4 The 'moral politics' of gambling is also complex. Brown's position was linked to his Presbyterian roots, as were those of James Purnell, who was, by then, Culture Secretary, as well as to Brown's interest in distancing himself from his predecessor, for example through more democratic parliamentary outcomes. As noted below, he approved tax concessions to the bookmakers when in control of the Treasury. The Labour Party still bears the imprint of its roots in Methodism, yet MPs representing northern constituencies, where economic regeneration and local authority support for casinos was extensive, were reported to be 'furious' about the abandonment of regional casinos.

5 41 local authorities had competed for the right to have such a casino licensed in their area, indicating the spread of support for such new initiatives. The successful include Blackpool which had also been a front-runner among the 27 authorities competing for the single regional site before it was abandoned.

6 The ban is voluntary (and does not apply to betting advertising around major sporting events) but is embedded in the social responsibility code (see note 3). Miers (2008: 589, note 50) notes the near total industry compliance with codes of practice on broadcast and print advertising.

References

Austrin, T. (1998) 'Retailing leisure: Local and global developments in gambling', in H. C. Perkins and G. Cushman (eds) *Time Out: Leisure Recreation and Tourism in New Zealand and Australia*, 167–81 (Auckland: Addison, Wesley, Longman).

Austrin, T. (2002) 'Symbolic technologies, surveillance and simulation: Embodied skills and programmed memories in casino gaming'. Paper presented at The 20th International Labour Process Conference, University of Strathclyde, Glasgow, April 2–4.

Austrin, T. and J. West (2004) 'New deals in gambling: Global markets and local regimes of regulation', in L. Beukema and J. Carillo (eds) *Globalism/Localism at Work*, 143–58 (Amsterdam: Elsevier).

Austrin, T. and J. West (2005) 'Skills and surveillance in casino gaming: Work, consumption and regulation', *Work, Employment and Society*, 19(2): 305–26.

BCP (Blackpool Challenge Partnership) (2000) *Blackpool Rejuvenated: Proposals to Bring Prosperity to Blackpool through Resort Casinos* (Blackpool: Blackpool Challenge Partnership).
Bowers, S. (2006a) 'Overseas operators make a play for the big money in British casinos', *The Guardian*, 25 October.
Bowers, S. (2006b) 'Online gambling firms' founders may face jail over French deal', *The Guardian*, 20 September.
Bowers, S. and A. Clark (2006) 'Sportingbet arrest threatens Internet gambling', *The Guardian*, 9 September.
Carmichael (1998) 'Foxwood's resort casino: Who wants it? Who benefits?', in K. Meyer-Arendt and R. Hartmann (eds) *Casino Gambling in North America*, 132–61 (New York: Cognizant Communications).
Casey, E. (2003) 'Gambling and consumption: Working class women and UK national lottery play', *Journal of Consumer Culture*, 3(2): 245–63.
Critcher, C. and J. Clarke (1985) *The Devil Makes Work: Leisure in Capitalist Britain* (London: Hutchinson).
Coren, V. (2007) 'Flush with cash', *The Guardian: G2,* 21 September, 12–15.
Coren, V. (2009) *For Richer, For Poorer: A Love Affair with Poker* (London: Cannongate Books).
Cosgrave, J. and T. Klassen (2001) 'Gambling against the state: The state and the legitimation of gambling', *Current Sociology*, 49(5): 1–15.
Crevelt, D., D. Crevelt, J. Gollehon and L. Crevelt (1989) *Slot Machine Mania* (Grand Rapids, MI: Gollehon Press).
DCMS (Department for Culture, Media and Sport) (2001) *Gambling Review Report* (Budd Report), CM5206 (London: HMSO).
DCMS (2002) *A Safe Bet for Success: Modernising Britain's Gambling Laws* (London: Department for Culture, Media and Sport).
DCMS (2004) *Draft Gambling Bill: Government Response to the First Report of the Joint Committee on the Draft Gambling Bill; Session 2003–4*, HMSO Cm 6253 (London: Department for Culture, Media and Sport).
Downes, D. M. et al. (1976) *Gambling, Work and Leisure: A Study across Three Areas* (London: Routledge & Kegan Paul).
Eadington, W. R. (1999) 'The economics of casino gambling', *Journal of Economic Perspectives*, 13(3): 173–92.
Earley, P. (2001) *Super Casino: Inside the 'New' Las Vegas* (New York: Bantam).
Farnsworth, J. and T. Austrin (2010) 'The ethnography of new media worlds? Following the case of global poker', *New Media and Society*, Prepublished 5, 4, 2010, DOI: 10.1177/1444809355648.
Gambling Commission (2009) *Industry Statistics 2008/09* (Birmingham: Gambling Commission).
Garnaut, J. (2009) 'Macau's seedy casino war turns to gold', *Sydney Morning Herald*, 22 September.
Goffman, E. (1967) 'Where the action is', *Interaction Ritual: Essays in Face to Face Behaviour* (Chicago: Aldine).
Goodman, R. (1995) *The Luck Business: The Devastating Consequences and Broken Promises of America's Gambling Explosion* (New York: Free Press).
Hobbs, D., S. Lister, P. Hadfield, S. Winlow, and S. Hall (2000) 'Receiving shadows: Governance and liminality in the night-time economy', *British Journal of Sociology*, 51(4): 701–17.
Judd, D. (1995) 'Promoting tourism in US cities', *Tourism Management*, 16(3): 175–87.
Kelly, J. (2008) 'New game revolutionizes roulette', http://gamingfloor/features/Rapid_Roulette.htm

Kingma, S. (1997) '"Gaming is play, it should remain fun!" The gaming complex, pleasure and addiction', in P. Sulkenen et al. (eds) (1997) *Constructing the New Consumer Society*, 173–93 (London: Macmillan).
Knights, D. and T. Tinker (1997) *Financial Services and Social Transformation* (Houndmills: Macmillan).
Lafferty, G. and J. McMillen (1989) 'Laboring for leisure: Work and industrial relations in the tourism industry: Case studies of casinos', *Labour and Industry*, 2(2): 32–59.
Lash, S. and J. Urry (1994) *Economies of Signs and Spaces* (London: Sage).
McMillen, J. (ed.) (1996) *Gambling Cultures: Studies in History and Interpretation* (London: Routledge).
Miers, D. (1983) 'Malpractices in British casino management', in M. Clarke (ed.) *Corruption: Causes, Consequences and Control*, 24–38 (London: Frances Pinter).
Miers, D. (2006) 'Implementing the Gambling Act 2005: The Gambling Commission and the casino question'. Paper presented at 13th International Conference on Gambling and Risk Taking, Lake Tahoe, Nevada, 22–26 May.
Miers, D. (2008) 'Gambling in Great Britain: Implementing a social responsibility agenda', *Gaming Law Review and Economics*, 12(6): 583–98.
Parker, R. (1999) 'Las Vegas: Casino gambling and local culture', in D. Judd and S. Fainstein (eds), *The Tourist City*, 107–23 (New Haven: Yale University Press).
Pratley, N. (2006a) 'The deck is stacked', *The Guardian*, 4 October.
Pratley, N. (2006b) 'They bet on weak regulation – and lost', *The Guardian*, 3 October.
Reith, G. (1999) *The Age of Chance: Gambling in Western Culture* (London: Routledge).
Richtel, M. (2005) 'Wall St. bets on gambling on the web', *The New York Times*, 25 December.
Ritzer, G. (1993) *The McDonaldisation of Society: An Investigation into the Changing Character of Contemporary Social Life* (Thousand Oaks, CA: Pine Forge Press).
Ritzer, G. (1999) *Enchanting a Disenchanted World: Revolutionising the Means of Consumption* (Thousand Oaks, CA: Pine Forge Press).
Rothman, H. K. and M. Davis (eds) (2002) *The Grit Beneath the Glitter: Tales from the Real Las Vegas* (Berkeley: University of California Press).
Sallaz, J. J. (2009) *The Labor of Luck: Casino Capitalism in the United States and South Africa* (Berkeley: University of California Press).
Skolnick, J. H. (1978) *House of Cards: Legalisation and Control of Casino Gambling* (Boston: Little Brown and Company).
Thompson, W. N. (2002) 'Nevada goes global: The foreign gaming rule and the spread of casinos', in H. K. Rothman and M. Davis (eds), *The Grit Beneath the Glitter: Tales from the Real Las Vegas*, 347–62 (Berkeley: University of California Press).
Volberg, R. and M. Wray (2007) 'Legal gambling and problem gambling as mechanisms of social domination?', *American Behavioural Scientist*, 51(1): 56–85.
Wilson, J. (2005) 'Millions at stake in huge Vegas poker tournament', *The Guardian*, Poker Column 7.
Wray, R. (2006) 'Washington's weekend ambush wipes £46bn off the value of on-line gambling shares', *The Guardian*, 3 October.
Yeager, H. and R. Blitz (2006) 'Gambling change that could be nothing but fantasy', *Financial Times*, 26 September.

8

Towards Web 3.0: Mashing Up Work and Leisure

Andy Miah

Chapter 8 is about the internet, an instrument now indisputably central to the politics of leisure and pleasure in Western societies and elsewhere. In mainstream political circles the internet is principally lauded as an instrument of creativity, free expression and democracy-enhancing communication. The advent of My Space, YouTube, Flickr, Twitter and other web facilities – collectively known as Web 2.0 has greatly enhanced this view. This chapter examines these claims, considering the tension between regulation and freedom in an online leisure world and evaluating the popular claim that the internet has transformed leisure.

Introduction

> Who are these people? Seriously, who actually sits down after a long day at work and says, I'm not going to watch *Lost* tonight. I'm going to turn on my computer and make a movie starring my pet iguana? I'm going to mash up 50 Cent's vocals with Queen's instrumentals? I'm going to blog about my state of mind or the state of the nation or the steak-frites at the new bistro down the street? Who has that time and that energy and that passion? (Grossman, 2006: html)

In 2006, TIME Magazine announced its Person of the Year as 'You', drawing attention to the boom of online user-generated content made available by new online environments such as YouTube, Facebook, Second Life, Wikipedia and MySpace, collectively described as social media for their capacity to share content across multiple platforms. Grossman claims that the collective achievement of social media users is no less than 'the many wresting power from the few' (ibid.: html). While the principles of these online environments have their roots in 1990s web architecture, with such platforms as eBay and Amazon, the mainstream utilisation of social media, as a more accessible form of publishing, derives from the proliferation of blogging platforms, mobile devices and the optical web – when anonymity shifted towards heightened transparency and visibility. The term that is now used to describe these digital environments is *Web 2.0*, a term coined by Tim O'Reilly in 2005 to denote a shift in how digital media content is generated and syndicated around the web. The concept describes a set of practices, protocols and interactions that have brought about a dramatic shift in how digital publishing and interaction takes place.

Two years later, TIME's Person of the Year for 2008 was US President-Elect Barack Obama, whose successful Presidential campaign was claimed as being due partly to his use of Web 2.0 environments. The legitimacy of this claim is difficult to verify. Yet, by the time of his inauguration on 20 January 2009, Barack Obama had over three million fans on one single page within Facebook, the membership of which had collectively posted over 500,000 comments in a few months.[1] In addition to this, there are countless other Barack Obama Facebook pages with similar numbers, the members of which will have been linked to numerous other platforms where their affiliation is syndicated multiple times, thus exponentially amplifying the impact of his personality.

Speaking in Liverpool in 2008 just after the US election result, life-long civil rights activist Jesse Jackson played down the role of the Internet in the election, explaining that it was won on the back of decades of campaigning for African American civil rights, rather than due to videos on YouTube. Yet, the Internet was an integral part of Obama's campaign discourse and received considerable prominence in the media, which, at least, might reveal something about how politics is reorganised in a Web 2.0 era. In this sense, even in the absence of evidence to clarify the Internet's role in determining the outcome of Obama's campaign, it is apparent that being part of the newly upgraded electronic superhighway was nevertheless an integral part of being newsworthy. Thus, being seen as an early adopter of new online environments offers some degree of cultural capital through which Obama was able to amplify his credibility.

Other governments have also invested into occupying Web 2.0 environments, to reach voters. For example, in the United Kingdom the home page of the 10 Downing Street website links to YouTube and Flickr accounts. Moreover, in 2007 at the height of YouTube's growth, UK opposition leader David Cameron set up 'WebCameron' as his way of speaking directly with the electorate. Most famously, when it launched, one of the first clips portrayed him in his kitchen talking direct to the camera about policy, while washing dishes and helping his child to clean her hands. In the camera shot, viewers also saw an indoor washing line in his living room, full of clothes, portraying a modest lifestyle and a relatively small home. While one might dismiss such performances as gimmicks calculated to humanise politicians, such exploits have now become a mainstay of political work in the West. A key reason for this has to do with how the Internet has changed as a mode of publishing. Thus, the principal value of WebCameron is its capacity to be shared across other platforms by the process of automated syndication. Were it just a static html website, like those of the Web 1.0 era, far fewer people would have been able to engage with the content. Coming to terms with these conditions of online communication has become an essential component of public relations campaigns.

The mainstreaming of the political web is one prominent example of how work and leisure is 'mashed up'[2] online. It is a clear case through which our consumption of leisure spaces becomes a subtle moment of leisure activism, if not leisurely activism. Even if we may resist labelling signing up to Barack Obama's Facebook page as a form of political engagement, this act of clicking a mouse button two or three times expresses something about what voters value. This performative act also becomes a form of social activity, a public performance of sorts, which reaches our peers and often, wider communities. In short, by performing such acts, we become social and political devices – literally forces of political labour. Our leisure time activity becomes part of the campaign.

As we begin to inhabit the period between Web 2.0 and Web 3.0, what can we expect from future interactions online? Over 20 years ago Clarke and Critcher's *The Devil Makes Work* articulated an emerging computer era in the home and the workplace:

> it is the home computer terminal which most clearly represents the potential of the new technology to overturn all our existing ideas of work, leisure and the home. The one piece of technology contains the worlds of work, entertainment, shopping, household responsibilities (accounts), and last, but not least, education. The whole world can be at our fingertips (1985: 192).

They highlight the possible 'decline of work, the terminal family, the information society, and the erosion of the social and geographical boundaries between work, leisure and the home' (1985: 192–3), as processes brought about by computer culture. They also consider some of the moral and political problems that could arise from computing, not least of which is the consequences of labour saving technology, the rise of a digital management elite, and the transformation of economic conditions.

However, in the 1980s there was little opportunity to foresee some of the major issues that now dominate debates about the Internet, such as privacy and surveillance, questions of identity, or indeed, what might be described as *leisure work*,[3] which is characterised by the kinds of enterprise that have built contemporary digital artefacts such as Wikipedia, YouTube (Gehl, 2009) or even eBay (Robinson and Halle, 2002). Moreover, there is little that could have been imagined about the intricacies of file sharing (Rojek, 2005), with its early manifestations through Napster through to the BitTorrent driven platforms such as the recently controversial Pirate Bay.[4] Each of these leisure spaces have found themselves locked into legal battles with those who hold the right to exploit creative media. Alternatively, while today the miniaturisation of media technology seems an obvious consequence of the microtechnology era, few could have foreseen the emergence of portable DVD players, Bluetooth headsets and the iPad. Each of these technologies has changed leisure experiences in some way, though the magnitude of this change is a matter of judgement.

These examples situate Internet politics firmly within the sphere of established leisure experiences. Yet, it is reasonable to ask whether any new leisure practices have emerged that are unique to the pervasive experience of new and mobile media and its emphasis on ubiquitous reporting. In short, is there anything new about leisure in an Internet era, or is it merely differences of degree rather than kind? After all, much of what happens online involves experiencing leisure activities that have been around for much longer: watching films, browsing pornography, listening to music and so on. Does, for instance, an Apple iPod – one of the many portable media players available now – constitute a fundamentally different personal music experience compared to the walkman stereo or even the record player? Alternatively, does the lifestyle appeal of an Apple notebook computer introduce a new aesthetic to computing which is qualitatively different to PCs? Each one of these devices, and many other nuances about the Internet era, has its own, distinct social history that deserves investigation, but what interests me here is to document the different trajectories that have emerged online throughout this 20-year period since Clarke and Critcher.

Perhaps one of the most defining dimensions of the Internet era is its having been accompanied by a series of new work and leisure practices. Recent examples of this include the rise of the new media entrepreneur, individuals who have a career history that crosses platforms (television, film, web, music, photography). It also encompasses the development of extreme programming as a new methodology of working within information technology industries, whereby new modes of collaboration have emerged. Thus, extreme programming describes a work force of computer programmers, where individual employees do not have a fixed role, but instead focus on tasks needing completion, thus allowing individual expertise to focus on tasks that they are most competent to handle. Additionally, open source technology has given rise to new communities of software development. Open source describes a form of software that, rather than protect the programming source code from outsiders, allows others to see how it is assembled, thus releasing the intellectual property of the proprietor. This allows others to collectively improve the code and the final product, whereby a large number of people within the user community become co-producers of the software, thus transforming the labour force that contributes to improving computing technology. These new forms of labour dominated popular discussions about the 'dot com' bubble and gave rise to new forces of capital and even new patterns of work. While the Internet has clearly not led to everyone working from home, as might have been imagined some years ago, new media work practices do often involve new conditions of office organisation. For instance, it is common for a new media developer – a programmer for instance – to also have their own creative practice, either as a photographer, film maker, or musician, for instance. Thus, one might argue that the distinct leisure activity that is afforded by the Internet is part of a broader shift towards what Florida (2002) calls a 'creative class', where people are able to experiment with computer programmes to make films, animate, compose, and so on. Thus, one might not be able to separate the rise of the Internet from the rise of the amateur digital photographer, or the YouTube film director. Of course, through these examples, one can also identify a crucial component of the Internet's leisure work practice: advertising. Perhaps the most prominent labour presence online is generated through advertising campaigns, which underpin many of the freely available tools used by millions of people in online communities. Again, our role in shaping how advertising functions online has become increasingly important, as programmes approach a form of collective intelligence through such innovations as Google AdSense.[5]

Anyone who has written about the Internet is conscious of how fast it changes and how quickly ideas date. Even the terminology one uses to describe certain digital practices can only be tentative since they change so quickly. Over the last 20 years, numerous dimensions of online activity have come and gone and the emerging redundancy or adoption of any platform is one of the most difficult aspects of the Internet to foresee.[6] These trajectories are also among the most interesting facets of online history, and they can easily be overlooked. Thus, it is tempting to talk about platforms as static environments, but actually they change constantly. Consider the recently prominent Twitter, which began as a microblogging device, but has become also an emailing and real-time search facility.

Thus, telling the story of any technological device – including the Internet generally – involves revealing a rich set of sociological configurations that characterise its use. Often, design intentions are usurped, rewritten and betrayed and unforeseen user cultures emerge, along with reactions to such use. For instance, who might have foreseen the rise of the Diana retro plastic camera, as a response to compact digital photography or, the growing prominence of camera mobile phones and photo blogging?[7] Alternatively, what happened to the Palm Pilot personal digital assistants (PDAs) of the early 2000s, the handheld computers that were all but wiped out by the Applie iPhone and Blackberry devices? These new devices offered integrated telephone and internet services, along with sophisticated software applications that provided, among other things, global positioning system (GPS) navigation devices and even compasses.[8] Who might have foreseen that free wireless at airports would lead to people walking around with their laptop under their arm at airports making free telephone calls via Skype, one of the early, free voice-over internet devices?

Other anecdotes from the Internet's history provide beguiling articulations of work and leisure innovation and integration. For instance, it is widely known that if one wants to see cutting-edge innovation online, then a good place to look is at pornographic websites (see Perdue, 2002), which, in the 1990s excelled in utilising 'pop-up' windows that would force users to shut down their computers completely or be drawn into an infinite loop of closing browser windows one after the other. Also, who might have predicted that, in the era of pervasive email, by 2009 around 94% of it would be spam? (Stone, 2009). Another rich layer of this computer culture is the way in which technologisation leads to the emergence retro cultures of creative media practice – consider the digital gaming environments of MiniClip.com or even ZX Spectrum games console emulators, and so on.

These introductory remarks offer a starting point for discussion about how the Internet has created a wide range of tensions between work and leisure practice. They also convey how a space of popular activity has been created somewhere between the two. Perhaps the single concept that best describes such activity is citizen media, a term that has emerged to recognise the way in which personal publishing online breeds the information economy and makes manifest the emancipatory potential of Castell's 'network society'. Yet, key questions remain about whether we should treat these practices as leisure activity or otherwise. Moreover, limitations to its popularity exist around the precise conditions that enable participation. Today, we speak more of a digital literacy divide than a digital divide, which implies accessibility differences. However, one cannot ignore the remaining technological barriers, which shift as the technology evolves. For instance, despite an advanced infrastructure, it was recently shown that broadband users in the United Kingdom have far less band width than they pay for, or would need to use many cutting edge platforms. Additionally, the original concerns about the digital divide remain, where there is much more work needed to bring the internet to most homes. Each of these matters, along with the question of how the Internet has emerged as a dominant media form over this period will be considered in this chapter, which will aim to articulate various periods of Internet development in the context of leisure experiences. Today, these eras may be conveniently characterised as Web 1.0, Web 2.0 and, soon, Web 3.0.

Web 1.0: The internet as a blurred boundary

In what ways do leisure and work activities occur within digital spaces? Is space even an appropriate metaphor through which to describe virtual information networks? Alternatively, how does our experience of being online compare to, say, watching television, listening to the radio or going to the theatre? These questions were asked when the Internet was first popularised as a publicly accessible environment. In some areas, debates focused on the subject of regulation, where the central question was (and remains) whether the Internet can (or should) be regulated. Perhaps, unlike any media form before it, our early experience of the Internet was as a loosely regulated space. Many of the regulatory devices that exist today emerged partly as a consequence of legal action arising from absences in how content was regulated, as for file sharing platforms or privacy invasions by web-based companies. This is not to say that internet service providers or, indeed, our own use of the Internet was not governed by rules or codes of practice. Rather, it is to draw attention to the fact that a great deal of user experience involved performative acts of freedom despite these rules (Cairncross, 1997; Castells, 1997; Jones, 1997; Turkle, 1995). For instance, Markham (1998: 35) describes how using a pseudonym online – a precursor to the use of an avatar[9] – provided a 'sense of freedom in a dislocated place where one can be anyone or anything simply by describing oneself through words and names'.

While it is difficult to claim that Web 1.0 was a wholly democratised environment, it was perhaps the most democratised form of publishing available which had a capacity to reach a wide and large audience. It also appeared to make geographical boundaries redundant (Cairncross, 1997). Nevertheless, the technical and financial costs were still significant obstacles to participation and, quickly, it became clear that a 'digital divide'[10] would emerge, as a defining feature of digital culture.

During this first Internet era, questions also emerged about how the broadly positive dimensions of the Internet compared to its darker side, perhaps as a vehicle for exploitation, anti-social behaviour and ill-health. What should we make of these kinds of activities and how should their existence affect our responses to the discussions about Internet regulation? One of the early episodes that drew attention in academic research was the virtual rape of a character within a text-based online game called LambdaMOO (MacKinnon, 1997). The event involved one character hacking into the other, taking control of it and then proceeding to violate the character sexually using text-based descriptions. The mythology surrounding the episode describes it as the event that turned a 'database into a society' (Dibbell, 1993 cited in MacKinnon, 1997), appealing to the idea that the community's subsequent reaction and self-organisation to establish laws of behaviour and an internal legal system and codes of conduct, denoted their becoming a tangible society, rather than just players of a fictional game.

In 2007 – deep in the midst of Web 2.0 – such questions arose again in the context of Second Life, the popular multi-player game from Linden Labs (Lynn, 2007). It transpired that a playground environment within this virtual world was being used by adults to covet young people in what the media called a 'paedophile's playground' (Associated Press, 2007). The episode – and its widespread reporting – led swiftly to the playground's removal from the virtual world, though the range of assumptions that are made about this action is deeply engaging. After all, isn't Second Life just a fictional environment, much like films, where the exploration of such behaviours should be treated as non-serious, trivial even? If we prosecute such behaviour, do we move closer and closer to prosecuting thought crimes? Thus, are we justified in treating such environments as fictional spaces, where only propositions of crime, rather than crime itself, take place? If the latter, then we should treat such environments as we treat other forms of fictional participation.

These episodes articulate one form of how boundaries become blurred when identity is played out online and, for this reason, they demonstrate the need to create new understandings of society and of governance. For instance, it requires explaining how behaviour in a virtual computer game can imply the same kind of criminal action as may take place in physical space, as it is apparent that some forms of social space permit the exploration of ideas and behaviours that would otherwise be considered immoral or even criminal. The way in which film and literature explore ideas are indicative of such exploration, though the question remains as to whether environments like Second Life are more like such fictional narratives or more comparable to life offline. They also demonstrate how similar issues pervade different eras of the Internet. Thus, discussions about what is public or private space, or how we make sense of such concepts as gender have been of interest since the early Internet years. Moreover, they continue to raise similar questions, specifically how we make sense of such concepts online. In the early 1990s, virtual leisure communities provoked various claims about the cultural politics of cyberspace. These new cyberspaces were seen to have activist and emancipatory tendencies, made manifest by such hacktivist[11] interventions as the Zapatistas, an almost mythical example of digital political activism (Cleaver, 1998; Russell, 2005), the modern equivalent of which may be the use of Twitter during the 2009 Iran elections.

From the mid-1990s to the early 2000s, the (first) *dot com* (.com) bubble grew and burst, provoking the creation of numerous spaces where consumption, labour and leisure were blurred. Prominent examples of this include *eBay*, perhaps the paradigmatic example of such blurring. Moreover, digital technology ushered in an era of *convergence*, whereby a number of leisure practices were oriented around the capacities of digital systems (e.g. gambling, film, television) which reconstituted the Internet as an integral component of leisure systems and their institutionalisation. Indeed, institutions with limited funds paid large amounts of money for websites, which did not allow owners the ability to manage their content. Instead, funds were allocated on the basis of, what seem today to be extraordinary principles – such as paying per hyperlink added to a website. During this period, sociological studies of the Internet, focused on identity (Miah, 2000). The internet was, primarily, a mechanism of self-expression. The backlash to this was a concern that we had become distracted from more prominent issues, such as the growing digital divide or, indeed, the increasing monopolisation of certain platforms over our online experiences. The user community became focused on specific search platforms to find information – AltaVista, Yahoo and finally stuck with Google and very little attention was given to how the Internet operated in different languages, or whether its emergence was to the detriment of linguistic diversity (though everyone did have to learn new programming languages).

Web 2.0: Social media as a distinct leisure form

In the mid-2000s, the concepts of social software, social networking sites, open source software mobility and their collective designation as 'Web 2.0' spaces became prominent. Their rise was accompanied by claims over the emergence of a new, digital intelligentsia led by the citizen journalist, who would hold traditional media to task, destabilising their power base and reconstituting mediation. Steadily, in key environments, the growing community of bloggers began to occupy some of the spaces that were traditionally reserved for the mass media, though almost concurrently, traditional media began to inhabit such environments themselves. A good example of this was the merger – and subsequent divorce – of America On-Line (AOL) and Time Warner. Their marriage for £163.4 billion in 2001:

> brought together two huge corporations involved in TV, film, magazines, newspapers, books, information databases, computers, and other media, suggesting a coming synthesis of media and computer culture, of entertainment and information in a new infotainment society (Kellner, 2004: 34).

Their subsequent separation in 2004 signalled the end of the convergence era, ushering in a period where new media needed to become newer, in order to retain its primary capital. The expansion of new, online leisure spaces in the era of Web 2.0 brought renewed optimism – a new bubble – and new playful leisure environments, which recentre knowledge hierarchies and leisure practices (e.g. Second Life, videoblogging, Wikipedia). This era also marked a shift from informationalism to ludology, as increasing numbers of platforms became playful environments, bringing together expertise from the communication technology and the creative design sectors. The mashing-up[12] of data that is made possible by open source programming is a critical dimension of this period, as is the transient quality of media.

Afterall, as was mentioned earlier, the way that any digital environment begins its life is nearly always different from how it is used or how it evolves. The rise of Twitter Search for instance, has the capacity to rival the omnipotence of Google search by providing real-time search results. Thus, while Twitter is often considered a means of instant messaging to people within one's social network, the function of 'Twitter Search' is to provide results about the most immediate and personal kind. This value is likely to have been the main incentive for Google to attempt a takeover of Twitter in 2009 and their eventual agreement (Mayer, 2009). Alternatively, the original design of a Web 2.0 platform only partially reveals its cultural significance and value. These *transient media* also convey the most significant contribution to how the Internet looks today. Consider Facebook, for instance, – one of the most prominent Western social networking websites, at least between 2005 (it began in 2004) and 2009, when this chapter was written. Its success is located in numerous dimensions of its capacity to generate ad-generating clicks, but it is centrally located in its capacity to work with other applications and through its promotion of application authorship. In short, one might des-cribe its core value as an application aggregator, an environment that works because of its openness to sharing programming script.

Facebook is also a helpful example of how the blurring between Web 1.0 and Web 2.0 endures. While there are few social studies of Facebook that can guide us, it is evident that individuals negotiate the Facebook space as both a professional device and as a leisure space (Ellison et al., 2007). It is also evident that this creates tensions, as stories emerge about people who have posted photographs on Facebook being watched by prospective employers. Facebook is also the most prominent manifestation of how public and private space is conflated online, as the privacy settings within the platform are often obscured from people's minds, until they encounter a moment that leads to some realisation of undesirable exposure.[13]

Digital gaming also features heavily in the web 2.0 era, but is notable for how it transforms the sociology of gaming cultures and, indeed its history. For instance, a prominent game within Facebook is 'Scrabulous' – essentially, a digital version of Scrabble™. Its existence on Facebook reconstitutes the history and contemporary status of the board game, but also brings into existence new populations of computer game players, who might otherwise have been alienated from the console generations of PlayStation, Microsoft X-Box, Nintendo and so on.[14] In addition to this, online platforms like MiniClip.com recreate early digital gaming experiences and, along with online games console emulators – they are providing additional chapters to the history of computer gaming.

As was noted earlier, leisure experiences within such spaces are often made possible because they are forms of leisure work. Each of the examples I have used is brought about by the support of advertising revenue, which underpins the business model of each environment. To this extent, one might describe leisure experiences online as consumerist practices, as they become part of market research or advertising campaigns. Consider another example from Facebook called 'Are You Interested'. This third-party application is probably best described as a dating application. It allows users to scan photographs of individuals and select whether they 'like' or 'dislike' them, a judgement based solely on a photograph and brief biography.[15] Once chosen, the other individual is notified and must reciprocate the 'like' in order for a 'match' to be created. From there, the couple can exchange messages and perhaps eventually meet. It is clear that the use of this space might be more for flirting than dating, or indeed, as a mechanism for some kind of sexual encounter. In any case, what interests me here is the mechanism of our engagement within such spaces. Within AYI, one of the principal advertising encounters occurs when scanning photographs, where every 11 clicks on the mouse – i.e. after clicking on 11 different photographs – an advert appears in the same space as the photos.

For a participant of this environment, it is apparent that the muscle memory involved in multiple repetitive clicking creates a high probability that the user will click the advert – which is sometimes also portrayed as a photo – and so the enabling intention of the platform is realised. In addition, other experiences emerge through the application, which work against its being neatly categorised as one type of leisure activity or another. First, it is apparent that some users within Are You Interested are, what might be called stooges, i.e. profiles generated by institutions which will take you to a subscription-based dating site. Alternatively, it is clearly populated by individuals who are, in some sense, operating as actors within the adult entertainment industry, though this is also a broader situation within Facebook.[16] A final component of this third-party application is how other applications link to it, generally with the intention of encouraging users to spend money on additional functionality, a common characteristic of Web 2.0 spaces (the basic functions are made available for free, but higher capabilities are licensed). On this basis, the collective intelligence that emerges from our shared online leisure experiences is not that of the user community, but rather is purposed intelligence, intending specifically to generate revenue or create new markets.[17]

Web 3.0: Where next for the Internet?

Opinions differ as to what era the Internet is currently in. There is considerable overlap between Webs 1.0 and 2.0 and it is likely that we are already at about Web 2.5. However, discussions have already emerged about the character of Web 3.0. This offers some opportunities to speculate on the third era of the Internet, though these speculations are not entirely without doubt. Nevertheless, a starting point should be those practices of online leisure activity that are just beginning to arise. For instance, if one looks at mobile computing, it is apparent that only a few are able to embrace the many new forms of digital leisure experience it provides. For example, the Apple iPhone, which has become the mobile device *par excellence,* remains too expensive for most people to afford while its functional capacities are also uneven. For instance, its camera is far inferior to even a modestly priced digital camera. Its storage capacities are much lower than a small notebook computer. It is, for want of a better term, a suboptimal product and will remain so until miniaturisation reaches the nano scale, at which point user needs will align with the technology. In comparison, we already see such alignment occurring in the development of notebooks. Consider hard disk space, for instance. Today, the hard disk space of a notebook, along with screen size, processing power and so on, are able to compete adequately with desktops, which, as a result, is becoming an increasingly redundant technological form.

Other indicators of issues as yet unresolved by Web 2.0 are questions of identity. Early research into cyberculture discussed its status as the post-modern culture, wherein identity is visibly fragmented. Today, such fragmentation takes various forms, but none so fundamental as the login name and password. Web 2.0 has still not solved the problem of maintaining a permanent identity online and individuals frequently create new profiles with complex passwords that they are unable to remember. Intimations of solving this problem are visible through OpenID, a standard device that should allow users to have one login for all platforms, though it has yet to bridge the 'trust' gap perhaps. Alternatively, it is for some years now that computers have had built-in finger print technology, as a security device.

Other prospects for the internet involve its trajectory towards what might be called the semantic web. If Web 1.0 was characterised by the perceived freedom brought about by anonymity and Web 2.0 celebrates and visualises identity through photo sharing, profile creation and so on, Web 3.0 might be best understood as a 3-Dimensional space infused with contextual meaning. Again, intimations of this are offered by Google Earth and Google StreetView, which allow us to virtually walk down streets seeing their precise visual form, thanks to the Google surveillance cars.[18] The layering of context that such spaces provide signals the end of the era of information saturation and a return to building narratives around our online social experiences. This is also what explains the rise of such platforms as Twitter, which provide us with certainty about there being a real person that informs our navigation through the internet, rather than an algorithm.

The way we experience places offline is also changed dramatically by new media. Consider the traveller arriving in a new city. Traditionally, this person might have bought a guide book, prebooked hotel and thought of important landmarks to visit before arriving. Today, through microblogging on Twitter, that individual arrives in the city, receives advice from the online community about where to go and what to do. Moreover, these users become the tourist's real-time guide and part of their extended social network. To this end, mobile devices reconstitute tourism experiences by providing the means for meeting up with others upon arrival. Here, again, we see the frustration of dubious traditional tropes about the Internet – that it isolates us from others. In this case, social experiences are enabled by the technology. Indeed, some of the most recent research into Facebook users demonstrates precisely these socialising effects of Web 2.0 (Ellison et al., 2007).

Conclusion: The rise of transient media

What would Clarke and Critcher have made of Web 3.0 back in the 1980s? What for them would have been the most profound transformations in how leisure is organised and experienced? Would it have been the way in which leisure consumers became complicit in their own surveillance by large corporations who seek to sell data about their leisure activity to any number of third parties? Would it have been the debates about censorship that rely heavily on highly questionable claims about media effects. Alternatively, would the way that internet labour is organised by communities through open source programming have been seen as the major transformation of our leisure practices? For example, consider how users of the blogging community Wordpress create templates and scripts, which can be freely used by many wordpress bloggers or how Twitter users redistribute information on behalf of news providers. Each of these mechanisms is redefining how information is shared over the Internet, though each also involves the use of informal labour brought about by volunteers who find value in these tasks, perhaps because it is a performative act of demonstrating knowledge or because it helps build reputation online.

Over the years, one overarching concern seems to have been the potential harm that such technology could create for society at large. Back in the 1980s, the connectivity and communication opportunities that would arise from the rise of the Internet were hardly evident. Tim Berners Lee had yet to type the note that became the first description of the Internet on November 12 1990. Indeed, the word Internet did not even appear in Clarke and Critcher's index. That said, speculations about the networked society were present in the lived reality of science fiction works, such as Neuromancer, the definitive cyberpunk romance novel. In the two decades since Clarke and Critcher, the Internet has more than one story to tell about how it has altered our leisure experiences.

Ten years after the publication of *The Devil Makes Work*, around 35 million people were online worldwide (Kitchin, 1998). Another ten years later in 2009, over 1.5 billion people are online and growth in all regions of the world remains high, though penetration varies considerably (5.6% in Africa versus 74.4% in North America) (see Internet World Stats, 2009). In terms of the global digital divide, change is also still occurring. For example, the Chinese online population exceeded the US in 2008 and its mobile 'phone population exceeds the entire UK population four times' (China Internet Network Information Center, 2007). We can observe how the Internet has evolved and answer tentatively what sort of space it has become. We know that people use the Internet for many kinds of pursuit, from watching movies to having 'cybersex', with many different practices in between. It is also clear that the Internet is an arena for work-based activity and that, with the rise of social media environments, there is an increasing level of leisure based interactions in the workplace as a result. In fact, one of the dimensions of the internet in the twenty-first century is how leisure activity can be construed as a kind of labour.

The Internet has not transformed leisure completely. Instead, its most dramatic effect has been its ability to create new questions about issues the culture industries had thought were resolved, such as the attribution of intellectual property or censorship. There is no clean break between the Internet and these other leisure experiences, though it is frequently clear how the emergence of some new online artefact creates catastrophic consequences for other leisure forms. The sharing of music and film through such platforms as the early Napster, the more recent Pirate Bay and the newest Spotify are exemplars of this temporary system failure.

At a time of financial instability and amidst considerable optimism within the online world, one might wonder when the Web 2.0 Internet bubble will burst. It seems far too early to predict, but the collapse of the first bubble seems to have brought a maturity of expectations to online entrepreneurialism, there is a different culture of risk taking evident in how collaboration takes place. However, perhaps the most defining dimension of computer culture is its transient character. After all, the era of Web 2.0 remains difficult to isolate from previous periods of computer culture and, over the years, academics have prematurely attempted to characterise paradigm shifts of Internet use. Even the notion of social media is contested as a way of distinguishing how today's digital technology should be characterised. Yet, the transient quality of media environments may best capture the way that populations move through different environments. It might also capture the way in which specific platforms evolve and become redundant as more compelling alternatives emerge. Thus, we cannot commit to the idea that any single digital platform we see today will be in use in five years from now. As noted earlier, Twitter's challenge to Google – which tried to buy it in 2008 for $500 million – is testimony to this idea. Consequently, the concept of *transient media* may be a reasonable way of describing today's media culture, as it draws attention to the fluidity of digital environments. It describes both the labour markets that underpin their development and the leisure communities that use them. The mashing-up of data described in the title also talks to this notion, since the relevant, enduring condition of the digital space will not rely on form, but on the cultural value attached to the performative act of mashing up. To this end, the prospect of Web 3.0 has its roots in coming to terms with the transient quality of media environments and the way that institutions orientate themselves around these mobile user communities.

Notes

1. The idea of *membership* is also made more complex in social media environments. To what extent does clicking on a page convey membership of a community?
2. The phrase 'mashing up' has become prominent in the Web 2.0 and describes the opening up of programming language, which allows different applications to integrate. A good example of this is the microblogging platform Twitter and the

Facebook status updates. Each of these involves a small piece of text – approximately the size of an SMS (text message) – conveying something about what is happening at any one time. By Facebook and Twitter opening up their application programming interface (API), users are able to combine the best facilities of each, so when, say, someone 'tweets', their Facebook status update is automatically changed as well. Here, I borrow the phrase to convey how work and leisure have begun to function similarly online.

3 Arvidsson (2006) develops a similar concept, 'fantasy work', though I wish to emphasise the location of this productive force within the leisure industries specifically. Moreover, he locates the concept in the contexts of dating sites, whereas my notion is broader. I also encompass the notion of 'gamework' offered by Ruggill et al. (2004).

4 Napster was one of the earliest examples of media sharing platforms, which created considerable legal controversy around 1999–2001 (McCourt and Buckart, 2003). BitTorrent amplified the potential of file sharing (Gardner and Krug, 2005), while The Pirate Bay is a major BitTorrent tracker, which gave rise to a major legal case into breach of copyright – the Web 2.0 equivalent of the Napster case (Kiss, 2009).

5 Google AdSense allows website hosts to integrate advertising that is tailored to their users, rather than using generic adverts to a broad audience.

6 However, it is useful to note that offline or traditional media assists the rise of any online platform remarkably. Consider the rise of Twitter in the first part of 2009. Within the UK, its popularisation by Stephen Fry – who, at that time, was in the top ten Twitter users in the world, where the highest tended to be major media feeds, such as the New York Times or CNN – along with other BBC presenters played a central role in its popularisation.

7 Gye (2007) provides a very useful case study of the camera phone, which is particularly relevant here because the mobile phone has become the ubiquitous device, always taken around with people.

8 For a detailed study of Blackberry users, which draws attention to its infiltration of non-work spaces, see Middleton (2007).

9 An avatar is a computer users virtual representation of themselves, often as a graphic or animation.

10 The phrase *digital divide* has been utilised by numerous authors, though particularly in Bolt and Crawford (2000), Compaine (2001) and Loader (1998).

11 The concept 'hacktivism' has been used to describe various forms of political action that utilise digital technologies as their primary device. A good example of this is the creation of 'Add-Art' (add-art.org) an art project aiming to hack Google, which transformed all commercials presented to browsers into art works.

12 Mashing-up is a phrase common in the Web 2.0 to describe how data from more than one source is combined to create a new kind of online experience, such as embedding photo content from Flickr into Google maps.

13 See Boyd (2008) for an analysis of one of Facebook's encounters with privacy settings. In 2009, a similar challenge arose around Facebook, though the user community rebelled and Facebook withdrew the proposed changes.

14 We can also look to such examples as moments of intellectual property debates. Scrabulous specifically encountered such difficulties.

15 Actually, the platform began by offering the chance to say you dislike a person, but evolved to simply choosing to 'skip' that individual, which tells us something about how the etiquette evolved within this space.

16 I do not wish to dwell too much on this, but it is apparent how some users have created profiles to draw in people to their online pornographic activities or,

indeed, for hacking purposes. Such individuals are easily identified by promiscuous profile photographs, along with having an unfeasibly large number of 'friends', most of which are male.

17 The common ground between this experience and other leisure pursuits is important to stress. For instance, consider how sponsorship operates around major sporting activities, such as the Olympic Games or professional soccer. Arguably, similar processes are at work.

18 Any number of issues might be discussed from this. For instance, should we think of Google as part of the governmental desire to increase the reach of CCTV. Can the Google cameras be used by the police to assist in fighting crime? Any number of issues has already emerged from Google StreetView, such as individuals protesting at their being published online or automobile licence plates being visualised (they are now blurred by Google).

References

Arvidsson, A. (2006) 'Quality singles: Internet dating and the work of fantasy', *New Media and Society*, 8(4): 671–90.

Associated Press (2007) 'Pedophile playground discovered in "second life" virtual world', FoxNews: http://www.foxnews.com/story/0,2933,306937,00.html.

Bolt, D. and R. Crawford (2000) *Digital Divide: Computers and Our Children's Future* (New York: TV Books).

Boyd, D. (2008) 'Facebook's privacy trainwreck: Exposure, invasion, and social convergence', *Convergence: The International Journal of Research into New Media Technologies*, 14: 13–20.

Cairncross, F. (1997) *The Death of Distance: How the Communications Revolution will Change Our Lives* (London: The Orion Publishing Group Ltd).

Castells, M. (1997) *The Information Age: Economy, Society and Culture: Volume II: The Power of Identity* (Oxford: Blackwell).

China Internet Network Information Center (2007) *Statistical Survey Report on the Internet Development in China*.

Clarke, J. and C. Critcher (1985) *The Devil Makes Work: Leisure in Capitalist Britain* (Basingstoke: Macmillan).

Cleaver, H. M. Jr. (1998) 'The Zapatista effect: The Internet and the rise of an alternative political fabric', *Journal of International Affairs*, 51(2): 621.

Compaine, B. M. (ed.) (2001) *The Digital Divide: Facing a Crisis or Creating a Myth?* (Cambridge, Massachusetts: MIT Press).

Ellison, N. B., C. Steinfield and C. Lampe (2007) 'The benefits of Facebook 'friends': Social capital and college students' use of online social network sites', *Journal of Computer Mediated Communication*, 12(4), http://jcmc.indiana.edu/vol12/issue4/ellison.html.

Florida, R. (2002) *The Rise of the Creative Class: And How It's Transforming Work, Leisure, Community and Everyday Life* (New York: Basic Books).

Gardner, S. and K. Krug (2005) *BitTorrent for Dummies* (New York, London, Sydney: John Wiley and Sons).

Gehl, R. (2009) 'YouTube as archive: Who will curate this digital wunderkammer?', *International Journal of Cultural Studies*, 12(1): 43–60.

Grossman, L. (2006) 'Time's person of the year: You', *TIME*, 13 December, 2006. Available online at: http://www.time.com/time/magazine/article/0,9171,1569514,00.html

Gye, L. (2007) 'Picture this: The impact of mobile camera phones on personal photographic practices', *Continuum: Journal of Media and Cultural Studies*, 21(2): 279–88.

Internet World Stats (2009) available: http://www.internetworldstats.com/stats.htm. Accessed: 21 May 2010.

Jones, S. G. (ed.) (1997) *Virtual Culture: Identity and Communication in CyberSociety* (London: Sage).

Kellner, D. (2004) 'The media and the crisis of democracy in the age of Bush', *Communication and Critical/Cultural Studies*, 1:29–58.

Kiss, J. (2009, April 17) 'The Pirate Bay trial: Guilty verdict', *The Guardian*. London: http://www.guardian.co.uk/technology/2009/apr/17/the-pirate-bay-trial-guilty-verdict. Accessed 22 April 2009.

Kitchin (1998) *Cyberspace: The World in Wires* (Chichester: John Wiley & Sons).

Loader, B. D. (ed.) (1998) *Cyberspace Divide: Equality, Agency and Policy in the Information Society* (London: Routledge).

Lynn, R. (2007) 'Virtual rape is traumatic, but is it a crime?', *Wired*, http://www.wired.com/print/culture/lifestyle/commentary/sexdrive/2007/05/sexdrive0504.

MacKinnon, R. (1997) 'Virtual rape', *Journal of Computer Mediated Communication*, 2: 4, http://www.ascusc.org/jcmc/vol2/issue4/mackinnon.html. Accessed June, 1999.

Markham, A. N. (1998) *Life Online* (London: AltaMita Press).

McCourt, T. and P. Burkart (2003) 'When creators, corporations and consumers collide: Napster and the development of on-line music distribution', *Media, Culture & Society*, 25: 333–50.

Mayer, M. (2009, October 21) RT @google: Tweets and updates and search, oh my! The Official Google Blog.

Miah, A. (2000) 'Virtually nothing: Re-evaluating the significance of cyberspace', *Leisure Studies*, 19(3): 211–25.

Middleton, C. A. (2007) 'Illusions of balance and control in an always-on environment: A case study of Blackberry user', *Continuum: Journal of Media and Cultural Studies*, 21(2): 165–78.

O'Reilly, T. (2005) 'What is Web 2.0: Design patterns and business models for the next generation of software', O'Reilly Net. Accessed 8 July 2006. http://www. oreillynet.com/pub/a/oreilly/tim/news/2005/09/30/what-is-web-20.html.

Perdue, L. (2002) *Eroticabiz: How Sex Shaped the Internet*, iUniverse.

Robinson, L. and D. Halle (2002) 'Digitization, the Internet, and the arts: Ebay, Napster, Sag, and e-books', *Qualitative Sociology*, 25: 359–83.

Rojek (2005) *Decentring Leisure: Rethinking Leisure Theory* (London: Sage).

Ruggill, J. E., K. S. McAllister and D. Menchaca (2004) 'The gamework', *Communication and Critical/Cultural Studies*, 1(4): 297–312.

Russell, A. (2005) 'Myth and the Zapatista movement: Exploring a network identity', *New Media and Society*, 7(4): 559–77.

Stone, B. (2009) 'Spam back to 94% of all e-mail', *New York Times*, March 31, http://bits.blogs.nytimes.com/2009/03/31/spam_back_to_94_of_all_e-mail/. Accessed July 8 2010.

Turkle, S. (1945) *Life on the Screen: Identity in the Age of the Internet* (London: Weidenfeld and Nicolson).

9
Tourist Bodies, Transformation and Sensuality

Annette Pritchard and Nigel Morgan

Chapter 9 *discusses the new* ***politics of pleasure, choice and desire*** *in relation to* ***tourism****. For previous generations, especially those on modest incomes, a gentle Sunday afternoon car ride or a few days on an English beach in the summer was the extent of their tourist ambitions and experience. Today* ***tourism*** *is a huge and diverse* ***leisure industry*** *and tourism studies a flourishing area of research in higher education. This chapter discusses the new* ***politics of leisure and pleasure*** *that have accompanied this expansion and argues that we should give more central consideration to* ***the body*** *and* ***the sensual*** *in our attempts to understand* ***contemporary tourism****.*

Introduction

As we approach the second decade of the twenty-first century now is a good time to reflect on the new politics of pleasure, choice and desire. For those who are able to engage in it (many are not, through material and socially constructed inequalities), the experience economy is designed to turn bodies on by sensual appeals. This is nowhere more evident than in the new kinds of tourism which are engaged in producing spaces which arrest and delight the senses (Thrift and May, 2001). Tourism, as it is commonly understood and practised, is a form of commoditised pleasures and these – whether tastes, touches, spectacles or sensations – are sensual and carnal. The warmth of the sun on the skin, the call of seabirds, the smell of unfamiliar food markets, the sight of a dazzling blue sea or the taste of warm fresh bread are all common memories of holiday times. We experience and enjoy the world through our senses or sense organs, we gratify and indulge our physical appetites and many seek out certain places for our holidays '....because there is an anticipation...of intense pleasures, either on a different scale or involving different senses from those customarily encountered' (Urry, 2002: 3). And yet, despite this fundamental connection between tourism encounters and our bodily senses, the corporeality of the holiday experience remains underexplored (Small, 2007). Whilst there is an increasing recognition that people engage with the social and physical world in a variety of different ways (Franklin, 2003), tourism researchers continue to emphasise the plethora of activities available to contemporary tourists without adequately exploring the inherently embodied nature of all tourist experiences.

Whether we describe society as in a modern (high or late) or postmodern phase, it is undoubtedly the case that our bodies are ever more central to, and have come to symbolise, our sense of self and identity. Moreover, these creative projects of managing self and identity are increasingly translated into a reflexive project of the tourist body as a vehicle of self and body management. Here in this chapter, we will see how the tourist body is constructed as both a site of freedom and of control, subject to discourses of hedonism and sensuality and of self-discipline and management. Essentially, we will explore the body as 'project' and to illustrate our discussion we will consider how tourists discipline their bodies through dieting, tanning and body transformation regimes as well as indulging in sensual gratification. Our world is obsessed by notions of the ideal body and the media are dominated by bodily fantasies and features, matched only by the volume of coverage devoted to transgressive bodies that challenge these hegemonic norms and subsequent concerns over eating disorders and warnings of the social, economic and individual costs of obesity. It is becoming more and more evident that 'real bodies and fantasised bodies are radically dissimilar' (Evans, 2002: 9) and whilst the fashion industry has been identified as a key driver of this dissonance, tourism (along with the media and the sport, leisure and fitness industries) is also a crucial discourse of bodily norms and deviations. Yet there has been little examination of the relationships between the fantasised body and tourism.

This chapter argues that embodiment is a conceptual necessity for understanding tourism experiences and, as such, tourism has a valuable contri bution to make to its study. The changing terrain of tourism is made up of the places, things and selves through which social processes are performed and negotiated in the context of space, time and the body. Building on our earlier engagements with tourism and embodiment (e.g. Morgan and Pritchard, 1998; Pritchard and Morgan, 2005; Pritchard and Morgan, 2007), we will show how the body is central to the pleasure arena that is tourism and how the human body is the ultimate site of tourism experiences. We begin by briefly discussing how, influenced by feminism, Foucault and queer theory, cultural studies researchers no longer regard the body as a simple material reality, but as a complexly constructed set of social discourses. This is followed by our examination of the sensuous tourism body and the regimes of discipline to which tourists subject their bodies.

The turn to embodiment

'The body' as a sociological concept is a much more complex entity than the biological form that defines the human person. Today, it is no longer considered to be an unproblematic natural, biological entity external to the mind but a complexly constructed object of social discourses, materialities and practices. The body is the site where our personal identities are constituted and social knowledges and meanings inscribed. As such it is no longer seen as a particularly sexed, raced or aged body; 'instead it is seen as an indeterminate potentiality' (Osborne, 2002: 51) inscribed with meanings 'that reflect the social, cultural, economic and political milieu of its experience' (Wearing, 1996: 80). And, just as the self has become a central concern of social theory in recent years, so has the body been 'discovered' by social and cultural theorists (Turner, 1984; Shilling, 1993 and 2003). The scholarship of Michel Foucault and Erving Goffman has been central to changing the way social science understands the body and Bryan Turner (1994) has identified three themes in the literature on social theory of the body. These are: the symbolic significance of the body as a metaphor of social relationships (an approach particularly pursued in social and cultural anthropology); work focused on the discourses of gender, sex and sexuality; and medical work on issues of sickness, disease and illness.

Foucault's work (1977a, 1977b) is particularly influential in examining the body as a socially constructed phenomena and he saw bodies as fluid and mutable, subject to various forms of power and providing the link between the everyday and prevailing discursive formations. In such contexts, the body and body identities emerge as mutable and changing entities invested with social meaning and control. Foucault's analysis has however been criticised for its essentialising and reductionist tendencies, whereby our active, physical and material bodies vanish and become forever lost in discourse. As a result of this *discursive essentialism* (Shilling, 2003: 71 italics in original) our 'authority, possession and occupation of a personalised body through sensuous experience are minimized in favour of an emphasis on the regulatory controls which are exercised from outside' (Turner, 1984: 245). To counterbalance this sense of the vanishing body, authors such as Chris Shilling have turned to the work of Erving Goffman, which emphasises that the body is central to human agency and social interaction. This scholarship, however, recognises that at the same time our bodies are constrained by and derive meaning and significance from societies and their prevailing social norms and values as evidenced through Goffman's (1963: 35) 'body idiom[s] including bodily behaviours, expressions, decorations and dress which help us make sense of embodied information. These body idioms dominate the interpersonal encounters which shape our social lives. The social roles we perform in these encounters (such as business traveller and cabin crew, diner and maître de, hotel guest and receptionist, traveller and tour guide) all involve our conscious and unconscious observance of particular bodily rules which indicate hierarchical social positioning and our locations within these unequal relationships of dominance and submission.

It is perhaps not surprising that there has been this upsurge in interest in the body in social science research. Pierre Bourdieu (1984) argued that the body as a form of physical capital underlines the all-consuming commodification of the body so that people's identities are linked to the social values accorded to the sizes, shapes and appearances of their bodies. In contrast Norbert Elias (1991) demonstrates how our bodies have become increasingly individualised and now act to distinguish us from each other. Wherever one chooses to place one's emphasis, the body has emerged as a socio-physical entity and is the ultimate site for the production of personal identities and for the experiencing of sensation and time. It is the threshold between the interior subjective self (the individual) and the exterior object world (society). Class, age, gender and sexuality are all inscribed on the body, contributing to the formation of self in terms of individuation, individuality and subjectivity. Moreover, with its multiple and diverse meanings, shapes and experiences, the body (as a concept and a focus of study) appeals to the emergent postmodern academy sceptical of modernist grand narratives and binary definitions.

Indeed, Moss and Dyck (2003: 61) have gone so far as to suggest that the body has become 'a central theoretical problematic' in the social sciences and certainly embodiment has become *en vogue* in many research fields. Complex and uncertain, no longer can the body be seen to be a biological given. Instead its parameters are much more fluid and hazy, dominated by wide ranging and deep-seated social expectations which are both culturally and historically specific (Butler, 1990, 1993; Bordo, 1993) and which have the power to reward and penalise. The body is not only a material entity, it is also a readable text constituted by bodily notions, ideas and inscriptions. As Foucault (1977a: 148) has suggested, the body is 'the inscribed surface of events' and societies 'invest it, mark it, train it, torture it, force it to carry out tasks, [and] to perform...' (Foucault, 1977b: 25). This is not to say that bodies do not have a real materiality – they do and it is this lived experience which is characterised as embodiment (the term in cultural studies and social anthropology which explains how culture is incorporated into the body). Bodies are always subject to interpretation and later in this chapter we explore the materiality of tourism bodies and their experiences. In this we will see that we live in a society which is dominated by body fantasies, yet which has an 'excessively regulatory...attitude to the body' (Evans, 2002: 2). Here again, Moss and Dyck (2003: 63) hit the right note when they say that bodies channel the dialectic between everyday activities and the larger organisation of social power and '...are deeply embedded within the ways people negotiate power through social relations. Regulatory mechanisms control the range of bodily activities as well as the bodies themselves thus producing bodies that are constantly under surveillance either by the self or by society'. As a result bodies present opportunities for freedom and escape and for control and regulation.

Western assumptions about the body have been dominated by the scientific gaze. It is interesting that, within this epistemology, men have been perceived to be largely disembodied, unencumbered by bodily needs, functions and desires, whilst by comparison women have been subject to a body tyranny. As a consequence, women's participation in the public sphere has only been within certain parameters, constrained by heterosexuality and traditional conceptions of womanhood and femininity. Gender is both embedded and embodied; embodiment refers to modes of being in bodies and gender is located not only on the surface of the body (in performance and doing), but becomes embodied (in being and becoming) and part of who we are physically and psychologically.

The overwhelming focus of body-oriented social science research has been on women's bodies and ironically the growing interest of feminist scholars in this sphere has actually 'shored up the dualism which links bodies and bodily matters to women and femininity' (Davis, 1997: 19). Indeed, it is only in the last decade that there has been significant interest in masculinity and men's lived experiences as work began focusing on the 'masculine crisis of identity' (Peterson, 1998). At the same time, heterosexual men also began to pursue their own work on masculinity – a subject which had hitherto been largely studied through the lens of feminist or gay scholarship. Whilst women have long been the object of study, relatively few scholars have questioned how Western societies have arrived at a conception of what constitutes 'normal', 'accepted' masculine identity and behaviour, and asked why certain research questions are articulated whilst others remain neglected, and what assumptions about male bodies and selves are embedded in everyday life (Peterson, 1998). More recently studies of media gender portrayals have begun to focus on masculinity, whilst earlier research concentrated on the social construction of femininity (Vigorito and Curry, 1998). As the privileged owner of the dominant gaze, a man had, in effect, no gender and was rarely subject to scientific scrutiny (Kimmel and Messner, 1992); only in the last decade has mainstream sociology recognised that men have bodies too and that the complex social meanings invested in and inscribed on them remains to be explored.

Tourism research's 'flirtation' with the body

The study of tourism is an inherently ill-defined and eclectic field of research and writing with its roots in the management studies, economics and geography of the 1960s. Against this philosophical backdrop, the field has long been heavily influenced by positivist approaches to intellectual enquiry, which sought to discover and measure universal, objective 'truths' about tourism. Specifically, the focus has been on the use of quantitative methods to model economic process, analyse effective resource deployment and explain and predict patterns of tourist behaviour. Since the later 1990s, however, following many similar fields of enquiry, tourism scholarship has also been influenced by the 'linguistic' and the 'cultural turn', leading some researchers to shift their focus away from issues of structure, economic determinism and supply and demand towards those of language, identity, meanings, representation and performance. As a result, recent discourses on tourism have begun to emphasise the interplay between tourism, landscape, representation and social structures, experiences and identities (see Ateljevic, Morgan and Pritchard, 2007 for a review of this turn). To such scholars, tourism spaces and sites are no longer objective physical entities with specific fixed characteristics, but rather subjective, culturally produced and consumed social constructs. This work also recognises that social categories such as 'gender', 'sexuality', 'race', 'dis/ability', 'age', and 'class' are no longer taken for granted or seen as exclusive and fixed, but rather understood to be socially constructed and performed. As such, these categories (and others) can be and are contested, resisted, affirmed and (re)negotiated.

Until very recently then, 'the body' has been absent from tourism enquiry, reflecting its masculinist, disembodied research traditions. In such traditions, the body has been considered to be 'an obstacle to knowledge in throwing up emotions, feelings, needs, desires, all of which interfered with the attainment of truth' (Alcoff, 1996: 15). In fact, the very methodologies promoted in mainstream tourism research have emphasised abstract, objective inquiry operating outside the body which has been seen to emotionalise and contaminate our understanding of social realities. Tourism is directly implicated in and concerned with the global circuits of bodies, body images and body servicing, the body occupies a central place in the tourist imagination and yet it remains taken-for-granted and under-theorised by tourism scholars. But as more reflexive and embodied researchers have engaged with tourism scholarship we have begun to see a shift as the field has started to address the new work on identity, difference, the body, gender and poststructural theories of language and subjectivity which has forced such a rethinking in the social sciences. Thus tourism work has now conceptualised: sensuousness and embodiment (e.g. Veijola and Jokinen, 1994; Veijola and Valtonen, 2007; Johnson, 2000; Franklin, 2003; Cartier and Lew, 2005), performativity (e.g. Edensor, 1998; Desmond, 1999), the senses (e.g. Dann and Jacobsen, 2002), and materialities and mobilities (e.g. Hannam et al., 2006).

The main theoretical concerns of this emerging recognition of embodiment in tourism have been mapped by Irena Ateljevic and Derek Hall (2007) as the: acknowledgement of our sensuous awareness in the experience of place and the performance of tourism; recognition of a context of representation in which culture is inscribed and with it power and ideology given spatial reference: identification of the body as our means of encountering the world as discursive experience; foregrounding of subjectivities and identities; emphasis on a reflexive situating of critical perspectives. Indeed, it could be said that one of the features of the 'new' tourism research (Tribe, 2005) is 'embodiment' – defined by Richard Osborne (2002: 51) as the term used 'to describe the way in which the bodily bases of individuals' actions and interactions are socially structured: that is embodiment as a social as well as natural process'. It must be said, however, that this recognition of the body and embodiment in tourism enquiry was a long time in coming and probably still remains marginal to most tourism scholarship. It is perhaps too soon to term it a turn to embodiment and as yet it remains more a flirtation by a small number of scholars whose work is firmly rooted in tourism studies rather than tourism management.

Corporeality and the tourist body

If one concept can be said to have gripped the imagination of the tourism academy more than any other, then that accolade must surely rest with the tourist gaze. John Urry's seminal work has shaped and directed many subsequent studies since its publication in 1990. Whilst Urry has himself (along with others) also explored issues of embodiment, the senses and performance since publishing that book, tourism studies has found it difficult to look beyond the gaze. Selanniemi (2001) has suggested that this occularcentrism is probably a result of the distinction between the 'higher' and 'lower' senses in the western philosophical tradition, since in the Aristotelian hierarchy of senses, sight, hearing and smell were classified as human senses, whilst taste and touch were regarded as animal senses. As a result, the passive tourist gaze has triumphed whilst the embodied tourist has been marginalised. As Soile Veijola and Eva Jokinen (1994: 149) highlighted over a decade ago, the tourist body has remained remarkably absent from tourism studies, despite their contention that the tourist gaze is far too abstract and remote from 'tourist events and encounters, in the duration of time and [the] sexed body'. For Selanniemi (2001) corporeality (sensuality and pleasure) is particularly significant and is articulated in what might be termed liminal, non-normal spaces and experiences such as tourism, where our senses are more stimulated than in everyday life.

Any tourism experience is a constellation of tiny, individual experiential moments, which build into an overall impression and, later, a memory of the event. In the few studies which have unpicked the intimate corporeal nature of the tourist experience, its emotional and sensual aspect is clearly apparent. Jenny Small's (2007) study of how women remember their holidays demonstrates how their memories are bound up with feelings of bodily states (e.g. feelings of relaxation or exhaustion), esteem, discomfort and appearance (e.g. feeling good or bad about themselves on the beach or in evening clothes). Holidays were seen as a time for treats and indulgence, a space where one could legitimately resist 'healthy' body discourses (and tuck into food and drink) and wear whimsical and attractive clothing (which perhaps contrasted with the women's more austere business attire). One of her participants, Carly, typified this with her comment that she enjoyed wearing '...the inciest, winciest, tiniest, weeny bikini' on holiday (Small, 2007: 83).

Other studies (e.g. Cockburn-Wootten, 2002; Dann and Jacobsen, 2002; Westwood, 2004) reveal the central role that taste and smell play in tourism experiences and today's postmodern tourist increasingly seeks meaning and pleasure from a variety of experiences (Ryan, 1995), often searching for encounters that delight and excite the senses. These senses come together in the consumption of food and drink, which are now considered to be major lifestyle activities (Abramovici and Ateljevic, 2003), defining, developing and demonstrating individuality and personality. Smells (both pleasant and unpleasant) are often greatly evocative of times and places as the sense of smell is the most ancient, the most basic of our senses; it requires no journey through the thalamus to be processed but is directly routed into our self. The sense of smell not only connects directly with the olfactory cortex in the medial temporal lobe but into all parts of the limbic system, plugging directly into the amygdale, our emotional centre and the hippocampus, the seat of memory. The smell and taste of (un)familiar foods becomes a significant tourism experience and subsequent memory for many people (see for example Small, 2007). Tourists recount how they remember the smell and sound of meat sizzling on a barbeque, the sweet taste of water drawn from an outside well, the flavours of freshly caught fish cooked over a beachside fire and many can identify with the anticipation of a romantic dinner in a local pizzeria or a fine dining restaurant. Typical of these were the couple who told Sheena Westwood in her study (2004: 6–44) 'Whenever we go away food is quite important to us, we're really interested in food as a general rule and so on holiday...a big part of it is traipsing around looking for the right little restaurant to try tonight....'

As well as satisfying bodily appetites for food and drink, many contemporary tourism experiences also offer solace, pampering and transformation for the stressed body in need of revitalisation. Spa, health and wellness tourism is now big business, offering mystical bodily relaxation and purification possibilities (frequently founded on Eastern philosophic traditions). These services engender particular 'body-and-soul experiences...couched in a mixture of descriptive, rhapsodic, metaphoric, medical and poetic style' (Meinhof, 2003: 199). Such philosophies and their manifestation within tourism offer the potential to fuse 'the inner with the outer, the self with the other, the tactile with the intangible, the lesser being with the supreme, the physical with the mental, and the ordinary with the extraordinary' (Singh and Singh,

2009: 139). In spa tourism we can thus see the material fusing with the spiritual.

> For Serefa, it was like prayer; she surrendered completely and was cured. Serefa closed her eyes as Vijaya massaged the warm golden oil into her back in slow, rhythmic movements. Around her flowers danced. The smell of the oil and the ethereal fragrance were soothing. This was the last day of the 14-day ayurvedic holiday. Fourteen delightful days of rejuvenating regimes, medical baths and herbal diets. Fourteen days that wiped away a five year old backache. Put the confidence back in her stride. The dance back in her limbs. Youth back in her life. And a smile back on her lips. As Serefa packed her bags she knew that Kerala had become an integral part of her life, where she would return year after year to revitalise her body, mind and soul (Conde Nast Traveller advertisement, 2007 February, p. 135).

Advertisements like this, linking body, mind and nature, emphasise the medical benefits of wellness tourism, benefits which can resolve long standing bodily discomforts. Indeed, surrendering to these traditional oriental medicinal systems which addresses the body, mind and spirit (in a way which western medicine does not) provides relief from modern day life and promises to allow us to reclaim youth, happiness and spiritual well being. We, just like Serefa, are invited to build this into part of our body maintenance and rejuvenation projects by this advertisement.

Transforming the tourist body

Although the only people we cannot escape from when we travel is ourselves (we bring our worries, concerns, anxieties and attitudes with us in our baggage), contemporary tourism carries powerful discourses of self-transformation. We have briefly seen how this has a spiritual dimension but it is also important to recognise that sensuous tourism bodies are also subject to body tyrannies; they are disciplined, scrutinised, evaluated, judged, valorised or sanctioned, depending on the extent to which they conform to 'desirable' bodily norms. Certainly, our contemporary media is saturated with body representations and body stories; our culture prizes young, slim and beautiful bodies, whilst simultaneously problematising and occluding ageing, overweight and disabled bodies. Our bodies are now objects for display (Featherston, 1982) and they are increasingly central to, and have come to symbolise, our sense of self and identity as people define their bodies as 'individual possessions which are integrally related to their sense of self identities' (Shilling, 2003: 28).

From an ever younger age, our bodies are in a constant state of becoming and we increasingly try to manage and maintain them, immersing ourselves in body projects through diet, fitness regimes, body building and sculpting, health plans, cosmetics and plastic surgery. Indeed, body modification techniques such as plastic surgery, and non-surgical procedures (such as botox and skin peels), are visible evidence of 'body nostalgia' whereby people try to stop or turn back the clock (Gullelte, 1997). Yet, as our ability to exert control over our bodies has grown, so have our body anxieties become ever more acute. Western women are particularly vulnerable to what Kim Chernin (1983) terms 'a slenderness tyranny' and are subject to a barrage of images of thin, young, air-brushed female bodies. In the emotional domain, sexualisation and objectification undermine confidence in, and comfort with, one's own body, leading to a host of negative emotional consequences such as shame, anxiety, and even self-disgust (Slater and Tiggemann, 2002). Today, society is increasingly concerned with obesity and its associated health consequences and the pursuit of the body beautiful is big business, with attempts to delay the physical and aesthetic signs of ageing fast developing into an obsession. There is a proliferation of 'health' and cosmetic products claiming to defy signs of ageing, and in North America the number of non-invasive cosmetic procedures increased by almost 800% between 1997 and 2005, so that over $12.5 billion is now spent there annually on such cosmetic treatments to counteract physical signs of age (Esfahani, 2006).

The beach is perhaps the archetypical territory of desire that embodies such processes of seduction, impression management, self-expression, and the construction, exchange, and interpretation of embodied sign-values. It is dominated by particular forms of masculinity and femininity based on the athleticism and beauty that is highly prized in contemporary society and replicated in countless media images of the perfect suntanned beach body. As a space of bodily undress it thus becomes an anxious place of display and body management, particularly amongst women. In 'beach cultures' there is a greater emphasis on needing to spend time at the beach and to be tanned as a result of greater bodily exposure and public scrutiny; for instance, a state-wide study of New South Wales adolescents' health concerns found that 'body image' was a concern for twice as many adolescents in beach and coastal locations as in inland locations (Quine et al., 2003). Before their tourist trip women are encouraged to physically prepare their bodies. Fiona Jordan (2007) has examined how women's magazines exhort women to aspire to, acquire and maintain the ultimate beach or bikini body, 'planned for and rigorously prepared for display' (p. 98). This is no brief encounter; instead it is more reminiscent of a carefully planned campaign involving dieting, waxing, plucking, working out, fake tanning, moisturising and much much more; at its most extreme bikini boot camps expressly link 'the body project and military discipline' (Jordan, 2007: 101). But once on holiday, the regime does not stop for many tourists, especially women.

Whilst Chris Rojek (1993: 193) sees tanning as 'the abandonment of work', for many sunbathing or sunbaking is a body project which requires an investment of skill, effort and dedication (Small, 2007). Each year millions of people purposefully expose themselves to harmful ultraviolet radiation to darken their skin and although most people are aware of the risks of tanning, many still engage in this practice because having tanned skin is often perceived as physically attractive and physical attractiveness is one of the ways in which people derive self-esteem. Moreover, a range of almost 200 studies suggests that high self-esteem helps defend people from the awareness of mortality (see Routledge, Arndt and Goldenberg, 2004) – thus people are tanning to gain self-esteem at the expense of their physical health. In western culture, tanned skin is one of many physical characteristics associated with beauty and physical appearance and since women are more likely than men to base their self-esteem and self-worth on their appearance, they are particularly likely to engage in frequent sun tanning (Pliner et al., 1990). Martine Abramovici's (2007) analysis of the sensual embodiment of Italian women emphasises how the sun gives pleasure, but it also generates hard work to regain the tan every year and to maintain it throughout the summer. It requires careful skin preparation through initial sun-bed use, the application of sun-creams, the careful realignment of bikinis and body positioning to guard against tan lines and social *faux pas* such as white necks or buttocks. Some bodies are more successful and sought after than others of course and bodily modification reflects this, as 'when walking down to the waterline ... only the young, slim, toned and tanned buttocks may remain unveiled. They are paraded for all to see' (Abramovici, 2007: 122).

Traditionally concern with body image has tended to be perceived as the preserve of women and in particular young women. Increasingly, however, men are also experiencing the pressure to look good. In a recent survey, around a fifth of British men in their early 20s admitted to having taken protein supplements to help bulk up their bodies and just under a quarter said they had considered having plastic surgery – with liposuction and nose jobs being the most popular procedures. Very muscular physiques are the most sought after body shapes for men, with almost four-fifths of men and two-thirds of women favouring this shape. Studies increasingly suggest that men are aspiring to the male celebrity physique seen in so many men's magazines, in much the same way as women respond to their female equivalents. One third of men hate their stomach and a quarter have issues with their entire body (www.bbc.co.uk men's beer bellies days are over, 16 March 2006). Recent research has also highlighted how body image issues are affecting boys and men and demonstrates how boys are under increasing pressure to conform to a muscular physical ideal (Smolak and Stein, 2006). Researchers and the media have drawn attention to an apparent crisis in contemporary forms of masculinity, marked by uncertainties over men's social roles and identities, sexualities, and work and personal relationships (Frosh et al., 2003).

Work on men and masculinities and tourism has been limited in both scope and scale and Chaim Noy's analysis of travelling for masculinity is one of the few explorations of masculine bodies in the tourism literature. He argues that 'the traveller's body is an ideal masculine body...founded on powerful romanticists and colonialist discourses of movement, expansion and exploration' (2007: 51). Valorised masculine bodily qualities include physicality, stamina, resourcefulness, self-discipline, adventurousness, asceticism, machismo and self-restraint. Thus, whilst the woman's beach body is subject to bodily discipline, qualification and bodily measurement (such as weight and inch loss, dress size, bust measurement and bodily transformation), so too is Noy's (2007: 67) 'militaristic' hiker's body (measured in terms of distances walked, weight carried in backpacks or the journey's duration and degree of hardship). The performance measures may differ but either offer tourist bodies' validation or approbation depending on the body's ability to shape up.

Conclusion

In this chapter we have tried to map some key aspects of embodiment, identity and tourism through our consideration of the corporeal, transformed and disciplined tourist body. If twentieth century tourism discourses were dominated by the ethnocentrically hegemonic gaze, then twentieth-first century tourism promises to be more concerned with being, belonging and becoming as peoples and communities use tourism to communicate, preserve, defend and engage their identities. Identity crises, embodied identities and identity politics have been the focus of key debates in recent years and the time is ripe for tourism researchers to take up the challenge and relocate the body to the centre of their work. The challenge this poses should not be underestimated since, just like subjectivity, the body is the real, the immediate and the experienced and 'in a sense one has to stand outside it to understand it, which makes theoretical thinking difficult' (Osborne, 2002: 51). This is further complicated for tourism scholars since, as an academy, tourism (perhaps in response to its need to be taken seriously as a research field) has long maintained a separation of body and mind, ignoring the body and the corporeal. Tourism's architecture of knowledge has been characterised by a remarkable rejection of body research and this is only slowly being countered as researchers recognise that it is only through our bodies that we experience tourism. Our call for the further embodiment of tourism research however, also asks for equal emphasis to be placed on male as well as female bodies. Ironically, much contemporary feminist research on the body has largely ignored the male body, thus paradoxically confirming the links between bodies and bodily matters and women.

At the beginning of the twenty-first century, the body has become 'the vehicle par excellence for the modern individual to achieve a glamorous lifestyle' (Davis, 1997: 2); it has emerged in consumer culture as the significant symbolisation of self, it is a work in progress, a key metaphor for identity and the media, advertising, fashion and tourism industries (amongst others) all contribute to the discourses within which we manage and locate our own bodies (Evans and Lee, 2001). Body projects 'promote an image of the body as an island of security in a global system characterised by multiple and inescapable risk' (Shilling, 2003: 5). It is through our bodies that we experience places and everyday interactions with people. The body symbolises the self, it connects us with other people and places but also marks us as different and 'out of place' (Cresswell, 1996). The corporeal is therefore the ultimate basis for spatial exclusion and inclusion since whether we are white or black, young or old, female or male, able-bodied or disabled, determines others' responses to us and at every scale from individual to nation, dictates what different bodies can or cannot do (Valentine, 2001). In the same way as we need to conceptualise masculine and feminine identities in tourism, we need to examine the cultural meanings and practices which surround embodied spaces, bodies in spaces, represented bodies, and the tourist as a liminal, metamorphic body, transposed from 'home' to 'holiday'. In the same fashion, as students and writers of tourism research, each of us needs to remember that we all participate in a plurality of identities – ethnicities, genders, sexes and races – and we need to turn our culturally diverse gazes to create alternative, inclusive and insightful ways of knowing and understanding tourism.

References

Abramovici, M. and I. Ateljevic (2003) What lies behind the romance of the grape? A case study of Wairarapa, New Zealand. Paper presented at the Social Science in the 21st century, Sociological Association of Aotearora (NZ), University of Canterbury, Christchurch.

Abramovici, M. (2007) 'The sensual embodiment of Italian women' in Pritchard, A., N. Morgan, L. Ateljevic, and C. Harris (2007) (eds) *Tourism and Gender: Embodiment, Sensuality and Experience*, 107–126 (Oxford: CABI).

Alcoff, L. M. (1996) 'Feminist theory and social science: New knowledges and new epistemologies', in N. Duncan (ed.) *Bodyspace: Destabilizing Geographies of Gender and Sexuality*, 13–27 (London: Routledge).

Ateljevic, I., N. Morgan and A. Pritchard (2007) (eds) *The Critical Turn in Tourism: Innovative Methodologies* (Oxford: Elsevier).

Ateljevic, I. and D. Hall (2007) 'The embodiment of the macho gaze in southern-eastern Europe: Performing femininity and masculinity in Albania and Croatia', 138–57, in Pritchard, A., N. Morgan, I. Ateljevic and C. Harris (eds) *Tourism and Gender: Embodiment, Sensuality and Experience*, 92–106 (Oxford: CABI).

www.bbc. men's beer bellies days are over, 16 March 2006 http://news.bbc.co.uk/1/hi/health/4812276.stm.

Bordo, S. (1993) 'Feminism, Foucault and the politics of the body', in C. Ramazanoglu (ed.) *Up Against Foucault: Explorations of Some Tensions Between Foucault and Feminism*, 179–202 (London: Routledge).

Bourdieu, P. (1984) *Distinction: A Social Critique of the Judgement of Taste* (London: Routledge and Kegan Paul).

Butler, J. (1993) *Bodies That Matter: On the Discursive Limits of Sex* (London: Routledge).

Butler, J. (1990) *Gender Trouble: Feminism and the Subversion of Identity* (London: Routledge).

Cartier, C. and A. Lew (eds) (2005) *Geographical Perspectives on Globalization and Touristed Landscapes* (London: Routledge).

Chernin, K. (1983) *Womansize: Tyranny of Slenderness* (London: The Women's Press Ltd.)

Cockburn-Wootten, C. (2002) *Gender, Grocery Shopping and the Discourse of Leisure*, unpublished PhD thesis (Cardiff: University of Wales Institute).

Conde Nast Traveller advertisement, 2007 February, p. 135.

Cresswell, T. (1996) *On the Move* (London: Routledge).

Dann, G. and J. K. S. Jacobsen (2002) 'Leading the tourist by the nose', in G. Dann (ed.) *The Tourist as a Metaphor of the Social World*, 209–35 (Oxford: CABI).

Davis, K. (1997) 'Embody-ing theory beyond modernist and postmodernist readings of the body', in K. Davis (ed.) *Embodied Practices: Feminist Perspectives on the Body*, 1–23 (London: Sage).

Desmond, J. C. (1999) *Staging Tourism: Bodies on Display from Waikiki to Sea World* (Chicago: University of Chicago Press).

Edensor, T. (1998) *Tourists at the Taj: Performance and Meaning at a Symbolic Site* (London: Routledge).

Elias, N. (1991) *The Symbol Theory* (London: Sage).

Esfahani, E. (2006) *The New Skin Trade* http://money.cnn.com/magazines/business2/business2_archive/2006/01/01/8368124/index.htm

Evans, M. (2002) 'Real bodies: An introduction', in Evans, M. and E. Lee (eds) (2002) *Real Bodies: A Sociological Introduction*, pp.1–14 (New York: Palgrave).

Featherstone, M. (1982) 'The mask of ageing and the postmodern life course', in M. Featherstone et al. (eds) *The Body: Social Process and Cultural Theory*, 371–89 (London: Sage).

Featherston, M. (1982) 'The body in consumer culture', *Theory, Culture and Society*, 1: 18–33.

Foucault, M. (1977a) *Language, Counter-Memory, Practice: Selected Essays and Interviews* (Ithaca, NY: Cornell University Press).

Foucault, M. (1977b) *Discipline and Punish: The Birth of the Prison* (London: Penguin).

Franklin, A. (2003) *Tourism: An Introduction*, (London: Sage).

Frosh, S., A. Phoenix and R. Pattman (2003) 'The trouble with boys', *The Psychologist* 16(2): 84–7.

Goffman, E. (1963) *Behaviour in Public Places* (New York: Free Press).

Gullelte, M. M. (1997) *Declining to Decline: Cultural Combat and the Politics of Midlife* (Charlottesville and London: University Press of Virginia).

Hannam, K, M. Sheller and J. Urry (2006) 'Editorial: Mobilities, immobilities and moorings', *Mobilities* 1(1): 1–22.

Meinhof, U. H. (2003) 'Bodies exposed: A cross-cultural semiotic comparison of the "Saunaland" in Germany and Britain', in Coupland, J. and R. Gwyn (eds) *Discourse, the Body and Identity*, 177–205 (Basingstoke: Palgrave).

Jordan, F. (2007) 'Life's a beach and then we diet: Discourses of tourism and the 'beach body' in UK women's lifestyle magazines', in Pritchard, A., N. Morgan,

I. Ateljevic and C. Harris (eds) *Tourism and Gender: Embodiment, Sensuality and Experience*, 92–106 (Oxford: CABI).
Kimmel, M. S. and M. A. Messner (eds) (1992) *Men's Lives*, 2nd ed. (New York: Macmillan Publishing Company).
Morgan, N. and A. Pritchard (1998) *Tourism Promotion and Power: Creating Images, creating Identities*, (Chichester: Wileys).
Moss, P. and I. Dyck (2003) 'Embodying social geography', in K. Anderson, M. Domosh, S. Pile and N. Thrift (eds) *Handbook of Cultural Geography*, 58–73 (London: Sage).
Noy, C. (2007) 'Travelling for masculinity: The construction of bodies/spaces in Israeli backpackers' narratives, 47–72, in Pritchard, A., N. Morgan, I. Ateljevic and C. Harris (eds) *Tourism and Gender: Embodiment, Sensuality and Experience*, 92–106 (Oxford: CABI).
Osborne, R. (2002) *Megawords: 200 Terms You Really Need to Know* (London: Sage).
Pliner, P., S. Chaiken and G. L. Flett (1990) 'Gender differences in concern with body weight and physical appearance over the life span', *Personality and Social Psychology Bulletin*, 16(2): 263–73.
Peterson, A. (1998) *Unmasking the Masculine: 'Men' and 'Identity' in a Sceptical Age* (London: Sage).
Pritchard, A., N. Morgan, I. Ateljevic and C. Harris (2007) (eds) *Tourism and Gender: Embodiment, Sensuality and Experience* (Oxford: CABI).
Pritchard, A. and N. Morgan (2007) 'Encountering scopophillia, sensuality and desire: Engendering Tahiti in travel magazines', in Pritchard, A., N. Morgan, I. Ateljevic and C. Harris (eds) (2007) *Tourism and Gender: Embodiment, Sensuality and Experience*, 158–81 (Oxford: CABI).
Pritchard, A. and N. Morgan (2005) 'On location: (Re)viewing bodies of fashion and places of desire', *Tourist Studies*, 5(3): 283–302.
Quine, S., D. Bernard and M. Booth (2003) 'Locational variation in adolescents' concerns over "body image"', *Health Promotion Journal of Australia*, 14(3): 224–25.
Rojek, C. (1993) *Ways of Escape: Modern transformations in Leisure and Tourism* (Basingstoke: Macmillan).
Routledge, C., J. Arndt and J. L. Goldenberg (2004) 'A time to tan: Proximal and distal effects of mortality salience on sun exposure intentions', *Personality and Social Psychology Bulletin*, 30(10): 1347–58.
Ryan, C. (1995) *Researching Tourist Satisfaction* (London: Routledge).
Shilling, C. (1993) *The Body and Social Theory* (London: Sage).
Shilling, C. (2003) *The Body and Social Theory* (London: Sage).
Selanniemi, T. (2001) 'Pale skin on the Playa del anywhere: Finnish tourists in the liminoid south', in Smith, V. L. and M. Brent (eds) *Hosts and Guests Revisited: Tourism Issues of the 21st Century* (New York: Cogizant).
Singh, S. and T. V. Singh (2009) 'Aesthetic pleasures: Contemplating spiritual tourism', in Tribe (ed.) *Philosophical Issues in Tourism*, pp.135–53 (Bristol: Channel View).
Slater, A., and M. Tiggemann (2002) 'A test of objectification theory in adolescent girls', *Sex Roles*, 46: 343–9.
Small, J. (2007) 'The emergence of the body in the holiday accounts of women and girls', in Pritchard, A., N. Morgan, I. Ateljevic and C. Harris (eds) *Tourism and Gender: Embodiment, Sensuality and Experience*, 73–91 (Oxford: CABI).
Smolak, L. and J. A. Stein (2006) 'The relationship of drive for masculinity to sociocultural factors, self esteem, physical attributes, gender role and social comparison', *Body Image*, 3(2): 121–9.

Thrift, N. and J. May (eds) (2001) *Timespace: Geographies of Temporality* (London: Routledge).
Tribe, J. (2005) 'New tourism research', *Tourism Recreation Research*, 30(2): 5–8.
Turner, B. S. (1984) *The Body and Society* (Oxford: Blackwell).
Urry, J. (1990) *The Tourist Gaze: Leisure and Travel in Contemporary Societies* (London: Sage).
Urry, J. (2002) *The Tourist Gaze* (London: Sage).
Valentine, G. (2001) *Social Geographies: Space and Society* (Oxford: Prentice Hall).
Veijola, S. and A. Valtonen (2007) 'The body in tourism industry', in Pritchard, A., N. Morgan, I. Ateljevic and C. Harris (eds) (2007), *Tourism and Gender: Embodiment, Sensuality and Experience*, 13–31 (Oxford: CABI).
Veijola, S. and E. Jokinen (1994) 'The body in tourism', *Theory, Culture & Society*, 11: 125–51.
Vigorito, A. and T. Curry (1998) *Marketing Masculinity: Gender Identity ... Race, Class and Gender on the Internet* (Westport, CT: Praeger).
Wearing, B. (1996) *Gender: The Pain and Pleasures of Difference* (Melbourne: Longman).
Westwood, S. H. (2004) *Narratives of Tourism Experiences: An Interpretative Approach to Understanding Tourist-Brand Relationships*, unpublished PhD Thesis (Cardiff: University of Wales Institute).

10
'They Can't Stop Us Laughing'; Politics, Leisure and the Comedy Business

Stephen Wagg

Chapter 10 *concerns the* ***comedy*** *business, another field of social endeavour central to the contemporary* ***politics of leisure and pleasure*** *and to the* ***politics of identity*** *that have characterised life in post-war Western societies. The rise of 'alternative comedy' in a range of societies during the last 30 years has brought perceptibly* ***greater freedoms in personal and public expression.*** *Indeed the emergence of new comedic discourses and styles is widely seen as synonymous with taboo-breaking and general iconoclasm. But 'alternative comedy' is rooted in identity politics, which carry their own proscriptions. The result has been still another* ***tension between freedom and constraint,*** *rendered in the comedy world as the battle over* ***'political correctness'*** *– the subject of close analysis in this chapter.*

Prelude: June 2007

'They can't stop us laughing' was a phrase often used on stage by Bernard Manning, a stand-up comedian from Manchester in the north of England. Manning lived all his life in Manchester and died there, aged 76, in June of 2007. The phrase, which Manning had used as punctuation between jokes, and the diverse commentary that followed his death (in addition to the numerous obituaries, Manning had been commissioned to make a television programme anticipating his own funeral) together evoke all of the themes of this chapter.

When Manning had begun performing as a comedian in the 1950s, 'comedy' had been a more or less taken-for-granted commodity: comedians knew the boundaries of public taste and the smaller ancillary markets – usually all-male gatherings at which 'you could do your blue material' – where these boundaries could safely be exceeded. Comics assumed a reasonably homogeneous audience. Manning perceived his constituency as 'grown men that work on building sites',[1] therefore, with shared assumptions about family life, sex, ethnic minorities and so on, although his actual audiences were predictably rather more diverse. Their comedy was resolutely local, with few global pretensions – although, interestingly, as he proudly boasted, Manning had played Las Vegas with great success in 1978 and refused further offers to perform abroad only because he didn't like being away from Manchester. In Britain, certainly, their typical *modus operandi* since the 1920s had been the telling of traditionally structured jokes ('Two blokes sat in a pub. One says....')[2] and not routines purportedly based on their own experiences, on current events, on shared experiences of television programmes or the like.

There had in the 1950s been a negligible relationship between comedy and politics, and comedy was, by and large, seen as a separate and sanctioned universe of discourse in which things said did not matter outside of that universe: facilitating the familiar comment 'It's just a joke'. This designation of comedy as 'non-serious' quintessentially places it as leisure. Manning's act from the 1960s until his death was replete with jokes at the expense of 'Pakis', 'Japs, 'niggers' and other ethnic groups but when in 2003 he was booked by the fascistic British National Party his agent offered the following reassurance to protestors:

> We have a contract with the Trafalgar Club, the social side of the BNP, for an appearance. He is doing them no favours. It is just another gig. We have done gigs for the Labour and Tory parties before. We are not politically motivated.[3]

Manning, like virtually every comedian of his generation, saw no connection between comedy and what he understood to be 'politics'. His death, nevertheless, drew a wide range of political and cultural responses. The website 'Funny.co.uk' showed little sorrow for the passing of 'Bernard Manning: Racist Tosser'.[4] The *Daily Mail* had asked him to provide his own obituary:

> 'Oh, I know there'll be a few who won't mourn my passing', he wrote, 'like mothers-in-law up and down the country. I'll never forget the day I took my own mother-in-law to the Chamber of Horrors in Madame Tussauds. Suddenly, one of the attendants whispered to me: "Please keep her moving. We're trying to do a stock take..." I don't think the Commission for Racial Equality will be holding a wake for me, either. Nor will the Lesbian and Gay Rights lot or the feminists. They were always banging on about how I was sexist or anti-gay.'[5]

The liberal *Guardian* had him as 'professionally racist and sexist but privately as tolerant as any man from his background and generation was likely to be'.[6] In the comparably liberal *Independent*, columnist Howard Jacobson defended Manning as some kind of social purgative:

> By what wilful misapprehension of his function did we censure him for being racist? The man was a comedian. Racist – in a race-jittery

> society – is precisely what a comedian is obliged to be. No racism, no offence. No offence, no comedy. No comedy, no removal of malignancy.[7]

And on a website emanating from a libertarian 'think tank' called the 'Institute of Ideas', writer Ed Barrett, while acknowledging Bernard to have been a racist, commented:

> Manning never made clever satirical points about the royal family. He preferred to insult them instead. At a recent charity dinner, he approached a friend of the late Queen Mother and said: 'One corgi turns to another and says, "Thank fuck the Queen Mum's dead, now we won't be blamed for the smell of piss"...Manning was the oldest – and truest – punk in town. He said and did what he liked, and didn't give a flying blue word what anyone thought.[8]

Introduction: Who would want to stop us laughing?

At some stage in the 1980s Bernard Manning had gone to bed as simply a comedian and woken up the next morning newly defined as a comedian of a certain stamp – 'racist', 'traditional', 'right wing', even 'iconoclastic' or 'transgressive'. He would probably have died in comfortable obscurity were it not for the cluster of structural, social and political changes that combined to produce the pervasiveness of comedy and comedians in contemporary life: they sell out theatres and on occasion larger venues (in 1993, for example, the comedians David Baddiel and Robert Newman performed to a capacity crowd of 12,000 at London's Wembley Arena; profane north of England comic Roy 'Chubby' Brown plays to an estimated 350,000 people a year);[9] they are columnists; they feature regularly in advertisements and on tongue-in-cheek TV quiz shows; their DVDs adorn our living room shelves; they are in some cases activists or *de facto* spokespeople; and they are, of course, celebrities – living, or dying, news items.

The first and most important change is the swelling, most prominently in North America, the United Kingdom and on mainland Europe, of two irreconcilable political tides: on the one hand, the growth from the late 1970s onward of free-market rhetoric and economic deregulation: on the other, the rise of a new politics, based on identity and deriving from the new social movements of the 1960s that campaigned for the rights of women, ethnic minorities, gay people and the disabled. Defence of the rights and sensitivities of these various minorities is now widely enshrined both in law and in popular consciousness. There are many who argue that free-market capitalism can not function without the exploitation of low-paid female and ethnic-minority labour, the scapegoating of 'Others' and so on. The specific implication for comedy practice is very clear: there cannot be *both* a free market in jokes *and* a protective concern for the dignity of these minorities – the butt, after all, of much mainstream comedy material before 1980. Those people – comedians, audiences, commentators – who favour discrimination in joke-making are now invariably styled by their opponents as 'politically correct'. This disturbingly durable political shibboleth expresses perfectly the irreconcilability of the two ideological currents, since it always carries two contradictory implications: (a) that a form of action is correct and (b) that it should nevertheless not be adopted.

It's reasonable to assume, then, that when, in the 1950s, Bernard Manning had begun to reassure his audience of imagined builders that 'they' couldn't 'stop us laughing', the 'they' doubtless referred to all the bosses and toffs and others who constituted the 'Them' in the 'Them and Us' world view prevalent in working-class districts. From the mid-1980s onward the same phrase was more likely to conjure images of a cadre of pious middle-class, college-educated comedians and po-faced cultural commentators trying to stop ordinary decent folk having a good giggle. That does not mean, of course, that the original implication of the phrase, or some approximation to it such as 'having a laff', can no longer apply.

Other important factors which bear on the emergence of the modern comedy business, and which therefore inform this chapter, are: the decline in deference in British society (and the corresponding waning of the power of the aristocratic element in the British establishment); the growing delegitimation of politicians and political processes; the decline in political activism and the membership of mass political parties – a factor, of course, particularly for the left; the deregulation and globalisation of the mass media; the residual influence of a 'counter-culture', originating in the late 1950s and 1960s; the corresponding (if paradoxical) increase in the control of public information, much of which has come now to be channelled through official spokespeople, 'spin doctors' and the like; the onset of Zygmunt Bauman's 'liquid modernity', in which individuals strive for stable identities in a world of increasingly transient social forms and institutions;[10] and the rise of what Bauman calls 'the confessional society' in which people increasingly feel the need to speak publicly about themselves.[11] As I have argued elsewhere,[12] the clear paradox of these new arrangements has been that, in the material sense, a large and growing section of the British population and others has had its freedoms reduced (through curtailments in public spending, increased surveillance at work, fear of redundancy and so on) while in the symbolic world greater and greater possibilities for individual expression are held out: in postmodern popular culture, people are encouraged to be who they want to be. It is this latter imperative that the 'alternative' comedian obeys; s/he is now, apparently, everywhere saying the unsayable, speaking the minds of some constituency of ordinary punters. As Jack Dee, one of the most popular of the 'alternative' comics to emerge in the early 1990s has said:

> Some of the stuff I do is releasing thoughts and ideas that the audience wouldn't usually be able to get away with and the audience need you for

> that reason. That's why they don't want you to be ingratiating. They want you to have a 'fuck you' attitude, because a lot of people wish they could have that 'fuck you' attitude all the time.[13]

One might add here the enormous popularity of mock TV documentary *The Office*, first shown by BBC TV in the UK in 2001 and since seen in over 80 countries;[14] an American version began in 2005.

This chapter will chart the emergence of comedy-as-leisure, as a playful discourse and/or state of mind, as well as an area of sanctioned truth-telling in an age of official blandishment, and trace the parallel development of the contemporary comedy business. Here it will have regard, in particular, to 'edgy' or self-consciously taboo-violating comedy and will draw intermittently on my previous work.[15] The chapter will then discuss several dimensions of contemporary comedy in more detail. These will be the relationships between: comedy and globalisation; comedy and the market; and comedy and politics. I conclude with a brief allusion to issues raised in the prelude when discussing comedy as heritage.

Define your terms: Some social roots of contemporary comedy

Many of the roots of contemporary comedic practice lie in the nightclubs of San Francisco and New York patronised by young, white, college-educated Americans in the late 1950s and early 1960s. Around this time a number of comedians began performing who eschewed jokes and their customary targets (usually wives) and instead delivered intellectualised routines – monologues that were often surreal, were likely to be derived from current affairs and sometimes had the feeling of a stream of consciousness. A pioneering figure here was Canadian-born American Mort Sahl (1927–) who seems to have been a template for the new comedy in several respects. He dressed for leisure – in slacks, sweater and open-necked shirt – in contrast to the conventional suit and bow tie. He disdained jokes and punchlines; instead, according to one biographer, he adopted a conversational style that dealt in 'free association, nervous digression, jazz-like improvisations, one-word references, parenthetical stream of conscious commentary and a mosaic of ad-libs and suddenly remembered set routines.'[16] One such routine 'was about a bank robber who comes up against an intellectual teller. The robber hands him a note that says "Act normal." The teller writes back, "Define your terms."'[17] 'Every comedian who is not doing wife jokes has to thank him for that', said the actor-comedian Albert Brooks in 2007.[18] 'I talked about social and political hypocrisy', remembered Sahl the same year;[19] this had made him a sharp contrast to more conventional American comics such as Bob Hope who merely teased successive presidents, usually about their golf swing.[20] Importantly, though, this did not prevent either man from befriending a number of American presidents and leading politicians. Sahl befriended both Adlai Stevenson, the Democratic candidate for the presidency in 1956 and John Kennedy, who ran (and won) for the Democrats in 1960;[21] he wrote jokes for the latter's campaign,[22] but happily told jokes against both men. On *The Next President*, his album of 1960,[23] he said 'Whoever the President is, I will attack him'. This is not simply a reminder that comedians must be seen as separate from their material; it assumes politics as *performance* and it distinguishes between politicians as *individuals* and as *texts*. Sahl became one of a number of performers who similarly departed from comedic orthodoxy and appeared to address sectional audiences beneath the all-American cultural radar: women perhaps, as with Joan Rivers (1933–) who began performing in New York clubs in the early 1960s or African Americans, as with Bill Cosby (1937–) and Dick Gregory (1932–). Of these comedians, the most influential, certainly in the short term, was Jewish New Yorker Lenny Bruce (1925–1966) whose satirical and profanity-ridden routines made him a lightning rod both for conservative cultural commentary and for liberal critique in the United States. He was arrested for obscenity several times in his short career but also drew some notable support – for instance, from the San Francisco columnist Herb Caen in 1959:

> They call Lenny Bruce a sick comic and sick he is. Sick of all the pretentious phoniness of a generation that makes his vicious humor meaningful. He is a rebel, but not without a cause, for there are shirts that need un-stuffing, egos that need deflating. Sometimes you feel guilty laughing at some of Lenny's mordant jabs, but that disappears a second later when your inner voice tells you with pleased surprise, 'but that's true'.[24]

Sahl, Bruce, Gregory and other emergent discursive comedians were promoted by *Playboy* magazine, flagship since the mid-1950s for a new hedonist leisure in the US, and performed at the *Playboy* club. When Bruce died prematurely in 1966 *Playboy* carried the eulogy (by sportswriter Dick Schaap) 'Finally, one last four-letter word concerning Lenny Bruce: Dead. At forty. That's obscene'.[25] Bruce posthumously became the defining symbol of what one writer has described as 'the vital Lenny Bruce era – nastiness, fuck-you, sick-comic daring'.[26] The wave-making comedians of this era, which may be understood as the historic moment of America's 'alternative comedy', helped to usher in a free market in American comedy, based upon artistic licence but in which ultimately the consumer was sovereign. As the African-American comic Richard Pryor once advised another performer, 'Whatever the fuck makes the people laugh, say that shit'.[27]

The influence of Sahl and Bruce was recognised by a coterie of young writers and media executives in Britain. Defined largely by their social class – most of them had attended private schools and/or elite universities – their focal figure was Peter Cook. Cook had been one of the writer-performers in *Beyond the Fringe*, a revue derived from the recent work of Cambridge

University's Footlights club and the Oxford [University] revue and first performed at the Edinburgh Festival in August 1960. Cook's roles in the show included an impersonation of the Prime Minister of the time, Harold Macmillan – something that startled some members of the audience. In 1961 Cook helped found both the satirical magazine *Private Eye* and the Establishment Club in London's Soho, both outlets for an iconoclastic and humorous commentary on public affairs. Cook was an admirer of Lenny Bruce and Bruce performed at the Establishment Club in 1962; the following year, booked for another appearance, he was refused admission to the UK, the Home Office classifying him as 'an undesirable alien'. Cook also wrote for *That Was The Week That Was*, a sketch show broadcast late on Saturday nights on BBC television between 1962 and 1963. *TW3*, as it was widely known, was arguably the first means of access to the British mass media for this new comedic form. Because of its newness observers struggled to interpret it politically. As I have argued elsewhere, most of the comedy in *Private Eye* and *That Was The Week*...seemed to be directed against the 'swineries of public life', all public figures being considered fair game. But some of *TW3*'s most swingeing parodies had been directed against American racism (following a recent murder by the Ku Klux Klan) and the persecution of homosexuals (after John Vassall, a gay civil servant, was found to have been blackmailed into spying for the Soviet Union in 1962).[28]

The creation of *Monty Pythons Flying Circus* for BBC television in 1969 is another milestone in the development of contemporary comedy. The *Python* project (1969–1983) also chiefly involved graduates from the elite English universities – three recent Cambridge graduates and two from Oxford – and is important to this narrative for two principal reasons. First, it moved the struggle for free expression away from stand-up, quasi-social commentary and toward the long-established comedic form of the sketch. In its second series, broadcast in 1970, the programme had included an 'Undertakers Sketch' in which undertakers casually discuss with a bereaved man how he might like to eat his recently deceased mother. This sketch had required some negotiation at management level and, four years later, *Python* was still being censored: Michael Palin, one of the *Python* team, wrote in his diary in November 1974 that the line 'Cold enough to freeze your balls off, freeze the little buggers solid in mid-air' had been excised from one of their scripts, 'as well as one "piss off" (we could keep the other)'. Later the same month he lamented that 'we no longer surprise and shock. We are predictable'.[29] Second, the *Python* programmes were an early example of the globalisation of comedy, initially through the endorsement of 'hip' and/or hedonistic sections of the leisure industry: their first film *And Now For Something Completely Different* (1971) was financed by Victor Lowndes, head of the London branch of the *Playboy* empire, and rock bands such as Led Zeppelin and Pink Floyd promoted them, financing their second film, *Monty Python and the Holy Grail* in 1975. George Harrison of The Beatles funded their most successful film *The Life of Brian* four years later.[30] The *Python* team performed stage shows in Canada in 1973 and made a promotional tour of the United States the same year. Their television shows began to be broadcast on Public Broadcasting Stations in the US in 1974[31] and the team performed at the Hollywood Bowl in 1980. They also had a considerable following outside the English-speaking world – in Germany, for example.[32] The term 'Pythonesque' has since entered many dictionaries to denote an absurdist, stream-of-consciousness or surreal humour, rejecting traditional structures and punch lines. Although rendered as 'very English' by some commentators and disparaged in some circles as 'undergraduate humour', its derivations and likely audiences were more complex and far-reaching. The Pythons acknowledged the influence of Sid Caesar (1922–)[33] and Spike Milligan (1918–2002)[34] – indeed Milligan claimed that the *Monty Python* writers had plagiarised his own creation *Q5*.[35] Caesar was a New Yorker, Milligan had been born in India to an Irish father and an English mother; neither was a graduate. Nor could a particular brand of humour, even if it did make undergraduates laugh, necessarily be confined within specific national boundaries.

A joke that feeds on ignorance....The arrival of alternative comedy

Trevor Griffiths' play *Comedians* was first performed at Nottingham Playhouse in 1975 and published the following year. In the play a veteran comedian has been tutoring an evening class in the art of stand-up comedy and each student is now preparing to do a turn for the external examiner, a comic of the old school. The tutor, however, urges the students to abandon the comedian's standard repertoire, with its assumptions about women, ethnic minorities, trade unionists and other familiar targets. He tells one: 'A joke that feeds on ignorance starves its audience'. The external examiner is unconvinced:

> Don't try to be deep. Keep it simple. I'm not looking for philosophers, I'm looking for comics. I'm looking for someone that sees what the people want and knows how to give it to them. It's the people pay the bills, remember....We're not missionaries, we're suppliers of laughter.[36]

Griffiths caught the mood of the time: ten years later a new, more diverse social world of comedy had been established in which the politics of both of these characters had been fully accommodated.

In Britain, what became known as 'alternative' comedy emerged around 1979. Two important social trends influenced its emergence. First, the progressive cuts in public spending, begun by the Callaghan government in the mid-1970s, withdrew subsidies from a lot of fringe theatre and such like experimental art work. This effectively destroyed much of the non-mainstream artistic environment that had sustained a large number of cultural workers. Many of these now sought some kind of self-employment. Several of the earliest 'alternative' comedians to perform – Jim Barclay, Tony Allen, Alexei Sayle, Keith Allen, John Hegley... – had worked previously in fringe, street or some kind of minority theatre. They now became what has been called 'the mauve economy' – what Helen Chappell called the 'sometimes rather *outré* services the western world will want if and when *making things* is as obsolete as sociologists predict'.[37]

Secondly, the first batch of 'alternative' comedians drew culturally on the punk movement and on older established traditions of libertarian comedy in the United States. This meant, first and foremost, an anguished political preoccupation with the 'sell out'. The first 'alternative' comedians, like many punk musicians[38] were concerned to capture an uncompromised working classness and a street authenticity in what they did. The initial paradigm for 'alternative' comedy was often fiercely anarchistic. Tony Allen and others performed what was advertised as 'alternative cabaret' at the Elgin pub in Ladbroke Grove, west London:

> the pub of preference for scruffy political extremists, bohemian arty types and those barred from everywhere else. On any given night you could meet up with feminist groups, plotting squatters and law centre workers.[39]

Malcolm Hardee (1950–2005), who with two mates made up The Greatest Show on Legs, a profane Punch and Judy Show formed in 1978 and featuring 'masturbation, sex and a punk rocker', spotted the general popularity of the show's bawdy humour and a growing political fragmentation in audiences – 'the feminists disapproved of our generally coarse attitude' he noted.[40] The comedy gigs of the late 1970s and early 1980s became an often chaotic crucible in which the working definitions of stand-up comedy were extended and reshaped to embrace a range of new political assumptions in relation to the body, the self and social difference. Keith Allen, who began doing stand-up in 1979, recalls:

> I loved being on stage without the constraints of either a recognisable act or the disciplines of a script. I could do anything – and, believe me, I did....Very early on in my career as a stand-up comic I learnt not to think about what I was going to say next. I developed a reputation for dealing with hecklers. Sometimes I would piss on them, other times pour their drinks over them.[41]

Allen recognised that comedians usually collude, and rarely collide, with their audience[42] and fiercely advocated the latter, more individualistic course:

> Comedians think they're brave, but they're not – all they're doing is joining that gang, which is the audience. Whereas my idea was always not to be one of them. That's what I didn't like about any of these other

> acts – because none of them were themselves. I would always go up as Keith Allen....[43]

Certainly the expectation grew that comedians would now be joker-jokewriters, writing their own material and basing it on their own experiences and philosophies: 'you never nicked other comics' jokes', said comedian Jim Tavare, adding that jokes seemed, by contrast, to be community property among northern club comics.[44] This remains an issue: in September 2008 comedian Lee Hurst broke the mobile phone of a member of his audience, believing the man to have been filming his act. In 2009, Stewart Lee, another comedian, added a page called 'Plagiarists Corner' to his website accusing a fellow comic of stealing one of his lines.[45]

Clearly, though, a nascent culture based on these principles and organised, by definition, around public declamation would have to address other questions: most of the performers accepted that the expression of their individuality could infringe the individuality of others. Racism and sexism, hitherto the staples of traditional joke-telling in a range of cultures, were widely outlawed by comedians and by proprietors of the new venues. With this customary *caveat*, the emergent comedy of the 1980s embraced all the previous innovations – Sahl's discursiveness, a widespread political iconoclasm and the surrealist tendency of Milligan and the Pythons: 'Revue was a dirty word', said Rik Mayall, who formed a double act in the late 1970s, 'and so was Oxbridge, we had a down on the Pythons, although we secretly all thought that the Pythons were great.'[46] Interest in 'alternative comedy' soon spread to high status, white-collar groups many of whom came to the Comedy Store, the most influential early comedy venue, which opened in the West End of London in 1979.[47] Someone who performed regularly there in its early months told me: 'You had to be good. The audience were mainly young professionals, estate agents from Kensington, and so on'. Peter Rosengard, the club's first proprietor, described the audience one night as 'sitting, standing, shouting and drinking champagne from the bottle'.[48] The Store was rowdy, in keeping with the prevailing ethos, but it was also governed by strict market forces: acts deemed unfunny or objectionable were required to leave the stage when the compere, interpreting audience reaction, banged a gong.

There was a growing sense through the 1980s, especially on the part of the most leftish and politically acute comics, that the new comedy was, for most people, no more than unpoliticised fun, with the audience happy simply to hear the piss taken, no matter who or what it was taken out of. Jeremy Hardy, who combined radio quizzes and a politically radical column in a broadsheet newspaper with his stand-up, said:

> I think there's something spreading in our society – people are just becoming hedonists, they don't want to think, they don't want to have

> ideas, they don't want to change, they just want to get into a groove and be debauched.

An audience that liked him, he reflected, might equally enjoy Bernard Manning or Jim Davidson, comedians for whom racist and sexist material was stock-in trade.[49] Laughing at a joke, perceived by its teller to be left-wing, was not necessarily itself a left- wing gesture, any more than buying William Morris wallpaper was. Comedy, it dawned on even the most reluctant of the 'alternatives', was a commodity; once people had received it, they were free to do what they wanted with it. This was reaffirmed in 1994 when , for the BBC2 television programme *Reportage*, the 'alternative' comedian Mandy Knight agreed to swap roles with the Northern club comic Jimmy Bright: Knight played a Northern club and Bright performed at the Bound and Gagged club in North London, an alternative venue; in the parlance of the circuit, Jimmy 'stormed' and Mandy, on her own admission, 'died'. There was in any event an ongoing *rapprochement* between 'alternative' and traditional comedians during the 1980s. Les Dawson (1931–1993), a northern comedian of the old school, who dealt heavily in mother-in-law and 'the wife' jokes, played the Comedy Store in its early days. '[He] turned up one night and did fifteen minutes and was brilliant', recalled Jim Barclay, a socialist and pioneer 'alternative' comedian.[50] In the late 1980s and early 1990s the term 'alternative comedy' drifted out of usage and was replaced, sporadically, by 'post-alternative comedy'.

The comedy business, following its 'alternative' moments in the United States (c1959) and Britain (c1979) respectively, has a number of perceptible characteristics: a largely homogeneous stable of comedians has been replaced by a diverse array of comedians and comedies, each with a recognisable constituency or demographic, but all more or less rooted in the politics of identity, of selfhood and what the social theorist Anthony Giddens has called 'reflexivity'.[51] By the 1990s in Britain there were left-wing comedians and right, self-consciously 'laddish' comedians and feminist ones, gay comedians and straight, comedians of various ethnicities, comedians who discussed politics, comedians who discussed sex, comedians who discussed football, and so on. Some told jokes; some did not. For the most part, comedians told the truths and spoke in the vernacular of their own constituency. Indeed, 'alternative' comedy's first empowerment had been in the validating of ordinary experience and everyday discourse. The comedian Frank Skinner commented in the early 1990s: '...it took me along time to realise that I could say things on-stage that I found funny and I'd say to my mates...'.[52] Likewise Jeremy Hardy remarked in 1985:

> 'When I started a year ago I couldn't think of anything interesting about myself. I come from this very straightforward uninteresting unethnic background in Surrey. So I thought, well, I'll make the most of that'.[53]

Similarly, Eddie Izzard, who sold out an extended run at the Ambassadors Theatre in the West End of London in 1994, saw his act primarily as an assertion of his own individuality:

> My agenda isn't to think what's in their [the audience's] minds, and where I've got to pull them....I can't be apologetic because it just doesn't work that way. You have to say 'This is what I am! This is what I think! This is how it is!'[54]

The general ethos on the comedy circuit of clubs, theatres, pubs and corporate entertainment that burgeoned after the early 1980s was libertarian, with periodic – often press-derived – squabbles over 'political correctness'. Most comedians did not make on-stage remarks of a sexist or racist inflection, either because they had no taste for them or because they knew their audience didn't care to hear them, or both. In other parts of the comedic forest, such material survived because it still made some people laugh. In 2000 Bob Monkhouse (1928–2003), doyen of right wing British post-war comedians, was asked by the journalist Lynn Barber about ' a routine he does about American Indian names':

> Only last week I ran into my first wife, Crazy Bitch. With her new husband, Nutless Wonder. On the way to visit her mother, Cow That Will Not Die.' Monkhouse replied: 'My entire act is a fiction. Jackie [his wife] once said years ago that for every thousand insults the fictitious wife gets on television, she gets another diamond.[55]

'Time Well Wasted':[56] The comedy business and postmodern society

The remainder of this chapter considers some of the implications of this reconstituted and now-pervasive comedy for the contemporary politics of leisure.

I begin with some thoughts on comedy and globalisation. The emergence in Britain of 'alternative comedy' coincided with the deregulation and globalisation of the mass media. In Britain the principal enactment here was the Broadcasting Act of 1990 which licensed satellite and cable broadcasting and applied a 'lighter touch' to requirements of 'quality' in programming. Comedians, with their talent to amuse, their often reflexive irony and the access they promised to particular taste groups were soon in great demand by media companies and advertisers. As the 1990s wore on, a growing number of performers made the transition from discrete stand-up comedian to multi-purpose media celebrities: they appeared in and recorded 'voice-overs' for advertisements; they presented documentary programmes; they took part in a myriad of TV game shows; they wrote novels, memoirs and newspaper columns; they chaired chat shows; they championed charities;

they had their own radio programmes; they were interviewed regularly across the media about their latest tour/DVD/book, and so on. In 2009, in a perfect evocation of this multi-faceted, multi-media existence, Frank Skinner, a stand-up comedian since the early 1980s, published a book about how in 2007 he had returned to stand-up comedy after ten years scripting situation comedies, hosting TV chat shows, writing an autobiography and appearing in a range of other playful television programmes.[57] Some, indeed, created characters for media purposes – as with the Manchester comedian Steve Coogan, whose various aliases included sports reporter-cum-chat show host Alan Partridge, unemployed Mancunian Paul Calf and Portuguese crooner Tony Ferrino; Caroline Aherne's fictitious TV personality Mrs Merton; and Sacha Baron Cohen's reincarnation as hip-hop parody Ali G,[58] maladroit Kazakhstani Borat and gay Austrian fashion reporter Bruno. Cohen's alter egos were especially cleverly wrought and media-wise because he never appeared as himself, thus heightening the mystery of the characters and affirming them as open texts. Moreover, the deregulation and globalisation of the media helped to expand exactly the niche markets that 'alternative comedians' had come into existence to serve. Television companies now had fewer obligations to produce programmes that were educationally or morally improving and a now overriding imperative principally to seek audiences: the 'phones of comedy agents began to ring. Furthermore, comedy that had tickled the ribs of growing student audiences in Britain might now very well, via the appropriate television deals, come to amuse similar audiences across a range of countries. By 1997, John Thoday, director of Avalon, one of the two major comedy agencies in London, was talking with pride of how many of his clients, most of them recently or currently on show at the Fringe of the Edinburgh Festival were receiving lucrative offers from the United States.[59]

The career of Sacha Baron Cohen provides a good illustration of the possibilities opened up for comedians around this time. Cohen, another Cambridge graduate, began his career in the newly established cable TV channels of the early 1990s. His character Ali G – an apparently witless white suburban youth affecting a black identity – first appeared on Channel Four's comedy programme *The Eleven O'Clock Show* in 1998. Much of the comedy generated by 'Ali G', and Baron Cohen's subsequent characters derived from the *faux* innocent questions they posed to people, many of them established celebrities who believed him to be real. In one or other of his guises the comedian gulled his interviewees into expressing uncomplimentary opinions of women, gay people or Jews, thus availing himself of the increasing global comedic paradigm of 'political in/correctness'. Neither the character comedy employed by Baron Cohen nor his 'candid camera' technique was new but the global media possibilities for the marketing of his characters were. All three of Baron Cohen creations had feature films built around them – the first, *Ali G Indahouse*, was released in 1999; in 2002 it was made available on DVD in Region 2 (principally most of Europe Western Asia, Japan and parts of Africa) and two years' later in Region 1 – effectively the United States and Canada. The second film, *Borat*,[60] opened in the USA (800 screens), Canada and 14 European countries in 2006. It was banned in all Arab countries, bar Lebanon. The culture minister of Dubai described the character of Borat Sagdiyev as 'vile, gross and extremely ridiculous' while liberal commentators saw this as evidence of the divide between irony-conscious, game-for-a-laugh postmodern Western culture and the socially conservative Persian Gulf – 'a region not known', as one put it, 'for its tolerance toward the arts'.[61]

Correspondingly, the capacity to appreciate comedy – satirical and/or stand-up – is increasingly seen as an index of cultural enlightenment in Western discourse. For example, in December of 2008, Dean Odeidallah, an American stand-up of Palestinian and Italian parentage, wrote euphorically about the recent Amman Stand-up Comedy Festival in Jordan, the first of its kind in the Arab world:

> While there is no history of stand up comedy in the Arab world, You Tube and American TV shows airing in the region have brought our comedy there and it's catching on fast. To give you a sense of how much Arabs love stand up comedy, I recently performed in Beirut with Middle Eastern-American comedians Maz Jobrani and Ahmed Ahmed and we sold over 5,000 tickets. Just a few weeks ago I co-headlined a show with comedian Aron Kader in Cairo and over 4,000 people attended. The material we perform is almost all in English and basically the exact jokes we tell in the comedy clubs in the US. (With a few local jokes thrown in as well.) The audiences in the Arab world – which are predominantly but not exclusively Muslim – have no problem laughing at themselves or jokes about relationships, politics, pop culture, or just standard US observational comedic material. Its been amazing to see these audiences laugh at the identical jokes we have told to US audiences. It makes you realize that we have a lot more in common than some would believe.[62]

Similarly, the emergence of India as a marketised economic power in the early twenty first century led some to suppose that she too had arrived at her moment of postmodern comedy: in April of 2009, it was announced that Don Ward, who had launched The Comedy Store in London in 1979, would open a similar club in Mumbai, beginning with some 'brand awareness' shows by British comedians. If successful, Ward intended similar ventures in other Indian cities. This 'brand awareness' would involve the importation of the now familiar mix of 'edge' and identity politics – as one journalist noted that one of the visiting comedians 'is openly gay, which is highly controversial in the sub-continent'.[63]

All this raises the question of the relationship between contemporary comedy and the market. What we have seen over the last 30 years or so (1979–2009) is the commodification of banter and the use of irony. By the mid-1990s a comedy industry had become established, built primarily on the now-burgeoning leisure pursuit of going to comedy clubs. Two agencies dominated the landscape – Off the Kerb (founded in 1983) and Avalon (f.1989) and, as one writer suggested, could promise clients:

> the sort of career structure the civil service might envy. Take Avalon. It has the resources to build up an act slowly but surely. It can test out newcomers on the National Comedy Network [the comedy club circuit], before sending them out as support on its bigger acts' tours of 2,000-seat venues and placing them on suitable radio shows. Finally, the comedians can start doing their own national tours, while the company's television wing develops neat little BBC2 [television] series for them.[64]

By 2009 Avalon had expanded to include a live events bookings agency, arranging 1,500 shows a year, a 'public relations arm' and a company making programmes for daytime television.[65] Other agencies and companies now thronged the market, many supplying the contemporary tendency of companies to merge notions of work and leisure at corporate events. Purple Cactus Productions, for example, counsel that 'Corporate comedians and live entertainment are some of the best ways to brighten up a company event.'; they offer a price range of comics for this purpose starting at 'Bronze' (£500 to £2,500) and graduating to 'Diamond' (£20,000 plus) and they add that 'if the name you search for is not here, don't worry. We guarantee that we are in touch with every comedian working today'.[66] The wide and lucrative availability of corporate work has naturally clashed with the politics of identity that has characterised much contemporary comedy. Ethnic personal reflection, for example, is particular and unlikely to be suitable for corporate audiences. Besides, as British-Iranian stand-up Omid Djalili commented in April of 2009: 'I don't want to be a role model. I'm just a comedian. If British-Iranian companies are keen for me to be a standard-bearer, they should show that in their corporate fees!'[67] A similar commercial pragmatism has gained a string of advertising contracts for leading comics from companies eager to associate their various drolleries and cheeky personae with their products. Understandably, this has often been in the realm of young male consumption – particularly drink. Since comedy clubs are invariably domiciled in pubs, comedy, like televised football, has been a major boon to the drinks industry since the mid-1980s. Stephen Frost and Mark Arden, who worked a double act called The Oblivion Boys on the 'alternative' circuit, were among the first of these when they did a series of notable TV advertisements for Carling Black Label in 1989; most recently, the Lancashire comedian Peter Kay (b.1973) has made a number of popular television adverts for John Smith's bitter and Mark Watson (b.1980) has endorsed Magners Pear Cider. (Red Stripe, Stella Artois and Boddingtons have all sponsored 'alternative' comedy). Here the political paradox of post-1979 comedy has continued to be played out: could a comedic voice still be called authentic once it was hired and scripted by a corporation? In 1997, Ivor Dembina, a veteran stand-up, publicly castigated the comedian Jack Dee (b.1961) for his recent 'No Nonsense' adverts on behalf of John Smiths bitter, Dee having been an alcoholic in the 1980s:

> he is allowed easily to deflect questions about his former drink problem – which is surprising, perhaps, when he spends so much of his time selling beer to others. Comedy should attack hypocrisy, not promote it.[68]

In September of 2009 this issue was still being debated with Dembina (and others) now reproaching stand-up Mark Watson for his recent endorsement of Magners Pear Cider. On the Chortle website he wrote:

> Mark's explanation for doing the advert is to support his family, but if he wants to do so by selling booze, he should open an off-licence. Sorry, when you operate as a comedian your first responsibility isn't to your family, it's to your audience....Whether you see comedy as art, entertainment or political activity, it's one of the few places left for sharing truth.[69]

Other comics took (almost certainly the majority) view that this was simply a symptom of life in the contemporary leisure industry, wherein comedians were both producers and consumers. 'Mark has never claimed to be a capitalism-bashing comic, he does sort of whimsical riffing on everyday nonsense, so how can he be selling out?' asked Lloyd Wolf, another stand-up. Besides, he added, 'a comedian endorsing alcohol is about as unhypocritical as it gets. I'm generalising of course. But a stroll around Edinburgh's Brooks Bar at 3am any night in August will pretty much back me up on that one.'[70]

The notion that comedy is one of the few remaining areas left for the sharing of truth raises the question of the relationship between comedy and politics.

In this relationship the paradox noted above is reproduced and rendered more complex. Comedy is play and politics is not. In Britain, the United States and elsewhere the cultural-comedic politics of the late 1950s and early 1960s long since became the basis of a paradigm in which politics and the people who practised it were the subject of ridicule and citizenship, by implication, a voluntary activity from which individuals might withdraw, in favour of social worlds perceived to be more playful, more pleasurable or more real.

In sanctioning *That Was The Week That Was* in 1961 Donald Baverstock, Assistant Controller of Programmes at the BBC, had suggested that 'late on Saturday night people were more aware of being persons and less of being citizens...'[71] By the 1990s 'satire' – the media mockery of politicians – was an entrenched part of the comedy business and its attendant philosophies were no longer confined to a particular period of the week. In 1995 one writer now spoke of 'assembly line satire'.[72] Periodically over its 50 odd year history a commentator would proclaim that 'satire' had lost its 'edge', its 'capacity to shock', and so on.[73] Several important points flow from this.

First, and crucially, 'satire' has, once again, affirmed *politics as performance*. In 2000 Christopher Booker, a founder in 1961 of the satirical magazine *Private Eye* remarked, that the *Eye*

> had emerged just when British life and our politics in particular were first passing under the fatally distorting spell of television. I cannot really imagine *Private Eye* could have come into being when Winston Churchill or Clement Attlee were still presiding over our national life. In those days our politics still had a substance now long departed, swept away into that all-embracing world of make-believe created by the telly.[74]

From this perspective, politics no longer mattered; 'substance' was elsewhere. And politicians were essentially in the same business as satirists and other comedians.

Second, the nature of pressing political issues has changed – particularly since the 1980s. This explains the pervading invocation of 'political correctness'. Britain and a range of other countries have, as the influential sociologist Stuart Hall pointed out, experienced 'a fragmentation of the political landscape' involving 'the erosion of the mass party as a political form, a decline in active participation in mass political movements and a weakening in the influence and power of the "old" social movements of the working class and industrial labour'. Political initiative has passed to the 'new social movements – "the soil in which PC has been nurtured"'.[75] These new social movements are now, tacitly at least, represented in comedy, with each political, gender, ethnic, sexual and disability group having its favoured jokers. This, to a significant degree, is where politics now resides – in a leisure space. A perfect illustration of this transition is the long-running TV panel game *Have I Got News For You?* which has a domestic audience of between six and eight million and was first made for the BBC by the independent production company Hat Trick in 1990. One team is led by Ian Hislop, Oxford-educated editor of Private Eye magazine, whose contributions are always politically informed and dyspeptic. Hislop, a career satirist, refers acidly to the week's hypocrisies; his manner is that of the thin-lipped English suburbanite, impatient with 'politicians'. The other team is captained by Paul Merton, a stand-up comic, formerly on the 'alternative' circuit. Merton's voice is working-class South London and his humour in the programme comes out of his political incomprehension. He doesn't know much about these things and he doesn't care to; with his 'dumb and puzzled logic'[76] he represents the Ordinary Bloke. The guests on the programme have often been 'alternative' or ex- 'alternative' comedians or (other) fellow-news items of the permanent cast, figures culled from elsewhere in the world of politics-publicity. Here left wing comedians, right wing or 'laddish' comedians, female comedians, black comedians, gay comedians and others may sit with the regular panelists and commune in their mockery of the week's parliamentary affairs. If there is a political issue here it may well, as elsewhere, simply concern *access*: "I've been going 10 and a half years and I haven't done any panel shows' complained Francesca Martinez, a stand-up comedian with cerebral palsy, in 2009.

> I've tried all of them – *Mock the Week, Have I Got News for You* – but my agent gets told 'we're worried she'll make the audience too nervous'. Ultimately, I'm not saying I should be on TV because I'm wobbly. I should be on TV because I've proved that I'm funny and I can be entertaining. Maybe the truth is that disability is the last remaining taboo. People are so nervous about it.[77]

Third, the increased centrality of 'political correctness' has increased the public salience of comedians. As professional arbiters of public taste and pushers-of-boundaries in an age of 24 hour news channels they promise a reliable news story, always likely to outrage some sections of a global audience and tickle others. Hence the mini-moral panics that now routinely attend comedians and their risqué remarks. At the time of writing the latest episode concerns Jimmy Carr who told audiences in late autumn 2009:

> Say what you like about those servicemen amputees from Iraq and Afghanistan, but we're going to have a fucking good Paralympic team in 2012.

Carr showed a shrewd appreciation of the resulting furore, arguing that it had been concocted by the media and was a symptom, merely, of the diversity of the media and of comedy markets:

> I played to 9,000 people that weekend. I did Manchester and Stockport, and two people complained. My audience aren't offended, but this other audience that reads the papers are offended.[78]

Fourth, in a number of countries, where the range of arguments, perspectives and/or rights has been narrowed in the official political arena, comedians have stepped into the breach. In Britain, with the demise of parliamentary socialism and the rebranding of the Labour Party as 'New Labour' in the mid-1990s, some comedians became spokespeople for the left. For example, the comedian and impressionist Rory Bremner moved to Channel Four in 1993 for his TV show *Rory Bremner – Who Else?* This and subsequent shows have carried much political argument and Bremner was dubbed 'the real leader of the opposition' by the writer Robert McCrum in 2001. Bremner himself said: 'Winston Churchill would be thrown out of New Labour for being too left wing'.[79] The comedian Mark Thomas has led numerous campaigns since the 1990s – particularly against the international arms trade, the international activities of Cola Cola and the erosion of civil liberties.[80] In the United States it is now, similarly, argued that the cultural inheritors of Mort Sahl and Lenny Bruce are among the most influential progressive voices in national life. Two of the most trenchant and popular critics of the Bush administration (2000–2008) were said to be the comedians Jon Stewart and Stephen Colbert. Stewart's *The Daily Show* (begun in 1993) and its spin-off *The Colbert Report* (a parody of the right-wing Fox News), both available on the Comedy Central channel, became the favoured news source for liberal and/or young America and led to accusations from the political right that they were hiding behind the designation of 'comedian'.[81] In Israel, *Eretz Nehederet (Wonderful Country)*, that country's equivalent of *The Daily Show* mocked the Israeli army's attack on Gaza in January of 2009 in a way mainstream politicians were disinclined to do:

> Parodying coverage of the Eurovision Song Contest – in which Israel competes – Wonderful Country correspondents report from different capitals on how many Palestinians the Europeans will allow Israel to kill. From Italy: 800. From Germany: 6,000.[82]

Later the same year Zargana, a popular Burmese comedian, was sentenced to 45 years in jail for criticising the response of the country's military government's response to Cyclone Nargis to foreign news media.[83]

Fifth, other comedians – now in many cases enjoying greater credibility than politicians – have agreed to endorse, and promote participation in, the orthodox political process. In 1996 British comedian Eddie Izzard spoke up for Rock the Vote, a campaign to persuade young people to register to vote: 'Apathy' he said, 'is never going to solve anything'.[84] More recently in 2008, in the United States, Sarah Silverman, an emergent comedian with a ditzy on-stage persona, whose thoroughly profane transgressions of political correctness invited comparisons with Lenny Bruce, made a video in support of The Great Schlep – an attempt to get young liberal-minded Jewish Americans to visit their elderly relatives in Florida and tell them to vote for Barack Obama in the presidential election. Remember, she told them sweetly, back in 2000 'Gore got fucked in Florida'.[85] Comedians have also lent credibility to politicians – as in 2007 when British Prime Minister Tony Blair, widely unpopular for his part in the American invasion of Iraq in 2003, took part in a TV sketch with comedy actress Catherine Tate. Tate played one of her characters, fast-talking Essex schoolgirl Lauren, and Blair had frequent recourse to her popular catchphrase 'Am I bovvered?'. The sketch was in aid of Comic Relief, a scheme inaugurated in 1985 by comedy writer Richard Curtis, wherein famous comedians help raise money for famine relief – seen by some comedians as obscuring the causes of poverty.[86]

Finally, a brief consideration of the continuing tradition of working-men's club or 'blue' comedy and of the argument that it serves a fractured proletariat and a vanishing working-class community. Although the best known exponent of this comedy was Bernard Manning, via the initial run of the ITV programme *The Comedians* (1971–1974), for which, of course, he 'cleaned up' his act, the most popular practitioner here in recent times has been the afore-metioned Roy 'Chubby' Brown, the stage name of Royston Vasey (b.1945) from Grangetown, Teesside in the north of England, who began playing the clubs in the 1960s. In 2007 Brown was the subject of a sympathetic analysis by the academic Andy Medhurst. This analysis attracted some *a priori* media curiosity because Medhurst is gay and Brown is wont to say during his performances 'If there's any poofs in, fuck off'. Similarly, he often remarks 'If there's any asylum seekers in, fuck off home'.[87] Brown trades heavily in self-consciously profane versions of the historic wife joke: 'My first wife died. I didn't notice for a week. The fucks were the same but the dishes piled up'.[88] His DVDs include a box set, issued in 2007, celebrating *Forty Years of Filth*. Medhurst suggests that Brown's popularity in certain circles is 'a flagrant testament to the persistence of class as a meaningful category'[89] and that his comedy, albeit hostile to minorities, is nevertheless:

> rooted in the life experiences and structures of feeling of another stigmatised group, those white working-class English, left behind by the turn toward hybridised, globalised culture.[90]

One cannot feel wholly comfortable with this view. It's hard, for instance, to see how the 'white working-class English' were ever a 'stigmatised group' – certainly they were never a comedy staple in the way that wives, mothers-in-law, gays and migrants have been. Nor can class be sensibly defined by its capacity to appreciate blue humour. Class is defined through economic situation and it is globalisation and deindustrialisation that have fragmented the working class not identity politics or 'political correctness'. Thus, there will be plenty of young working-class people – in the north of England and elsewhere – who cannot see their experience reflected in Brown's material – young women, for example, who would rather listen to the cheerful sexual voraciousness of Jo Brand or Jenny Éclair than to one liners about sexless wives-as-dishwashers or instructions to asylum seekers to 'fuck off home'. Brown's audience and Manning's are unlikely to have been exclusively working class in any case – when in 1995 Manning was secretly filmed by ITV's *World In Action* making offensive remarks to a black member of the audience, the occasion was a police charity dinner, with senior officers present. Chubby and Bernard's public was/is probably made up, like the current membership of the British National Party, of the most unprogressive elements in the working class and in the status-anxious lower middle class, who seek an ordered world, unpolluted by black people, queers and feminists. This is a male world – there could never be a female Brown or Manning – and it abhors social change: women are wives who do or don't want sex and children are, on the whole invisible. For all his on-stage expletives, Brown relies on the survival of Edwardian social mores in order to transgress them in his performance. Indeed, in 2003 he was fined by Blackpool magistrates for hitting a man with an umbrella. He told the court: 'I just wanted the man to stop swearing and being abusive in front of women and children who were on the pier.'[91] Such comedians resent the onset of new social movements – parodied as 'political correctness' – because it threatens the very assumptions on which their acts rest; they do not hope, for instance, for more openness about sex because then it ceases to be 'filth' and becomes desire. This undermines the argument that these comics are transgressive. They have no politics of the self; rather than rattle or remove taboos, they seek to preserve them. So Bernard Manning was certainly not a punk – a word which has connotations of social revolt; his Queen Mother joke mocked not royalty, but the aged.

Conclusion: Comedy as heritage

Although Brown and Manning have each made a good living and attracted big audiences in the time since the diversification of comedy in the late 1970s, their comedy is essentially comedy-as-heritage – a celebration of Comedic Times Past. Many people clearly either regret the diversification or in some way hanker for past comedic times – times when most public comedy was apparently uncontentious and comedians helped to promote a sense of the national family. Central to this nostalgia are the artfully clumsy magician-comedian Tommy Cooper, who died on stage in 1984 and, most especially, the popular double act Morecambe and Wise who performed together from 1941 until Morecambe's death, also in 1984. The cultural importance of Cooper and Morecambe and Wise is twofold. First, they recall a time when comedy could be both innocent and universal, somehow beyond politics and 'edge'. The old school of comics and the spectrum of post-alternative comedians have united in affection for Morecambe and Wise; no memorable comedic voice has been raised against them. Moreover, the estimated audience of 28,835,000 who watched their Christmas Day TV show in 1977 (six million more than the audience for the Queens Speech)[92] is the ultimate symbol of prefragmentation comedy. When Mark Watson was rebuked for his cider advertisements and for not using comedy to 'shine a light on how power is used to maintain the status quo', he replied 'Well, I don't remember Morecambe and Wise doing much of that.'[93] Second, Morecambe and Wise achieved their greatest success in an incipiently post-modern age wherein people were beginning to define themselves through their leisure – through what they liked, rather than what they did for a living. Thus, 25 years after they last trod the boards, a mini-Morecambe and Wise heritage industry continues to thrive. Amazon.com, for example, has a web page designated as The Morecambe and Wise Shop[94] where DVDs of their TV and live performances, biographies (both learned and intimate – Morecambe's son Gary has written three books about him), radio tapes and Morecambe's edited diaries can all be purchased.

This pattern, however, will not be confined to the lovers of 'national' or inoffensive humour. Satirical comedy is old enough now to have become a theatre of memory: for example, at least five biographies or intimate reminiscences of Peter Cook have been published since his death in 1995 and a range of newspapers and news websites in Britain and the United States celebrated 40 years of the 'silliness of Monty Python' in October of 2009 and a sell-out celebratory show – *Not the Messiah (He's a Very Naughty Boy)* – featuring several of the original performers was staged at the Albert Hall. Comedy in the twenty-first century will continue to be an important way in which people tell themselves who and what they are, and who and what they've been.

Acknowledgement

Many thanks to Holly Bentley at the *Guardian* archive, Pete Bramham and Anne Mitchard for their help in the preparation of this chapter.

Notes

1 *Daily Telegraph* 19th June 2007 http://www.telegraph.co.uk/news/obituaries/1554926/Bernard-Manning.html Access: 2nd November 2009.
2 See Roger Wilmut (1985) *Kindly Leave the Stage: The Story of Variety 1919–1960*, pp.26–35, 116–125 (London: Methuen).
3 http://www.funny.co.uk/stand-up-comedy/art_64-966-Bernard-Manning-Racist-Tosser.html Posted 7th August 2003. Access: 2nd November 2009.
4 Ibid.
5 Bernard Manning 'Bernard Manning: His own obituary, in his own words', *Daily Mail* 20th June 2007 http://www.dailymail.co.uk/news/article-462884/Bernard-Manning-His-obituary-words.html Access: 2nd November 2009.
6 Stephen Dixon Obituary of Bernard Manning http://www.guardian.co.uk/news/2007/jun/19/guardianobituaries.obituaries1 Posted 19th June 2007. Access: 2nd November 2009.

7 Howard Jacobson 'Arise Sir Salman, and goodbye Bernard, those two experts at stirring things up' http://www.independent.co.uk/opinion/commentators/howard-jacobson/howard-jacobson-arise-sir-salman-and-goodbye-bernard-those-two-experts-at-stirring-things-up-454301.html Posted 23rd June 2007. Access: 2nd November 2009.
8 Ed Barrett 'Bernard Manning: The oldest and truest punk in town' http://www.spiked-online.com/index.php/site/article/3500/ Posted 19th June 2007. Access: 2nd November 2009.
9 See Andy Medhurst (2007) *A National Joke: Popular Comedy and English Cultural Identities*, p.188 (London: Routledge).
10 Zygmunt Bauman (2000) *Liquid Modernity* (Cambridge: Polity).
11 Zygmunt Bauman (2000) *Society Under Siege*, p.166 (Cambridge: Polity).
12 See Stephen Wagg (1996) 'Everything else is propaganda: The politics of alternative comedy' in George, E. C. Paton, Chris Powell and Stephen Wagg (eds) *The Social Faces of Humour*, pp.322–3 (Aldershot: Arena).
13 Quoted in William Cook (1994) *Ha, Bloody Ha: Comedians Talking*, p.195 (London: Fourth Estate).
14 http://en.wikipedia.org/wiki/The_Office_(UK_TV_series) Access: 7th January 2010.
15 Stephen Wagg (1992) 'You've never had it so silly: The politics of British satirical comedy' from *Beyond the Fringe* to *Spitting Image*', in Dominic Strinati and Stephen Wagg (eds) *Come on Down? Popular Media Culture in Post-War Britain*, pp.254–84 (London: Routledge); Stephen Wagg (1996) 'Everything else is propaganda: The politics of alternative comedy', in George C. Paton, Chris Powell and Stephen Wagg (eds) *The Social Faces of Humour*, pp.321–47 (Aldershot: Arena Press); Stephen Wagg (1998) 'They already got a comedian for governor: Comedians and politics in Britain and the United States', in Stephen Wagg (ed.) *Because I Tell a Joke or Two: Comedy, Politics and Social Difference*, pp.244–72 (London: Routledge); and Stephen Wagg (2002) 'Comedy, politics and permissiveness: The 'satire boom' and its inheritance' in *Contemporary Politics*, 8(4): 319–34. With the permission of the publisher, which I gratefully acknowledge, I have re-worked some of the material from 'Everything Else…' for this essay.
16 Biographical notes adapted from Ronald L. Smith 'Who's who in comedy' http://mortsahl.tripod.com/id15.html Access: 8th November 2009.
17 Ibid.
18 John Rogers 'Mort Sahl, the angry young comic, at 80' http://www.sfgate.com/cgi-bin/article.cgi?f=/n/a/2007/06/26/entertainment/e104540D01.DTL Posted 26th June 2007. Access: 8th November 2009.
19 Ibid.
20 See Wagg 'They already got a comedian for governor'.
21 Rogers 'Mort Sahl, the angry young comic…'
22 Mort Sahl (1976) *Heartland*, p.80 (New York: Harcourt Brace Jovanovich).
23 Verve Records MGV 15022–3.
24 Quoted in Albert Goldman (1976) *Ladies and Gentlemen Lenny Bruce!!*, p.210 (London: Picador).
25 http://www.hollywoodusa.co.uk/biographies/lbruce.htm Access: 8th November 2009.
26 Bob Woodward (1984) *Wired: The Short Life and Fast Times of John Belushi*, p.55 (London: Faber and Faber).
27 Quoted in Gerri Hirshey (1989) 'Vicious boys', *Time Out* 29th November to 6th December, p.18 and subsequently in Wagg 'They already had a comedian…'. See the latter for a fuller account.

28 For a fuller analysis, see Wagg 'You've never had it so silly'. See also Roger Wilmut (1985) *From Fringe to Flying Circus* (London: Heinemann) and Humphrey Carpenter (2000) *That Was Satire That Was* (London: Victor Gollancz).
29 Michael Palin (2007) *Diaries 1969–1979: The Python Years*, pp.215, 221 (London: Phoenix).
30 Ibid., p.166.
31 Geoff Hammill *Monty Python Flying Circus: British Sketch Comedy/Farce/Parody/Satire Series* http://www.museum.tv/eotvsection.php?entrycode=montypython Access: 10th November 2009.
32 Palin *Diaries...* pp.97–102.
33 See Sid Caesar (and Bill Davidson) (1982) *Where Have I Been? An Autobiography*, p.173 (New York: Crown Publishers).
34 Palin *Diaries...* p.15.
35 Channel Four 22nd May 1995. Quoted in Wagg 'Everything else....' p.325.
36 Trevor Griffiths (1976) *Comedians*, pp. 23, 33 (London: Faber and Faber).
37 Helen Chappell (1983) 'The mauve economy', *New Society*, p.123, 28th July.
38 See John Street (1986) *Rebel Rock: The Politics of Popular Music*, pp.147–9 (Oxford: Basil Blackwell).
39 From Tony Allen's book *Attitude*, an extract of which can be found at http://www.newagenda.demon.co.uk/redpepper.html Access: 11th November 2009.
40 Malcolm Hardee (with John Fleming) (1996) *I Stole Freddie Mercury's Birthday Cake and Other Biographical Confessions*, pp.90–1 (London: Fourth Estate).
41 Keith Allen (2007) *Grow Up: An Autobiography*, p.230 (London: Ebury Press).
42 I'm grateful to George Paton of Aston University for passing this thought on to me some years ago.
43 See Roger Wilmut (1989) *Didn't You Kill My Mother-in-Law*, pp.34–5 (London: Methuen).
44 See William Cook (2001) *The Comedy Store: The Club That Changed British Comedy*, p.72 (London: Little, Brown). See Chris Green (2009) 'Plagiarism is no laughing matter for comedians', *The Independent*, 11th November, p.17.
45 See Chris Green (2009) 'Plagiarism is no laughing matter for comedians', *The Independent* 11th November, p.17.
46 Wilmut *Didn't You Kill...* p.95.
47 For an exhaustive history of the Comedy Store, see William Cook *The Comedy Store....*
48 Wilmut *Didn't You Kill.....* p.6.
49 Ibid., p.276.
50 Cook *The Comedy Store* p.71.
51 Anthony Giddens (1991) *Modernity and Self-Identity* (Cambridge: Polity Press).
52 Cook, *Ha, Bloody Ha*, p.298
53 Colin Shearman 'Weed that pulls you up', *The Guardian* 31st May.
54 William Cook 'Going straight' [Profile of Eddie Izzard] *The Guardian* 4th February 1994.
55 Lynn Barber 'Beneath my underpants I'm a riot of polka dots and moonbeams' [Interview with Bob Monkhouse] *The Observer Magazine* 20th August 2000, p.14. Available at http://www.guardian.co.uk/theobserver/2000/aug/20/features.magazine27 Access: 13th November 2009.
56 'The slogan of the comedy network', a Canadian, English-language cable TV channel.
57 Frank Skinner (2009) *Frank Skinner on the Road* (London: Arrow Books).

58 See Sharon Lockyer and Mike Pickering (2009) 'The ambiguities of comic impersonation', in Sharon Lockyer and Michael Pickering (eds) *Beyond a Joke: The Limits of Humour*, pp.182–99 (Basingstoke: Palgrave).
59 John Thoday 'What ITV must learn from the fringe', *The Guardian* 16th August 1997, p.6.
60 Full title: *Borat: Cultural Learnings of America for Make Benefit Glorious Nation of Kazakhstan*.
61 'Arab countries ban *Borat*' http://www.guardian.co.uk/film/2006/dec/01/film-censorship Posted 1st December 2006. Access: 15th November 2009.
62 Dean Obeidallah 'Middle East sees "explosion" of comedy, first stand-up comedy festival in Arab world' http://www.huffingtonpost.com/dean-obeidallah/middle-east-sees-explosio_b_148357.html Posted 4th December 2008. Access: 15th November 2009.
63 Arifa Akbar 'An Indian walks into a comedy club...', *The Independent* 22nd April 2009, p.9.
64 James Rampton 'Top of the bill (less 15%)' http://www.independent.co.uk/arts-entertainment/comedy–top-of-the-bill-less-15-1568567.html Posted 18th January 1995. Access: 18th November 2009.
65 See James Ashton 'Comedy firm Avalon find fun way to earn £4m' http://business.timesonline.co.uk/tol/business/industry_sectors/media/article6814958.ece Posted 30th August 2009. Access: 18th November 2009.
66 See http://www.purplecactus.co.uk/comedians.php Access: 18th November 2009.
67 James Rampton 'Consider yourself a Muslim, Fagin', *The Independent* [*Independent Life* Section] 23rd April 2009, p.14. Available at http://www.independent.co.uk/arts-entertainment/theatre-dance/features/consider-yourself-a-muslim-fagin-1672544.html Access: 18th November 2009.
68 Ivor Dembina 'Who's laughing now?', *The Guardian* [*The Week* section] 7th June 1997, p.6.
69 Ivor Dembina 'Doing adverts. It's a crime against the truth' http://www.chortle.co.uk/correspondents/2009/09/14/9604/doing_adverts%3F_its_a_crime_against_the_truth Posted: 14th September 2009. Access: 18th November 2009.
70 Lloyd Wolf 'All this tosh about adverts' http://lloydwoolf.blogspot.com/2009/09/all-this-tosh-about-adverts.html Posted 14th September 2009. Access: 18th November 2009.
71 Quoted in Alasdair Milne (1988) *DG: The Memoirs of a British Broadcaster*, p.32 (London: Hodder and Stoughton).
72 Edward Pearce 'Spitting out the bile', *The Guardian* 14th March 1995.
73 See, for example, Michael Collins 'You are naughty!', *The Observer* [*Screen* section] 5th November 2000, pp.6–7.
74 Christopher Booker 'How our schoolboy humour became a way of life – and libel', *Daily Mail* 14th April 2000, p.13.
75 Stuart Hall (1994) 'Some "politically incorrect" pathways through PC', in Sarah Dunant (ed.) *The War of the Words: The Political Correctness Debate*, pp.166–7 (London: Virago).
76 Hugh Hebert 'Three men and a joke', *The Guardian* 21st April 1995.
77 Emily Dugan 'A wobbly girl's battle against the last taboo', *The Independent on Sunday* 8th November 2009, pp.26–7. Available at http://www.independent.co.uk/news/people/profiles/francesca-martinez-a-wobbly-girls-battle-against-the-last-taboo-1816861.html Access: 19th November 2009.

78 Stephen Moss 'Jimmy Carr: I thought my Paralympics joke was totally acceptable', *The Guardian* 5th November 2009. Available at: http://www.guardian.co.uk/culture/2009/nov/05/jimmy-carr-paralympics-joke Access: 19th November 2009.
79 Robert McCrum 'The real leader of the opposition', *The Observer* 1st April 2001 p.19.
80 See, for instance, Mark Thomas (2006) *As Used on the Famous Nelson Mandela: Underground Adventures in the Arms and Torture Trade* (London: Ebury Press); Mark Thomas (2008) *Belching out the Devil: Global Adventures With Coca Cola* (London: Ebury Press).
81 See Theodore Hamm, *The New Blue Media*, pp.155–90 (London: The New Press).
82 Ben Lynfield 'Israeli comedy show satirises Gaza violence', *The Independent* 10th January 2009, p.28. Available at http://www.independent.co.uk/news/world/middle-east/israeli-comedy-show-satirises-gaza-violence-1297579.html Access: 19th November 2009.
83 Phoebe Kennedy 'Burma jails comedian for 45 years', *The Independent* 22nd November 2009, p.38.
84 Alex Bellos 'Votes campaign off to flying start', *The Guardian* 11th April 1996.
85 See http://www.thegreatschlep.com/ Access: 19th November 2009.
86 See, for example, Mark Steel 'If the poor are hungry, send them arms' *The Independent* 18th June 2008. Available at: http://www.independent.co.uk/opinion/commentators/mark-steel/mark-steel-if-the-poor-of-africa-are-hungry-send-them-arms-849170.html Access: 19th November 2009.
87 Medhurst *A National* Joke, pp.190–1. Chris Arnot 'Making sense of humour: Chris Arnot discovers what a gay academic finds to enjoy and admire about a homophobic northern comic', *The Guardian* 6th November 2007. Available at: http://www.guardian.co.uk/education/2007/nov/06/highereducationprofile.academicexperts Access: 19th November 2009.
88 Ibid., p.189.
89 Ibid., p.199.
90 Ibid., p.197.
91 http://icteesside.icnetwork.co.uk/0300entertainment/showbiz/2003/08/07/chubby-couldn-t-stomach-swearing-50080-13265120/ Access: 19th November 2009.
92 Graham McCann (1998) *Morecambe and Wise*, p.268 (London: Fourth Estate).
93 Mark Watson 'Why I made those Magners ads' http://www.chortle.co.uk/correspondents/2009/09/11/9586/why_i_made_those_magners_ads Posted 11th September 2009. Access: 20th November 2009.
94 http://astore.amazon.co.uk/themorecwisehome/detail/0007234651 Access: 20th November 2009.

11
Noughties Reading

Nicole Matthews

Chapter 11 *is about the contemporary* ***politics of leisure and pleasure*** *as they concern another historically established leisure activity:* ***reading.*** *Historically, reading has been regarded as an inherently civilising activity, albeit that reading matter was subject to moral regulation, especially with regard to social class and gender. A perfect illustration of this came in the trial of Penguin Books in 1960. Penguin were prosecuted under the Obscene Publications Act for publishing D. H. Lawrence's Lady Chatterley's Lover. Prosecuting counsel Mervyn Griffiths Jones memorably asked members of the jury if Lawrence's novel was the sort of book 'you would wish your wife or servants to read'. The book, in any event, was thought to be under threat from the popularisation, successively, of film, television and the internet. But the book trade has thrived and, as this chapter notes, publishers now strive to meet demand, rather than to educate their readers or to guide public taste. The chapter explores the emergence of these new* ***cultural politics*** *in the* ***leisure world*** *of* ***reading*** *with particular reference to women's fiction.*

What's naughty about reading?

Reading seems the most demure of leisure activities. While writers around leisure have turned in recent times to 'the dark side' of leisure (Rojek, 1999: 126), the privacy and silence we associate with much modern reading would seem to be at odds with the kinds of activities Chris Rojek describes as 'wild' leisure – imbued with the passion of the disorderly crowd. Reading has often been associated with virtue, progress and reason and viewed as a key route to learning and wisdom. Historically, public libraries, for example, have been viewed as places to promote the social good through educative reading, distracting working-class readers from vices like the 'demon' drink (Brophy, 2007: 34). Libraries, one recent author comments, 'are one of the marks of civilization' (Brophy, 2007: ix). Even in newly popular social spaces for reading and discussing books, such as the book clubs popularised by Oprah Winfrey and the UK's *Richard and Judy*, have been framed as therapeutic communities organised around ethical principles (Driscoll, 2008; Fuller, 2008).

In contrast, Rojek's account of edgework emphasises the way in which leisure draws on human capacities for play and excesses of energy unable to be contained in the regulated activities of work and the rational-legal values that accompany them. Rejecting 'moral fastidiousness and judgementalism' and a medical framework for understanding abnormal leisure practices (1999:125), Rojek seeks to understand 'decadence and deviance' within leisure in terms of its relationships to a carnivalesque inversion of everyday hierarchies and regulations.

Media cultures play a marginal role in Rojek's account of 'dark' leisure. He sees engagement in a fantasy life through the products of the cultural industries as part of the complex of emotional withdrawal and privatisation he describes as 'reservation culture' (1999: 135). While setting the mass media aside in considering pathological forms of leisure – like drug use, violence and alcoholism – Rojek borrows a popular, medicalising metaphor to describe the contemporary relationship of individuals to the media: 'we depend upon them as addicts depend upon their next fix' (1999: 138). He notes the continuities between dimensions of escape in routine activities like shopping and travel and more transgressive forms of leisure, and certain types of reading – of pornography for example (1999: 164) – are included in his discussion of 'abnormal' leisure. The status of reading in Rojek's account of the darker side of leisure, then, seems to be ambivalent – a commercialised and thus routinised part of the industries against which much leisure edgework rebels, while at the same time offering potential for transgression and illegality. So does Rojek's account of contemporary leisure – not simply an unquestioned social good, but a complex activity including destructive, risky and anti-social elements – offer a way of understanding the contours and pleasures of contemporary reading?

While reading has often been viewed as a hallmark of a civilised culture, historical accounts of reading underscore the fact that this uplifting character of book culture has been a consequence of exclusions, categorisation and regulation of readers and reading material. Certain types of readers – particularly women, working-class readers and children – and certain types of books – novels and comics for example – have been the focus of social concern and control. Those stocking lending libraries in the nineteenth century attempted to prompt working-class readers towards instructional works and away from 'dangerous literature with tendencies towards socialism, excessive superstition or obscenity' (Lyons, 2003: 334). Folk stories were reformulated for children with educative and moral endings. Social anxieties around women as readers focused on the dangers of reading novels, of which women were, and still remain, the primary consumers. A novel could 'excite the passions, and stimulate the female imagination. It could encourage romantic expectations...it could make erotic suggestions which threatened chastity and social order' (Lyons, 2003: 319). These concerns about the seditious, blasphemous and sexually exciting dimensions of fiction informed its legislative regulation throughout the twentieth century (Caso, 2008).

However, the idea that leisure reading should uplift, educate and civilise has been much less frequently heard in recent times. Laura Miller's ethnographic work with booksellers in *Reluctant Capitalists* documents a shift over the course of the twentieth century in the way book shop owners and staff imagined their role. She points out that book professionals have become less invested in the idea that they might be cultural mediators of taste, helping book buyers make informed and cultured judgements about what might count as a good book. Miller describes this as a shift from an 'educational mission' to a 'service mission' (2006: 66). She quotes a book wholesaler, for example:

> I order a lot of different things, there are many of which I find personally repugnant. But I'm not there to pass judgement on other peoples's tastes by saying 'No, I don't like these things'. That's not for me to do (Miller, 2006: 66).

This shift to a pluralism about taste and affirmation of the notion of the 'sovereign consumer', capable of making independent judgements about their reading, suggests that changes in the book world may have made space for forms of leisure which are unconstrained by the aesthetic judgements of cultural elites.

A number of writers on the publishing industry in the UK and US have mapped similar shifts: away from a view of publishing houses as having a mission to educate and guide public taste, towards an understanding of their role is simply to provide the public with reading material they will buy, or that might be profitable (e.g. Moran, 1997: 242–3; Schiffrin, 2000). Such shifts are viewed with great concern by editors and publishers of a previous era who continue to consider the educational mission of book publishing as critical (e.g. Schiffrin, 2000). It has been argued that a similar shift from guardians of public taste to warriors for free expression has taken place in the professional practice of librarians (Finan, 2007: ix; see also Jenkins, 2006: 195–6). Interestingly, high cultural tastes were also promoted by the state in the very different context of Soviet Russia. The fall of the Berlin Wall, it has been argued, has led to the easier availability of 'pulp fiction' (Zavisca, 2005).

There are good reasons, then, to suggest that hierarchies of value supported by elitist cultural practices, which affirm the significance of high cultural tastes, have been eroded in, and since, the latter part of the twentieth century. Reading for pleasure, escape, arousal and the satisfaction of desire rather than for information or education is perhaps more practised, but certainly more acceptable, now than at many times in the past (e.g. Flint, 2002). Armando Petrucci, for example, discusses with some concern the rejection by ordinary readers of systems of canonisation and classification of books and their arrangement in hierarchies – a refusal that might be associated with what is sometimes called 'postmodern culture'. He notes that many contemporary readers deploy 'openly consumeristic reading practices and reject all systems of values and all pedagogical attitudes in the name of absolute freedom' (1999: 359) including 'trash' genres like mysteries, science fiction and westerns.

Guilty pleasures (1993)

In thinking through the value of Rojek's assertion of the significance of leisure 'edgework' for understanding contemporary reading, I want to turn my attention to the novels of Laurell K. Hamilton. These books, I think, offer a neat, if unexceptional, example of the purchase of the idea that reading might be considered a liminal leisure activity (Rojek, 1999: 148). Hamilton is a prolific writer of novels which might variously be dubbed horror-romance or gothic romance. She has written two bestselling series in this vein, with 16 books in the *Anita Blake, Vampire Hunter* published since 1993, while seven *Merry Gently, PI* books to date bring together romance, the supernatural and to a very limited extent 'hard boiled' detective fiction conventions. The Anita Blake series has also spawned a comic published by Marvel (Reid, 2007).

While my main purpose here isn't an aesthetic evaluation of these novels, the series nature of these fictions, their repetition of very similar sexual scenarios – in short, their recursive banality as well as their commercial success – makes them all the more indicative of shifts in what constitutes acceptable themes for leisure reading for adults in the West. These books point towards a number of trends in contemporary publishing: they are part of a commercially successful revival of gothic-inspired fiction (Punter and Byron, 1999; Lutz, 2007; Wisker, 1998), and a tendency towards more sexually explicit romantic fiction (Moody, 1998: 144). As popular romances, they offer an opportunity to explore the shifts in social regulation of reading which aims to 'excite passions' and 'make erotic suggestions' (Lyons, 2003: 319).

Kath Albury cites Gayle Rubin's discussion of the distinction between 'good' and 'bad' sexual practices in what Rubin describes as the contemporary 'sexual value system'.

> Any sex that violates these rules is 'bad', 'abnormal', or 'unnatural'. Bad sex may be homosexual, unmarried, promiscuous, non-procreative, or commercial. It may be masturbatory or take place at orgies, may be casual, may cross generational lines, and may take place in 'public', or at least in the bushes or in the baths. It may involve the use of pornography, fetish objects, sex toys, or unusual roles (Rubin, 1992, cited in Albury, 1998: 57).

With one or two exceptions – sexual acts between men, and between women, for example, play a minor role in these series – this list of 'bad' sexual practices reads like a check-list of the types of sexual encounters staged with Hamilton's novels. The combination of sex and violence is a rich source of plot lines within these books, while sexual acts between (were) animals and humans and between the living and the (un)dead are also a prominent feature. Most of these encounters include the series protagonists, Anita and Merry, who narrate their sexual experiences, pleasures and conundra as first-person narrators. As a number of writers have pointed out, gothic fiction and especially the character of the vampire offer a longstanding set of metaphors for sexuality outside the bounds of marriage and other conventional restraints (Wisker, 1998: 63). Indeed, a number of writers have interpreted the figure of the vampire as an embodiment of queer desires (Krzywinska, 1995). Significantly, perhaps, despite their dazzling display of 'bad' and arguably queer sexual practices, Hamilton's novels in contrast rarely represent men having sex with other men, or women with other women.

These books stage moral decision making around sexual practices, and to that extent, Hamilton's novels fit Rubin's description of heterosex as having a morally liminal status:

> Heterosexual encounters may be sublime or disgusting, free or forced, healing or destructive, romantic or mercenary. As long as it does not violate other rules, heterosexuality is acknowledged to exhibit the full range of human experience (Rubin cited in Albury, 1998: 56).

Certainly, moral dilemmas about who to love and/or have sex with battle for space in Hamilton's books with what *Publishers' Weekly* describes as the 'lusty detail' (Schindler, 2004: 42) of sexual encounters and otherworldly battles. Across the Anita Blake series, the ground on which these moral decisions moves considerably. In the first book of the series, *Guilty Pleasures* (1993), Anita is a gifted and hard working zombie raiser, who struggles to restrain her sexual desires for her fiancé, a clean-living school teacher. Later books have her wrestling with the morality of 'dating' two men, or more accurately, a werewolf and a vampire; of being part of a vampiric ménage-a-trois; of establishing a sexual relationship with a much younger man with a history of being sexually victimised; of engaging in consensual rough sex; of having sex with someone who, on occasion, eats people. By the end of the series, Anita lives with or has regular sex with more supernatural men than you can count on both hands. The sexual content is ramped up, book by book – reviewers commenting on *Incubus Dreams*, noted that it has 'more mating than murder', for example ('It's Getting Hot', 2004).

A significant part of both series is also the detailed description of social and ethical practices within supernatural societies. Merry and Anita ruminate frequently about the relationship between their sexual practices and the codes of the otherworldly societies of which they are part. Here we can see again the staging of heterosex as a space for moral decision. In each series social rules, diplomatic negotiations and life-and-death emergencies make it necessary for Anita and Merry to engage in plenty of varied and transgressive sexual activity. As a review of the 14th in Anita Blake series, *Dans Macabre*, in *Publishers' Weekly* put it her 'sexual magic powers require multiple lovers' (2006), while in volume three of the Merry Gently series, the half-fairy royal can only inherit the throne if she becomes pregnant by one of her large bevy of personal guards. The protagonists are not just drawn into (usually pleasurable) sexual encounters because of their own desire, though the protagonists' lust is foregrounded in both series. Rather the rules of the game – commitment to a were-animal pride or pard; the obligations to vampiric servants or masters; the rules of fairy succession – obligate frequent, casual, orgiastic, cross species, public sexual encounters. If Hamilton's heroines move beyond passivity in their sexual encounters, they do have an alibi, beyond their own pleasure, for doing so.

Rojek provides an approving account of Weber's discussion of the orgy in his consideration of 'wild' forms of leisure. 'Weber argues that the orgy is an attempt to find greater intensity to life by treating prohibitions as alienable moral and physical restrictions' (Rojek, 1999: 159). Anita or Merry's orgiastic encounters with multiple sexual partners might well be experienced as an imaginary violation of sexual norms by these novels' readers. Within the world of the novels, such sexual acts are indeed described as channelling excessive magical energy in a way that neatly parallels Rojek's account. However, rather than violating the moral restrictions of the non-human social norms described in the books, these kinds of sexual acts are depicted, on the whole, as part of the maintenance of social order – allowing its protagonists to claim power in an otherworldly battle, negotiate with hostile clans of fey creatures, or revive a mortally wounded were-animal.

The backstory to the Anita Blake series points towards the ambivalent status of sexual danger and violence in the series. The milieu of these novels is one in which vampires have, despite the protests of religious conservatives and a police force reluctant to view vampires on a par with humans, gained civil rights. While the political metaphor implied by the notion of vampire rights recedes from the narrative foreground as the series progresses, this doubling – vampires both as a threat and as a fully paid up part of civilian life – parallels the insertion of eroticised acts as part of the upholding of vampiric and were-animals' power relations and social behaviours. As Wisker points out, such stories 'reuse the disgust and rejection of conventional vampire narratives [offering instead] an exciting double rejection and transgression' (Wisker, 1998: 65). In Hamilton's fictions, vampires, or at least well behaved ones, are protected by the law, and promiscuous, orgiastic and animalistic sex is part of the otherworldly law that governs their own social practices. This incorporation of 'dark leisure' into schemas of moral regulation provides an excuse for the erotic pleasures of protagonists and readers. It can also be used as a metaphor for the ways in which the 'wild' might be incorporated into mainstream, commercial fiction.

Noughties novels: Deregulated culture?

Hamilton's novels are very much in the publishing mainstream: successful genre-based products, capitalising on their series characters across a range of formats to offer reliable pleasures to their readers and predictable profits for their publisher, the multi-national Random House. While they are not marketed as such, they might easily be described as soft-core or erotica. Nonetheless, they appear to be readily stocked by libraries (Rothman, 2006), and although subject to some challenges within US libraries (e.g. Denver Public Library, 2008) are generally widely available. What does this tell us about 'noughties reading' and the level to which it has been deregulated?

One way of framing the success of Hamilton's novels, and the very ordinariness of their relentless representations of polymorphous sexuality, is through a story about the deregulation and liberalisation of culture across the course of the twentieth century and into the twenty-first. This account might have two variants: a conservative one bemoaning hyper-sexualisation in the contemporary era and the loss of social and religious values; or a liberal one, celebrating openness and freedom of expression in the once heavily censored domain of popular literature. Both of these accounts fall short of describing adequately the context in which Hamilton's novels sit unmolested on bookshop shelves. I'd like to highlight three limitations shared by both lapsarian and whig accounts of a deregulated literary culture.

First, this narrative of deregulation neglects what Christopher Finan in his discussion of censorship and its enemies in the US calls 'the counterattack' on civil liberties: the response by social and religious conservatives to the liberalisation of legal and social norms in the 1960s (Finan, 2007). Challenges by parents to the books stocked by US libraries, for example, actually increased between the 1980s and the noughties (Caso, 2008: 48). The infamous debates over pornography within western feminism during the 1980s brought together anti-porn feminists and social conservatives to create legislative frameworks like the 1992 Canadian Supreme Court decision *Regina v. Butler* which has been used to seize what are now viewed as obscene works from bookshops and at the Canadian border (Califia, 1995). More recently, in the wake of the 9/11 attacks, laws such as the US 'Patriot Act' gave government agencies sweeping new powers to engage in secret surveillance, including surveillance of the books borrowed by individuals from libraries (Finan, 2007: 268; Caso, 2008: 47).

In some respects, the fact Hamilton's novels have gone relatively unchallenged in the US is something of a surprise, given the concerns of these conservative forces. Griffin and Iball (2007) have argued that there has been a shift in the flashpoints for censorship, away from sexuality and towards controversial representations of religious figures and religious differences. Most public attention and discussion has been devoted to cases in which Muslim and Sikh believers have taken offence at fictional representations in books like *The Satanic Verses*, or more recently *The Jewel of Medina*. In the latter case, the home of the publisher was bombed by those outraged by the book's romance narrative, which focused on Mohammed's favourite wife (Flood, 2008). However, many challenges have been made in the US to libraries stocking books which include non-Christian, particularly pagan or Wiccan, religious ideas or symbolism (Caso, 2008: 45). Fantasy novels, most particularly children's books, have been a particular focus of challenge by parents and religious groups. J. K. Rowling's immensely successful Harry Potter novels, for example, have been one of the most challenged books in the US during much of the late 1990s and early noughties (Jenkins, 2006: 192; Falk-Ross and Caplan, 2008). If Hamilton's novels fall foul of conservative norms of both sexual behaviour and religion, they are directed not at children but at adults. While feminist-inspired anti-pornography legislation may frame women as vulnerable to violent and demeaning representations, the conception of children as readers who need to be protected from dangerous books continues to have more purchase in the West.

Dawn Sova in her work identifying examples of books which, in a range of national contexts, have been subject to censorship for what she describes as 'social' reasons comments:

> Most books challenged for social reasons in the past century have suffered because of the language used or the racial relationships portrayed within. In contrast, the new definition of 'obscene' which emerges in the last decade of the twentieth century and continues to the present, appears to emphasize the discussion or the depiction of homosexual relationships (Sova, 2006: 12; see also Falk-Ross and Caplan, 2008).

One reason, then, for the presence of Hamilton's novels, relatively untouched by censorship, on library and bookshop shelves may be because of the way they avoid what could be described as 'gay' or 'lesbian' encounters or relationships. A whig narrative of the sexual liberation of print can't fully account for the significance of these kinds of ellipses and their consequences.

A second limitation of narratives of cultural deregulation is their ethnocentricity. Much of the academic work on the publishing industry has focused on the Anglophone world. However the global pattern of regulation of publishing and reading is far more complex than this account acknowledges. While both discourses of civil rights and of neo-liberal deregulation circulate globally, national regulation of what can be published, imported, stocked by bookshops and various kinds of libraries, bought, borrowed and read by different categories of readers inevitably differs from context to context. For example, even within the countries of Europe, the extent of government support for authors, translators and publishers and the available funding for libraries and international marketing of local authors and publishers varies significantly (Smith, 2004). In other parts of the world, Egypt, Turkey and Algeria are among countries which have arguably experienced moves away from the liberalisation of what can be published and read (Karolides, 2006; Gafaiti, 1996: 80).

Talking meaningfully about reading across all these national settings is simply an impossible task. While my discussion will soon return to the well-trodden ground of the circulation and use of texts within the Anglophone world, I would first like to explore some of the implications of even momentarily considering forms of regulation of reading outside of this US and British-centred frame. Perhaps ironically, a number of writers analysing the regulation and circulation of print in authoritarian or single-party regimes have offered an account of censorship which exceeds some of the commonplace ways in which these processes are often imagined. Hafid Gafaiti, for example, has argued that a Manichean vision of the relationship between the author as creative producer and the repressing censor, derived from a Western context, fails to capture the complex negotiations between writers and the political and social circumstances under which they write. Gafaiti argues that even when writers are liable to assassination by regimes such as those prevailing in Algeria or Iran under Khoumeini, whether a writer publishes or is censored, or even lives or dies, is in part dependent on a complex social 'ecosystem' (Holquist cited in Gafaiti, 1996: 79) that 'relies, to be effective, on implicit and explicit cooperation by artists and workers who engage in ongoing negotiation with this system of permissions and prohibitions' (Gafaiti, 1996: 79). Similarly, Jirina Smejkalova, writing about Czechoslovakia during the post-1968 communist era calls for an analysis which considers the diversity of sometimes disjunctive mechanisms that 'institutionally limit... access of the reader to the text' (Smejkalova, 2001: 100).

One productive concept for thinking through the circulation and consumption of texts is Pierre Bourdieu's notion of 'structural censorship' (cited in Billiani, 2007: 6). This notion underlines the way in which a field of cultural production works shapes the kinds of things it is possible, advisable or likely for people in that field to say or do. Processes we could cite as part of such structural forms of censorship might include a list generated by Richard Burt:

> cutbacks in government funding for controversial art, boycotts, lawsuits, and marginalisation and exclusion of artists based on their gender or race to 'political correctness' (Burt, 1994: xiv).

The very breadth of this list turns Burt away from the term 'censorship' to broader conception of the 'administration of aesthetics' (Burt, 1994: xvi).

This shift in terms points to the productive, not simply repressive, character of the regulation of what is said and what is read. We could add still further to this list of processes shaping readers' access to books. Are books bought by libraries, made openly available on shelves, set as school textbooks? Which books are translated, and into what languages, with distribution (Billiani, 2007)?; how are they reviewed, marketed and distributed? All of these processes limit the access of readers to a diverse range of leisure reading. Here we can see the third limitation to claims of a general liberalisation and deregulation of book culture in the period leading up to the noughties: the range and complexity of processes that shape readers' access to books.

Markets and 'market censorship' (Schiffrin, 2000: 106)

How, then, to trace the impact of some of the more significant of these processes shaping readers' experience of leisure reading? One key debate has been the impact of neo-liberal regulatory regimes on the book world. As we have seen, Laura Miller has identified shifts in the underpinning discourses of the book world towards an acknowledgement of the sovereignty of the consumer. Embedded in this discourse is an understanding of the marketplace as a space of free choice – a shibboleth which Miller is at pains to critique. Nonetheless, the question of whether a deregulated market for leisure reading provides great diversity of choice for readers, driven by an increasingly open market, or a more restricted diet of reading, is still hotly debated.

Major transformations of the book business have taken place during the last years of the twentieth century. Since the 1960s, many independent and smaller publishing houses have been bought out by major publishers. A trend beginning in the same period but accelerating in the 1980s and 1990s brought these larger publishers into the fold of multi-national media corporations with interests in a range of communications industries (Moran, 1997; Greco et al., 2007). Liberalisation of cross and foreign media ownership legislation in a number of countries has, in part, underpinned these developments. Paralleling these changes in publishing has been the emergence of large chains of bookstores, including very large superstores, and a decline in the number of independent booksellers (Miller, 2006). Restrictions on the places and prices at which books can be sold, such as the UK Net Book Agreement, have also been liberalised in a number of countries, although some European nations have imposed fixed price schemes (Stockman, 2004; Hollier, 2007). Elsewhere, there has been erosion of subsidies and other forms of financial support for parts of the publishing industry – for example, in Australia, where sales tax exemptions for books and financial incentives for Australian publishers to print their books locally have been terminated (Hollier, 2007).

What has been the effect of such deregulation and liberalisation of the book world? One popular position amongst left and liberal critics of the corporatisation of book publishing has been to claim that 'by the end of the twentieth century, corporations did the censoring and the state merely facilitated the practice' (Vaidhyanathan, 2004: 186). There are a number of variants of this argument. At one end of the spectrum are examples of direct interference in the kind of books commissioned by publishers. A famous example would be HarperCollins' decision not to publish a book by Chris Patten, former governor of Hong Kong, because of Patten's status as a prominent critic of the Chinese government (Schiffrin, 2000: 133) and the desire by Rupert Murdoch's News Corporation, who owned HarperCollins, to make inroads in the Chinese media market. More broadly, Andre Schiffrin argues that fewer and fewer left-of-centre books are published in the US, in part because of potential conflicts with the financial interests of other arms of large media conglomerates. Conversely, it has also been argued that the decision to publish particular books has been shaped by the potential profits to be made from spin-offs in a range of media within, although, at least during the 1990s, such synergies seemed to be less profitable than commentators and multi-nationals themselves might have predicted (Moran, 1997).

Albert Greco, an economist of publishing, takes on such arguments that consolidation within the publishing industry has diminished readers' choices. Drawing on empirical evidence from the publishing industry and economic models of monopolistic trade practices, he and his colleagues critique the idea that US publishing has become oligopolistic, undermining free trade. His work points out that there are many more publishers in the US now than there were in the post war period, releasing a much large number of new titles (Greco et al., 2007: 3–4, see also Miller, 2006: 57).

It is interesting to note that claims by Schiffrin and other left critics of a narrower and less politically diverse output from the major publishing houses are not directly contested by Greco and colleagues – indeed, it is hard to imagine how their quantitative research methods could be turned to this question. Most writers on contemporary publishing agree that consolidation, the emergence of large book chains and the introduction of new computer technologies to track sales, ordering and returns have changed the types of books that are sought by publishers and the way they are marketed. Schffrin quotes a small publisher: 'Big houses think in terms of big numbers...new, strange, crazy, intellectually innovative, or experimental books are published in small to medium size print runs. That is the task of the smaller houses' (Klaus Wagenback, cited in Schiffrin, 2000: 146). At the same time, writers like Jason Epstein claim that blockbuster titles have become an increasingly prominent feature of the book world, with mid-list titles generating more modest sales less likely to be published or actively marketed (Epstein, cited in Horvath, 2001: 88. As

Laura Miller has pointed out, the great array of titles available makes the mechanisms which bring books to readers' attention all the more important (2006: 57). Larger publishers have larger budgets for marketing and for developing relationships with the large chain bookstores.

With ongoing sales able to be dissected more closely through the use of technologies like BookScan, quick sellers have become more important to publishers and booksellers (Hollier, 2007; Miller, 2006). 'High turnover, short shelf life and tighter stock control... make for a more unforgiving market' (Lee cited in Hollier, 2007: 67). Computerised ordering leads to a 'retrospective' form of shelf stocking (Hollier, 2007: 66) in which the successful sellers of last week are reordered again for next week. It is easy to see how series novels, especially those within a recognisable genre, like Hamilton's, are winners from the point of view of publishers and booksellers – their 'brand' recognisability enabling the reliable ongoing sales. If Hamilton's novels themselves relate and celebrate 'wild', excessive pleasures, as saleable items they offer reliable, workaday profits in an industry now much more clearly organised according to rational profit-making principles.

Reading beyond scarcity?

If consolidation, centralisation and technologisation of book marketing and book selling may well have homogenised the offerings on bookshelves, it is often argued that information technologies offer an alternative diverse source of reading material (e.g. Young, 2007). If processes of 'market' or 'structural' censorship can prevent readers accessing a diverse range of books in bookshops and libraries, has the web made a diversity of books more widely available to readers? Online bookshops, for example, can make a wide range of backlisted material available, at the same time making a critical difference to independent publishing houses' bottom line (King, 2007b: 109). Publishing on demand – perhaps consumer-requested, on-demand production of print copies of digital books – may in the future enable a greater diversity of reading material from both the past and the present to be available internationally. Certainly the logic of shifts in public funding for libraries in the UK, US and Australia towards funding computers and digitally accessible material, in preference to books, suggests that policy-makers are convinced of the significance of the web in accessing leisure reading (Brophy, 2007: 35; Buschman, 2003: 64–5; Greco, 2005: 44).

Don Slater, in his ethnographic analysis of online-communities based around an exchange of sexually explicit photographs, describes the internet as a 'culture beyond scarcity', though paradoxically one in which processes of exchange and relations of ownership are foregrounded. Similarly, writers about queer culture on the web have pointed to the way the Internet makes available texts and social networks that would otherwise be prohibited in publicly homophobic cultures (e.g. Khoo, 2003; Offord, 2003). Rojek himself comments that 'the rise of the information society has produced a quantum leap in the opportunities to go "beyond limits"' (Rojek, 1999: 190) to enjoy porn or violence.

This account of otherwise regulated material being readily available on the web is borne out in academic discussion of fan fiction, particular fan fiction with sexual or sexually transgressive themes. One notable example is Slash or YAOI fiction. These fan fictions, which reinscribe male characters from mainstream television, games or books in stories of sex and romance between men, were available before the advent of the internet via magazines and self-published zines (McLelland, 2000; Sabucco, 2003). However, the availability of this fiction has proliferated online. This kind of sexually explicit rewriting of media narratives, largely by female authors, has been a longstanding topic of fascination for media studies authors (Penley, 1997; Jenkins, 1992; Jenkins, 2006) because of the way these fictions underline the power of the marginalised reader, drawing on her own fantasy life and countercultural community to make their own kind of sense as 'textual poachers' of the meanings intended for the text by its original makers.

While the case of slash exemplified a paradigm of media studies stressing the transgressive power of readers to resist the limitations and homogeneity of commercial cultures, it may be that considering more everyday forms of reading on the Internet challenges the terms through which Rojek frames his conception of 'dark leisure'. In this chapter, my focus has been on reading for pleasure, and I have taken advantage of commonsense oppositions – which also underpin Rojek's discussion – between pleasure on the one hand, and acquisition of education and information which shape activity as a citizen and as a worker on the other. This set of binaries has led me to focus especially on fiction. In some respects this seems most appropriate – the terrain of press freedom and an informed citizenry on the one hand and the territory of stories of 'decadence and deviance' on the other appear to be so far apart as to be incommensurable sets of terms. Questions of freedom of the press evoke the image of the rational consumer of information, the voter, the activist while 'wild' leisure evokes the reveller and the hedonist.

However, as researchers on the publishing industry have pointed out, reading conventional print fiction it is increasingly an endangered practice, competing for leisure moments with a variety of electronic mediated leisure activities, including exchanging gossip on social networking sites, reading online newspapers and magazines, reading and writing blogs and engaging in internet chat. Such online spaces incorporate gossip, conspiracy theories, links to all kinds of other sites, outings, comments and flames alongside more formal and informational forms of textual exchange (e.g. Finkelstein and McCleery, 2005: 127; Young, 2007). Reading for leisure and relaxation and reading for information and instruction, the lines between fiction and non-fiction, may well be being blurred in these genres of online reading (e.g. Young, 2007). Many of these on-texts swirl uneasily in the space which

Sherman Young maps out between book culture proper and 'functional' reading. Perhaps in order to understand the contours of contemporary reading we need to challenge a way of understanding 'wild leisure' as that which is by necessity and definition excluded from the rational, public world. Rather we might need to consider that the infectious misdemeanours of crowd behaviour or moments of sexual excess, impropriety or disclosure might emerge adjacent to or even in response to local history, government pronouncements or political punditry in the new spaces for reading online.

Acknowledgements

Thanks to Derrick Cameron, Sherman Young and Nickianne Moody for their advice on drafts of this chapter.

Bibliography

Albury, K. (1998) 'Spanking stories: Writing and reading bad female heterosex', *Continuum: Journal of Media and Culture,* 12(1) April: 55–68.

Billiani, F. (2007) 'Assessing boundaries – Censoring and translation: A discussion', from F. Billiani (ed.) *Modes of Censorship and Translation: National Contexts and Diverse Media* (St. Jerome Publishing).

Brophy, P. (2007) *The Library in the Twenty First Century* (London: Facet Publishing).

Burt, R. (1994) 'Introduction: The "new" censorship', from Richard Burt (ed.) *The Administration of Aesthetics: Censorship, Political Criticism and the Public Sphere* (Minneapolis: University of Minneapolis Press).

Buschman, J. (2003) *Dismantling the Public Sphere: Situating and Sustaining Librarianship in the Age of the New Public Philosophy* (Westport: Libraries Unlimited).

Califia, P. (1995) 'Dangerous tongues', from *Forbidden Passages: Writings Banned in Canada* (London: Cleis Books).

Caso, F. (2008) *Censorship* (New York: Infobase Publishing/Facts on File).

Driscoll, B. (2008) 'How Oprah's book club reinvented the woman reader', *Popular Narrative Media,* 1(2): 139–50.

Falk-Ross, F. and J. Caplan (2008) 'The challenge of censorship', *Reading Today,* 25.5 April–May: 20(1).

Finan, C. (2007) *From the Palmer Raids to the Patriot Act: A History of the Fight for Free Speech in America* (Boston: Beacon Press).

Finkelstein, D. and A. McCleery (2005) *An Introduction to Book History* (New York: Routledge).

Flint, K. (2002) 'Reading practices', from D. Finkelstein and A. McCleery (eds) *The Book History Reader,* Second edition, 416–23 (London: Routledge).

Flood, A. (2008) 'Publication of controversial Muhammad novel delayed', *The Guardian,* October 10.

Fuller, D. (2008) 'Reading as social practice: The "beyond the book" research project', *Popular Narrative Media,* 1(2): 211–18.

Gafaiti, H. (1996) 'Between God and the President: Literature and censorship in North Africa', *Diacritics,* 27(2).

Greco, A. (2005) *The Book Publishing Industry,* Second edition (London: Laurence Erlbaum Associates).

Greco, A. N., C. E. Rodriguez and R. Wharton (2007) *The Culture and Commerce of Publishing in the 21st Century* (Stanford: Stanford Business Books).

Griffin, G. and H. Iball (2007) 'Editors introduction to gagging – Forum on censorship', *Contemporary Theatre Review*, 17(4): 516–56.
Hollier, N. (2007) 'Between denial and despair: Understanding the decline of literary publishing in Australia', *Southern Review*, 40(1): 62–77.
Horvath, S. (2001) 'Publishing's brave new world – according to Jason Epstein', *Logos*, 12(2): 87–93.
Jenkins, H. (1992) *Textual Poachers : Television Fans & Participatory Culture* (London: Routledge).
Jenkins, H. (2006) *Convergence Culture: Here Old and New Media Collide* (New York: New York University Press).
Karolides, N. J. (2006) *Literature Suppressed on Political Grounds* (New York: Facts on File).
Khoo, O. (2003) 'Sexing the city: Malaysia's new "cyberlaws" and Cyberjaya's queer success', from Chris Berry, Fran Martin, and Audrey Yue (eds) *Mobile Cultures: New Media in Queer Asia*, 222–44 (Durham: Duke University Press).
King, N. (2007b) 'The main thing we book publishers have going with our books is the books themselves', an interview with Peter Ayrton (London: Serpent's Tail Press, Islington), 12 July 2006, *Critical Quarterly*, 49(3): 104–19.
Krzywinska, T. (1995) 'La Belle Dame Sans Merci?', in P. Burston and C. Richardson, *A Queer Romance: Lesbians, Gay Men and Popular Culture*, 99–110 (London: Routledge).
Lutz, D. (2007) 'The haunted space of the mind: The revival of the gothic romance in the twenty first century', in S. Goade (ed.) *Empowerment versus Oppression: Twenty First Century Views of Popular Romance Novels*, 31–92 (Newcastle: Cambridge Scholars Publishing).
Lyons, M. (2003) 'New readers in the nineteenth century: Women, children, workers', in G. Cavallo and R. Chartier (eds) *A History of Reading in the West*, 313–44 (Cambridge: Policy).
McLelland, M. (2000) 'No climax, no point, no meaning? Japanese women's boy-love sites on the internet', *Journal of Communication Inquiry*, 24(3): 274–91.
Miller, L. J. (2006) *Reluctant Capitalists: Bookselling and the Culture of Consumption* (Chicago: University of Chicago Press).
Moody, N. (1998) 'Mills and Boon's *Temptations*: Sex and the single couple in the 1990s', in L. Pearce and G. Wisker (eds) *Fatal Attractions: Re-scripting Romance in Contemporary Literature and Film* (London: Pluto Press).
Moran, J. (1997) 'The role of multimedia conglomerates in American trade book publishing', *Media, Culture and Society*, 19(3): 441–55.
Offord, B. (2003) 'Singaporean queering of the internet: Toward a new ford of cultural transmission of rights discourse', in Berry, Chris, Martin, Fran and Yue, Audrey (eds) *Mobile Cultures: New Media in Queer Asia*, 133–57 (Durham: Duke University Press).
Penley, C. (1997) *Nasa/Trek: Popular Science and Sex in America* (London: Verso).
Petrucci, A. (1999) 'Reading to read: A future for reading', in G. Cavallo and R. Chartier (eds) *A History of Reading in the West*, trans. Lydia G. Cochrane, 345–67 (Cambridge: Polity Press).
Publishers Weekly (2006) '*Danse Macabre*', 00000019, 5/22/2006, Vol. 253, Issue 21.
Punter, D. (1999) 'Introduction: Of apparitions', in Glennis Byron and David Punter (eds) *Spectral Readings: Towards a Gothic Geography* (London: Macmillan), 1–10.
Punter, D. and B. Glennis (eds) (1999) 'Special readings: Towards a Gothic geography' (Basingstoke: Macmillan).
Reid, C. (2007) 'Comics bestsellers', *Publishers Weekly*, 00000019, 9/3/2007, 254(35).

Rojek, C. (1999) *Leisure and Culture* (London: Palgrave).

Rosen, J. (2003) 'Booksellers help Harry Potter'. *Publishers Weekly*, 00000019, 5/12/2003, 250(19).

Rothman, D. (2006) 'How sexy can a library e-book be?', from TeleRead 'Bring the E-Books HomeNews & views on e-books, libraries, publishing and related topics', http://www.teleread.org/blog/2006/01/17/how-sexy-can-a-library-e-book-be/

Sabucco, V. (2003) 'Guided fan fiction: Western "readings" of Japanese homosexual-themed texts', in Chris Berry, Fran Martin and Audrey Yue (eds) *Mobile Cultures: New Media in Queer Asia*, 70–86 (Durham: Duke University Press).

Schiffrin, A. (2000) *The Business of Books: How International Conglomerates Took Over Publishing and Changed the Way We Read* (London: Verso).

Schindler, D. T. (2004) 'Laurell K. Hamilton: Underworld seductress', *Publishers' Weekly*, September 20, 2004, 42.

Smejkalova, J. (2001) 'Censors and their readers: Selling, silencing, and reading Czeh books', *Libraries & Culture*, 36.1 (Wntr 2001): p.87(17), (6908 words).

Smith, K. (2004) 'Publishers and the public: Governmental support in Europe', *Javnost – The Public*, 11(4): 5–20.

Sova, D. (2006) *Banned Books: Literature Suppressed on Social Grounds*, revised edition (New York: Infobase Publishing/Facts on File).

Stockman, D. (2004) 'Free or fixed prices on books – Patterns of book pricing in Europe', *Javnost – The Public*, 11(4): 49–64.

Vaidhyanathan, S. (2004) *The Anarchist in the Library: How the Clash Between Freedom and Control is Hacking the Real World and Crashing the System* (Cambridge: Basic Books).

Wisker, G. (1998) 'If looks could kill: Contemporary women's vampire fictions', in L. Pearce and G. Wisker (eds) *Fatal Attractions: Re-scripting Romance in Contemporary Literature and Film*, 70–86 (London: Pluto Press).

Young, S. (2007) *The Book is Dead, Long Live the Book* (Sydney: UNSW Press).

Zavisca, J. (2005) 'The status of cultural omnivorism: A case study of reading in Russia', from *Social Forces*, 84.2 (December 2005): p.1233(23).

12
Sport and Lifestyle

John Horne

Chapter 12 *relates the* ***new politics of leisure and pleasure*** *to the world of* ***sport*** *– an activity constituted in its modern form in the latter half of the nineteenth century, but established as a key definer of individual identities across many societies in the last 50 years. This period has seen a huge growth in the* ***consumption of sport****, principally via television and mega-events such as the Olympics, and also in* ***participation****, as seen in the proliferation of 'lifestyle sports', fitness clubs and the like. Once again, an important* ***political tension*** *is explored – in this case, the tension between the* ***commercial exploitation*** *of sport by corporations and the* ***health- and pleasure-giving****, and thus emancipatory,* ***properties of sport****.*

Introduction

In the three decades since *The Devil Makes Work* was published many changes have occurred that have influenced social and cultural life. There are also some important continuities; not least the importance of capitalism for understanding developments in the position of sport and leisure lifestyles. This economic system is an economy of time as Clarke and Critcher (1985: 239–40) concluded: time is a 'resource that structures many of the conflicts and inequalities of leisure'. Such a temporal economy continues to have profound implications for the social distribution of rewards and disadvantages, poverty and wealth, and the creation of social divisions, and hence alternatives to it have to engage with a politics of time.

In these circumstances commodifying trends in sport and leisure culture have continued to unfold. Business-oriented commentators on sport have provided a compelling list of developments in the past 25 years – the blurring of what is sport and what is entertainment, the vertical and horizontal integration of sport enterprises by entertainment and media companies, an increase in venture capital and investment in transnational sports and sport properties, and the integration and consolidation of sport, leisure, recreation, television, film and tourism into elements of the entertainment industry.

In the present economic climate it might be thought that these developments were about to come to a grinding halt. Whilst we might disagree with some futurological studies, it is undoubtedly the case that sport has become and will continue to be more commercialised. The bursting of the protective bubble surrounding sport – in which some could argue it was *hors commerce* (outside the market) and therefore not for sale – happened at least 30 years ago. Since then we have seen commercial interests in sport – especially media and corporate sponsorship arrangements – become the main drivers of leading professional sports and flagship sports mega-events.

In circumstances where there has been a reduction in the amount of time actually in work or paid employment the relationships between sport and leisure activities and between social and personal identities have also come to take on greater significance. It has led to the idea that leisure lifestyles may have come to supplant work as the major shaper of personal identity. This is partly affirmed by the growth of interest in so-called 'lifestyle sports' (Tomlinson et al., 2005; Wheaton, 2004). As Tomlinson et al. (2005: 2) suggest such new forms of sport 'have commercial and competitive dimensions, but are essentially understood by participants as bodily experiences – about "doing it"'.

A second major development since the 1980s has been the growth of markets for sports goods and services on a global level. This in turn relates to the growth of media audiences and potential consumers for sports-related products (including competitions and matches), the growth of the commodity chains involved in the cultural division of labour (Miller et al., 2001), and the increasing spread of elite athlete migration.

Referring to their social and historical contexts, this chapter will review five broad issues: first, economic trends and their impact on the viability of sport sponsorship; secondly, the changing format of sport and its relationship to the mass media; thirdly, the relationship between athletes and the institutions and organisations that enable sports to operate; fourthly, developments in policy – both of government and sports bodies; finally, the social relations and the associated lifestyles and identities which frame sport. Before tackling these themes, I look briefly at the development of consumer culture as an object of study that helps to frame these developments.

Sport in consumer culture[1]

Debates about consumption, consumer society and consumer culture have developed from a minority academic issue into a public concern. Part of the debate about postmodernity is the suggestion that social scientists, especially sociologists, have focused too much on the experience and effect of paid work and production and not enough on consumption. Are people increasingly addicted to spending and shopping? Does affluence create happiness? Is shopping the ultimate freedom? One suggestion is that affluence

– like poverty – is relative and hence, as opportunities to spend increase, so too do the comparisons with others. People may compare their position in society with celebrity lifestyles – of David and Victoria Beckham, for example. The fact that a professional football player (and his pop-star wife) has become for some a central indicator of the 'good life' is a development that has been considered by several sociologists of sport (Cashmore, 2002; Whannel, 2002). Aside from individual sports stars, other recent attempts to discuss aspects of sport in consumer culture have focussed on fandom and fans (Crawford, 2004), advertising (Jackson and Andrews, 2005), individual teams (Andrews, 2004) and mega-events (Horne and Manzenreiter, 2006).

The relationship between sport (and active lifestyles) and consumer culture has been noted by several social commentators. One of the first was John Hargreaves (1987: 151) who argued 'What links consumer culture with sports culture so economically... is their common concern with, and capacity to accommodate, the body meaningfully in the constitution of the normal individual'. Sport can be seen as central to the economics of late capitalist modernity. In these economies the body is more than an instrument for producing material goods and getting things done. The body, including the sporting and physically active body, is now portrayed as an object of contemplation and improvement, in the spectacular discourses of the mass media, the regulatory discourses of the state and in everyday practices (or 'body projects'). Contemporary advertisements for both commercial sport and leisure clubs in the UK combine discourses of both medical science and popular culture in such phrases as 'fitness regime', 'problem areas like the bottom or the stomach', 'consultation' and 'fix'. By exhorting potential consumers/members to 'Flatten your tum and perk up your bum' and reassuring us that 'Gym'll fix it', regulatory control of the body is now experienced through consumerism and the fashion industry. As times get tougher economically, one health club recommends 'Investing in your health' as a means to 'leave the pressure of everyday life behind'. Sport has thus become increasingly allied to the consumption of goods and services, which is now the structural basis of the advanced capitalist countries through discourses about the model, (post-) modern consumer-citizen. This person is an enterprising self who is also a calculating and reflexive self. Someone permanently ready to discipline her- or himself – through crash diets, gymnastics, aerobics, muscle toning, tanning, strip-waxing, and cosmetic ('plastic') surgery (including breast enlargement and cellulite reduction) as well as sporting physical activity – in order to fit in with the demands of postmodern consumer culture.

How are we to make sense of these developments? There have been three main approaches to understanding consumption and consumer culture: the production of consumption approach, the modes of production approach and the pleasures of consumption approach (Featherstone, 1991). Each has been subject to criticisms – the first because it underplays human agency and the second and third because their emphasis on the active and (apparently) powerful consumer ignores the wider structural forces of exploitation and injustice. Hence in the study of sport in consumer culture the familiar debates between sociologists over theoretical approaches that emphasise agency and those that emphasise structure loom as large as ever.

The three main concerns about contemporary consumer society relate to growing inequality, increased commodification and globalisation (Schor and Holt, 2000). Capitalist economics and the associated ideology of consumerism have been spreading for decades but, since the end of the 1980s and the collapse of the Soviet Union, they have taken off globally. In this context there are two main reasons why commercial sports have also become global in scope and impact. The first is that driven by the profit motive the economic owners of sport are always looking to expand their markets and increase their profits. The second is that transnational companies, as sponsors, can use sport to promote their goods and services in new markets. That business strategies developed in one society may work well in other parts of the world is an argument that underpins George Ritzer's 'McDonaldization thesis' (Ritzer, 1993). Recently Alan Bryman has suggested that in twenty-first century marketing and branding the 'ludic ambience' developed by the Disney Corporation (Bryman, 2004) increasingly acts as the ideal template for the spread of contemporary consumer cultures of advanced post-Fordist capitalism throughout the world. However, efforts to use successful sports clubs to make money may come up against stiff opposition if they conflict with the social and cultural values and meanings associated with teams. This partly helps to explain the protests staged by fans and the initial reticence of the board of Manchester United to embrace the offer made by Malcolm Glazer, the fish-oil and property tycoon ranked 278th richest American by *Forbes* magazine, in 2005. What worked reasonably well in the USA was not always received in the same way on the other side of the Atlantic Ocean.[2]

At the end of the first decade of the twenty-first century we are faced with a new political environment in the midst of an old type of economic crisis. The financial system that helped fuel the consumer culture of the past 25 years is in chaos. Yet recession, deflation, rising unemployment, declining purchasing power, were all features of the capitalist Britain that John Clarke and Chas Critcher surveyed in the early 1980s. The impact of this socio-economic situation on specific social groups and the responses of the government and private sector are as of much interest today as they were a quarter of a century ago. The next section considers recent economic trends and their impact on the viability and currency of sport.

The credit crunch meets the sponsorship squeeze

Writing shortly after the Hillsborough football disaster in 1989 the late Ian Taylor identified six features of the state of football in England. Football grounds were treated as shrines by fans and they still acted as an emblem of locality for them. However, aspects of what Taylor was later to call 'market football' were developing. Then the sport was beginning to be discussed by politicians and administrators using the rhetoric of modernisation and normalisation, whilst it was still pervaded with violence and the theme of disaster. Market football has prospered in the 20 years since with regular coverage of the excesses that vastly increased revenues from broadcasting and sponsorship have brought. The vast salaries, exploits and lifestyles of top-flight players in the English Premier League (EPL) are regular features of the news, celebrity and entertainment pages as much as the sports pages. Yet in the current economic circumstances with reduced attendances for many teams, sponsors and advertisers cutting back on marketing budgets and broadcasters having less discretionary money to invest, it is likely that news of the indebtedness of clubs, especially in the lower leagues, will also start to appear with more frequency. Small clubs are the most vulnerable in these circumstances, although it is already apparent that there will be few billionaires prepared to bail out bigger clubs. New building projects are likely to be put on hold and excessive salaries will either come down or be subject to a form of capping, as already exists in Rugby Union and League.

In 2008 Tiger Woods lost his sponsorship with the US car manufacturer Buick, Johnson and Johnson, Kodak, Manulife and Lenovo all withdrew from sponsoring the International Olympic Committee (IOC) through its TOP (The Olympic Partnership) Program, and some people expected 'US Treasury' to replace the logo for insurance company 'AIG' on the shirts of Manchester United when the company received billions of dollars from the US government. Early in 2009 AIG formally announced that it was quitting as shirt sponsor.[3] If some of the most marketable athletes, teams and organisations in the world are finding sponsorship sources difficult to sustain, it is understandable why some see the corollary of the credit crunch for sport as the sponsorship squeeze. In such a climate it is likely that events-based sports – such as golf and tennis – will shrink their schedules. Marginal, what might be called 'relevance challenged', events will wither away. Teams in successful events will either fold altogether – such as Honda in Formula One (F1) motor racing – or need to seek new sponsors – such as the English cricket team whose sponsor Vodafone announced it would not renew after January 2010.

There will still be some anomalies. At the end of the week in which the Royal Bank of Scotland (RBS) recorded the biggest loss in British corporate history – £28 billion – journalist Simon Hattenstone (2009) went in search of the former RBS chief executive Sir Fred Goodwin to obtain a verbal apology. He encountered something else when he arrived at Edinburgh airport:

> Every wall is plastered blue and white with RBS logos and sponsorships. 'Sport and RBS. Both global and driven by competition and success,' Sir Jackie Stewart, RBS global ambassador and three times formula one

> world champion, tells me in one ad. 'The principles behind sporting success an (sic) business success are identical' Jack Nicklaus, RBS ambassador and golfing legend, tells me in another (Hattenstone, 2009: 28).

Despite gaining substantial support from the UK taxpayer in order to remain in business the bank also found time to agree an extension to its deal to sponsor the Six Nations rugby union championship in November 2008.

The following quote from Sir Bobby Charlton, former Manchester United and England forward, reflects the feeling that he was one of the lucky few to benefit from doing something he loved and being rewarded handsomely for it:

> When I look back on my life and remember all that I wanted from it as a young boy in the north-east, I see more clearly than ever it is a miracle. I see one privilege heaped upon another. I wonder all over again how so much could come to one man simply because he was able to do something which for him was so natural and easy, and which he knew from the start he loved to do more than anything else (Bobby Charlton cited in Hayward, 2009: 11).

In contrast with this heartfelt expression of amazement at his good fortune, at the start of 2009 it was announced that Andy Murray, the Scottish tennis player, and his brother Jamie were to switch to the same sports agency that helped promote the Beckhams; CAA Sports and 19 Entertainment. This would assist in turning Murray's image into a more global one at a time as he began to reach the apex of the top ten of the Association of Tennis Professionals (ATP) rankings.

Other means to obtain wider media exposure and hence sponsorship funding usually reserved for the more televisual sports – such as football, rugby, tennis, and F1 – include moves to attract sponsorship and audience interest in the Far and Middle East, and to create shortened versions of sports and sports events. Time is a crucial feature of sport for the media, after visual attractiveness and simplicity of rules, as we will see in the next section.

Express yourself – faster: The changing format of sport and its relationship to the mass media

The sport media industry is at a crossroads as the global economic crisis impacts on media companies, slowing revenue growth and narrowing profit margins. As Owen Gibson of *The Guardian* writes, 'neither Pay-TV nor the sports rights market has ever known a recession' (Gibson, 2009a: 6; see also Gibson, 2009b). This will make them more concerned to select the most attractive sports to cover. As we have noted, even the top tier of sports competitions – such as the EPL and the IOC – face a sponsorship squeeze. In the UK with respect to broadcasting deals the leading sports – cricket, rugby, tennis and track and field athletics – have recently agreed medium-term deals that should see them out of the current negative economic situation. That EPL football is looking to expand its coverage and revenues – from a previous deal worth £1.7 billion over three years – when it receives bids for broadcasting rights from 2010 onward is indicative of the convention, in the UK at least, that rights for coverage of football are different from other sports. Yet broadcasters are still likely to want, and to achieve, more control in the scheduling, structure and marketing of the sports they cover. Hence sports and events that have the ability and share the ambitions of broadcasters more closely – to build global audiences in order to offset the impact of domestic conditions – are likely to be the most wanted. In the UK whilst ITV and pay-TV company Setanta began coverage of the English FA Cup competition in 2008, ITV announced that it would not be renewing its coverage of the Oxford and Cambridge boat race and F1 racing.[4]

Another way in which television will drive the development of sport is the move toward faster, more spectacular, entertainment, through the packaging and presentation of events for audiences with short attention spans. Hence there has been the creation of new forms of 'express sport' – shortened versions of sports such as golf, tennis, squash, eventing and most spectacularly cricket.

Here is the advertisement for 'express eventing' staged at the Millennium Stadium in Cardiff in November 2008:

> Eventing comes to the Millennium Stadium for the very first time this coming November. Express Eventing is one of the most exciting developments in the sport today. It is set to do for eventing what Twenty20 has done for international cricket.[5]

What exactly Twenty20 cricket has done for international cricket, is a very good question. Mike Selvey (27 October 2008: 10) writing in *The Guardian* referred to Twenty20 as cricket's 'dot.com' boom; an economic bubble that is likely to explode after making some people – including a few players – very wealthy. Twenty20 is fast – 20 overs per side – which means that matches can be finished in a few hours in an evening, and/or several matches can be played on the same day. Three major championships have developed since the idea for Twenty20 was developed in Britain – two in India and one in the Caribbean funded by the (now discredited) billionaire Allen Stanford. The Stanford Super Series 20/20, the Indian Premier League (IPL) and the Twenty20 Champions League (postponed for a year after the Mumbai terrorism in 2008) are all examples in cricket where media relations are vitally important for the future of the sport.

Mobile commodities

Another feature of contemporary sport is the changing relationships between athletes, and the institutions and organisations that enable sport to operate. One way in which clubs and leagues can seek to maximise revenue – both during a crisis and as a means of expanding the audience for a sport – is to play more games. Additional fixtures, and fixtures played in different places than usual, are two ways forward for sports. Leading American sports – baseball, basketball, ice hockey and gridiron football – have pursued such an 'outsourcing' strategy for several years. Major League Baseball (MLB) began an All-Stars tournament against a Japanese selection drawn from the Japan Professional Baseball league (NPB) in 1986 and has staged three regular season openings in Japan. The National Basketball Association (NBA) staged a preseason match in London in October 2008 between Miami Heat and the New Jersey Nets, and has the overriding ambition of developing interest in the game in China. The National Hockey League (NHL) played games in London in 2007 (Anaheim Ducks v LA Kings) and two regular season games in Stockholm and Prague in 2008. The National Football League (NFL) has been developing a global strategy for the past three decades, but in 2007 and 2008 staged regular season games at Wembley Stadium in London for the first time. This meant that one of the eight home games for Miami Dolphins in 2007 (v New York Giants) and New Orleans Giants (v San Diego Chargers) in 2008 was played thousands of miles away from base. Such developments have led to suggestions that EPL teams might play a '39th game' somewhere in the world other than England, but this has not so far met with much support.

Whilst the increasing mobility of sports talent has been an issue for researchers in the past 20 years, a novel twist is that the family of Novak Djokovic, at the time world number three in ATP tennis rankings, bought the organising rights of the bankrupt Dutch Open (Amersfoort) in 2008. Hence in May 2009 the first Serbian Open would be played near Djokovic's home town in Belgrade as a part of the ATP World Tour.

Governing bodies

Government policies surrounding sport and lifestyles have demonstrated the importance of governing somatic developments in late modern society. Obesity has developed as a, if not *the*, major public health concern. Strategies for health and active living have begun to revolve around tackling obesity and increasing involvement and engagement in sport. Rates are estimated to have doubled in England in the 14 years between 1993 and 2007. Hence targets have been set for reducing obesity and promoting healthy living – the 'Change4Life' strategy costing the government £75 million over three years was launched at the beginning of 2009. The government's sports event strategy includes the ambitious legacies projected for London 2012 for increasing participation in active sport. Driving up participation in sport was a recurring theme of the 'New' Labour Governments from 1997 until 2010. This policy of promoting active lifestyles was linked to a strategy of attracting the staging of sports mega-events. Other events include the 2014 Commonwealth Games in Glasgow, also staged in Manchester in 2002, the Rugby Union World Cup (2015), and bids for the Football World Cup (2018) and the Cricket World Cup (2019).

Despite concerns that the London 2012 Olympics might not have been bid for if the economic conditions at the end of the decade had existed at the beginning of it, the British Government Olympic Executive still sought to reassure the public that the London 2012 Games will have a positive impact during an economic downturn at the beginning of 2009:

> The 2012 Games will play an active role in meeting the challenges of the current economic environment by:
>
> - Creating 75,000 supply chain opportunities – there will be opportunities for businesses across the country to get involved.
> - Creating opportunities for the tourism, media, sport and many other sectors.
> - Increasing the potential for inward investment and export.
> - Reducing unemployment – with a target of 70,000 people across London, including 20,000 in the five host boroughs.
> - Creating temporary jobs – with 10,000 jobs on site at the peak and 30,000 contract jobs to stage the Games.
> - Long-term job creation, with up to 12,000 new jobs in the Olympic Park and up to 50,000 in the Lower Lea Valley as a whole.
> - Opening up skills and employment opportunities for most, if not all, sectors, extending well beyond the Olympic Park and London.[6]

It remains debatable whether in stable economic or crisis conditions hosting a sports mega-event is perceived as a good thing for promoting government policies even in the rhetoric of metropolitan or civic boosterism.

Exactly how these strategies impact on individual people's engagement in sport and active leisure may revolve around a better appreciation of the perceived value active participation in sport and physical activity has for different sections of the population. This relates to the connection in consumer culture between lifestyle and personal identity.

The creativity and construction of physically active lifestyles

Booklets to help readers 'Get fit the Olympic way' were given away with a national newspaper at the start of 2009 and offered advice from British

Olympic medallists on how to train. They joined the usual exhortations at that time of the year to create 'A New Year. A New You' and 'Burn Fat. Get Fit. Look Good'. Lifestyle has become one of the most over used words of the past three decades, confirming the prediction of the first social scientist to systematically review it as a concept. In 1981 Michael Sobel (1981: 1) wrote, 'If the 1970s are an indication of things to come, the word 'lifestyle', will soon include everything and mean nothing, all at the same time'. Understood broadly as 'the distinctive pattern of personal and social behaviour characteristic of an individual or a group' (Veal, 1993: 247) lifestyle contains within it meanings of both the possibilities of individual and collective creativity and the potential for imposed consumerist conformity and control. Which of these it offers is an empirical question, but in the past 30 years it has underpinned conceptual and theoretical challenges to the prevailing social scientific orthodoxies in the study of sport and leisure.

Coakley (1993) for example summarised research into involvement in sport derived from this conceptual shift in three ways: that it involved a process of identity construction and confirmation; that the establishment of personal identity was a primary factor in decision making over participating or not participating in sport; and that individuals and groups create and negotiate their involvement in sport. This emphasis on active individuals making choices is symptomatic of the positive qualities attributed to lifestyle. Chaney (1996: 86) concluded that as a concept lifestyle was able to capture 'processes of self-actualization in which actors are reflexively concerned with how they should live in a context of global interdependence'. It has been suggested that lifestyle sports, such as windsurfing, skateboarding and snow boarding, offer a challenge to high-performance, achievement orientated, sports. 'In contrast to capitalism's temporal production, lifestyle sport time is immediate and discontinuous' (Wheaton, 2007: 298). But are lifestyle sports subversive, resistant or merely just another way of playing under capitalism?

Alternatively it can be argued that the use of lifestyle images in popular culture and official strategies and policies to create depictions of particular ways of life not only differentiates and fragments social groups, it also acts as a political concept. For example, by linking lifestyle to health on the basis of solving social problems by individualising them serves to deflect people's attention away from other 'forms of collective, social action' (O'Brien, 1995: 193). Lifestyle can be used in the construction of ways of life and create rather than resolve social problems.

Whilst there appears to be a greater degree of choice of leisure and sport than ever before, non-communicable diseases, such as obesity, have increased. Is there a relationship between these things? One answer is to argue that people are not sufficiently concerned about obesity and their health. Another is that people are not sufficiently able to make healthy choices given wider sets of social relationships, institutions and processes. Hence, despite occasional healthy eating campaigns run by retail supermarkets, during popular sports events such as the Football World Cup many of the related deals offer 'snack foods' – beer, soft drinks, pizza, chocolate, biscuits and potato chips – in exchange for entering sport event related competitions. The largest sports mega-events accept sponsorship from makers of calorie-dense beverages and food, and equipment that fosters sedentary activities such as watching television and using motorised transport. In this way it could be said they actually promote mortality, morbidity and disability as they contribute to the global spread of obesity or 'globesity' (Dickson and Schofield, 2005).

Veal (1993: 248) identified four key questions for research into lifestyles:

1. What are the processes by which people adopt lifestyles or, alternatively, have lifestyles thrust upon them?
2. What is the meaning and importance of actual or desired lifestyles to individuals – are they as important as some people believe?
3. Are lifestyles expressions of freedom or a contrived tool of consumer capitalism – are people heroes or dupes, victims or playful communicators?
4. Has lifestyle replaced traditional social variables, such as social class, gender, age, etc, as the key differentiating variable in society, and what might the implication be for the analysis of leisure and sport?

By extension, in his framework for a radical politics in contemporary society/late modernity, Anthony Giddens (1994) identified life politics as having a more central role to play. If identity has become a 'life project', under constant (re)construction, does this mean that lifestyle sports will play a much larger part in the development of and the politics of identity recognition? These remain important questions as we consider whether lifestyle is now more salient than class, gender, race, ethnicity and other social categories in the formation of personal identity. Evidence exists that points to the relative instability of lifestyle identities compared to social class identities – they are more subject to change across the life course (Runnymede Trust, 2009). Hence lifestyle will remain an essentially contested concept.

Conclusion

Underpinning this chapter, and in keeping with the analysis by the authors of *The Devil Makes Work*, is an emphasis on the importance of capitalism for understanding developments in the position of sport in society. More business-oriented commentators on sport provide a compelling list of associated developments (Westerbeek and Smith, 2003: 48–9). The consequences include a growth in the economic effects and impacts of sport, the ongoing increase in the value of genuinely global sport properties, including athletes and players themselves, and the convergence of economic power in sport ownership. These effects link up with the defragmentation of sport governance and the simultaneous professionalisation and marginalisation of smaller sports and leagues – 'the gap between the sport enterprises that are globally successful and those which remain domestically viable, will grow' (Westerbeek and Smith, 2003: 48).

Whilst we might disagree with some of their futurological study, especially in the light of the financial crisis stimulated by the subprime lending crisis, it is undoubtedly the case that sport has become more commercialised in the past 25 years. Each of the trends identified by Westerbeek and Smith relates to the development of capitalism as the dominant economic system throughout the world. In order to make better sense of their list of developments I have argued elsewhere that it is necessary to adopt a 'production of consumption' approach to consumer capitalism (Horne, 2006).

In the midst of consumer society class, 'race' and gender continue to influence life trajectories. The social divisions between rich and poor continue to grow. The production of consumption approach is therefore most useful to help explore questions about sport and lifestyle for three main reasons. First, to understand the social environment of contemporary advanced societies adequately it is necessary to come to terms with the emergence of a radically transformed global capitalist free market economy. Secondly, although society may have moved beyond the (Protestant) work ethic as its major disciplinary force, consumption increasingly operates in the service of production. Thirdly, consumer culture is simply capitalistic. As Smart (2003: 163) suggests 'hardly anyone's life is unaffected by the direct or indirect consequences of the global diffusion of capitalist forms of economic life'. In these circumstances corporate brands become experiences, offering lifestyles and identities. Ironically, rather than being a threat to the established social order, the identity politics of the 1990s may have been 'a gold mine' to corporations as Klein (2000: 115) suggested since they revealed diverse market niches, without the need for additional, expensive, and difficult, consumer research.

This chapter's overall message is that to understand sport in consumer culture it is necessary to have an understanding of what creates the conditions within which consumer culture flourishes and that is the economic system. Rather than draw on outdated and outmoded general models of society the task of social science is to provide nuanced accounts of the specifics of social and historical contexts, or conjunctures. Despite the convulsions that the global economy went through at the end of the first decade of the twenty-first century, the major issues in the study of sport in consumer culture will remain. These offer a research agenda for the sociology of sport and lifestyles that follows in the tradition established by John Clarke and Chas Critcher:

1. Identifying trends in the global sports goods and services market.
2. Monitoring the growth of sports coverage in the media, especially their role in the process of creating consumers out of sports audiences and fans ('consumerization').

3. Examining the importance of sponsorship and advertising for contemporary sport.
4. Assessing the changing role of government – concerning regulation, consumer protection and sports promotion.
5. Researching how much, as a result of increasing consumerisation, has the role of sport in the construction, maintenance and challenging of lifestyles and identities altered.
6. Revealing how consumerisation is reflected in social divisions in patterns of involvement and participation in sport.

Notes

1 The next six sections of this chapter are informed by Horne, 2009.
2 However since then American and other foreign ownership has increasingly become the norm in elite English football.
3 Replaced by Aon Corp, another American insurance and financial enterprise, from 2010 for a four-year deal reported widely in the press to be worth £20 million per year.
4 In the summer of 2009 Setanta went into administration and its coverage of English football was taken over by ESPN.
5 http://www.millenniumstadium.com/301_9998.php , accessed 21 January 2009.
6 http://www.culture.gov.uk/what_we_do/2012_olympic_games_and_paralympic_games/5602.aspx, accessed 21 January 2009.

Bibliography

Andrews, D. (ed) (2004) *Manchester United* (London: Routledge).
Bryman, A. (2004) *The Disneyization of Society* (London: Sage).
Cashmore, E. (2002) *Beckham* (Cambridge: Polity).
Chaney, D. (1996) *Lifestyles* (London: Routledge).
Clarke, J. and C. Critcher (1985) *The Devil Makes Work: Leisure in Capitalist Britain* (Basingstoke: Macmillan).
Coakley, J. (1993) 'Sport and socialisation', in J. O. Holloszey (ed.) *Exercise and Sport Sciences Reviews, Volume 21*, 169–200 (New York: Williams and Wilkins Publishing).
Crawford, G. (2004) *Consuming Sport* (London: Routledge).
Dickson, G. and Schofield, G. (2005) 'Globalisation and globesity: The impact of the 2008 Beijing Olympics on China', *International Journal of Sport Management and Marketing*, 1(1/2): 169–79.
Featherstone, M. (1991) *Consumer Culture and Postmodernism* (London: Sage).
Gibson, O. (2009a) 'Where will the big bucks go when the bubble bursts?', *The Guardian*, Sport section, 22 January, 6–7.
Gibson, O. (2009b) 'Olympics are the jewel in the crown but cricket is the flashpoint', *The Guardian*, Sport section, 23 January, pp.6–7.
Giddens, A. (1994) *Beyond Left and Right* (Cambridge: Polity).
Hargreaves, J. (1987) 'The body, sport and power relations', in J. Horne, D. Jary and A. Tomlinson (eds) *Sport, Leisure and Social Relations*, 139–59 (London: Routledge and Kegan Paul).
Hattenstone, S. (2009) 'Sir Fred, just say sorry', *The Guardian*, 24 January, pp.28–9.

Horne, J. (2009) 'Sport in a credit crunched consumer culture', *Sociological Research Online*, Volume 14, Issue 2/3, http://www.socresonline.org.uk/14/3/7.html.
Horne, J. (2006) *Sport in Consumer Culture* (Basingstoke: Palgrave).
Horne, J. and W. Manzenreiter (eds) (2006) *Sports Mega-Events* (Oxford: Blackwell).
Hayward, P. (2009) 'United again: Me, Jack and the meaning of brotherly love', Sir Bobby Charlton Interview, *The Observer*, Sport section, 11 January, pp.10–11.
Jackson, S. and D. Andrews (eds) (2005) *Sport, Culture and Advertising* (London: Routledge).
Klein, N. (2000) *No Logo* (London: Flamingo).
Miller, T., G. Lawrence, J. McKay and D. Rowe (2001) *Globalization and Sport* (London: Sage).
O'Brien, M. (1995) 'Health and lifestyle: A critical mess?', in R. Bunton, S. Nettleton and R. Burrows (eds) *The Sociology of Health Promotion*, 189–202 (London: Routledge).
Ritzer, G. (1993) *The McDonaldization of Society* (Newbury Park, Ca: Pine Forge).
Runnymede Trust (2009) *Who Cares about the White Working Class?* (London: The Runnymede Trust).
Schor, J. and D. Holt (eds) (2000) *The Consumer Society Reader* (New York: New Press).
Selvey, M. (2008) 'This mercenary match will resolve none of England's dilemmas', *The Guardian*, Sport section, 27 October, p.10.
Smart, B. (2003) *Economy, Culture and Society* (Maidenhead: Open University Press).
Sobel, M. (1981) *Lifestyle and Social Structure* (London and New York: Academic Press).
Taylor, I. (1989) 'Hillsborough, 15 April 1989: Some personal contemplations', *New Left Review*, 177: 89–110.
Tomlinson, A., N. Ravenscroft, B. Wheaton and P. Gilchrist (2005) *Lifestyle Sports and National Sport Policy: An Agenda for Research*, Report to Sport England, March.
Veal, T. (1993) 'The concept of lifestyle: a review', *Leisure Studies*, 12: 233–52.
Westerbeek, H. and A. Smith (2003) *Sport in the Global Market Place* (Basingstoke: Palgrave).
Whannel, G. (2002) *Media Sport Stars* (London: Routledge).
Wheaton, B. (2007) 'After sport culture: Rethinking sport and post-subcultural theory', *Journal of Sport and Social Issues*, 31(3): 282–307.
Wheaton, B. (ed) (2004) *Understanding Lifestyle Sports* (London: Routledge).

13

Between the Devil and the Deep Blue Sea:[1] Music and Leisure in an Era of *X Factor* and Digital Pirates

Brett Lashua

***Chapter 13** discusses two particularly salient aspects of the **new politics of leisure and pleasure** in the realm of **popular music**. TV talent contests such as* X Factor, *begun in 2004, and the downloading without payment of music from the internet. Television programmes such as* X Factor *are extremely popular but their critics contend that, as with **sport**, they are destroying the human essence of the activity, swamping it with commercialism and corporate mass marketing. From time to time there are campaigns to stop a talent show winner reaching the top of charts and there is much popular support for those who download their music free from the internet, in apparent defiance of the record companies. Free downloads are widely perceived to be a blow for freedom in the leisure politics of music. Once again a complex **political tension** is exam-ined – in this instance between the **corporate contrivance** and **marketing of contemporary popular music** and the **attempts to subvert** these processes in the name of **individual freedom**.*

Introduction

> It is hard to listen to a programme of pop songs ... without feeling a complex mixture of attraction and repulsion. (Richard Hoggart, 1970, in Storey, 1997: 77)

This chapter embraces the paradox that Hoggart expressed (above) in relation to two prevailing twenty-first century music and leisure phenomena: *Pop Idol* reality TV talent programmes and digital music piracy. Certainly more is at stake in popular music than the aversion and compulsion these phenomena provoke; however, at the end of 2009, two news items about these issues caught my attention. The first was an online campaign attempting to thwart the apparent certainty that the Christmas Number 1 song in the UK charts would come from the winner of the immensely popular TV singing contest *X Factor*. The second was a series of newspaper reports on the continuing threat (or resistance) to the music industry posed by digital piracy, including one claim that 95% of music downloads are illegal (Allen, 2010). Both issues lead me to consider further the relations between music and leisure, culture and consumption, legitimacy and illegality.

These popular phenomena – digital music piracy and talent contest chart dominance – are widely considered oppositional. On one hand, the '*Pop Idol* phenomenon' (Fairchild, 2008) epitomised by *X Factor* arguably represents the penultimate music production line, churning out musical ready-mades (Adorno and Horkheimer, 2000[1944]) on an annual basis, powered by a massive marketing machine that ostensibly involves viewers' participatory votes, while garnering millions in revenues for the corporations and music moguls behind the artists. In the last decade, reality-style singing contests (e.g. *Pop Idol, X Factor, American Idol,* and similar programmes around the globe, see Dann, 2004) have proven enormously successful, producing massive popularity for themselves and the artists they have showcased. In terms of sheer numbers, more people voted during a recent season of *American Idol* than in any US presidential election (Sweney, 2006), and more voted during *X Factor* than in the previous UK General Election (Goodman, 2009). Ytreberg (2009: 470) reported that a colossal 78 million text message votes were sent during the spring 2008 season of *American Idol*. The most recent final of *X Factor* (in December 2009) was viewed by 19.1 million people, roughly one-third of the total UK population. At the same time, the UK music charts were dominated by albums that were products of *X Factor* artists (or contestants from related talent programmes, such as *Britain's Got Talent*). In the run up to Christmas 2009, the popularity and success of the show and its format appeared both inescapable and unassailable.

On the other hand, one could download – illegally – many of the songs from these hit albums, and millions of other tunes, seemingly for free via internet file-sharing sites. Despite continuing attempts by organisations such as the Recording Industry Association of America (RIAA) and the International Federation of the Phonographic Industry (IFPI) to stop illegal downloads, digital music pirates continue to take advantage of the murky regulatory waters of the internet. Music piracy also taps into deeply-rooted mythologies of heroic cultural rebels (e.g. down-loaders) taking from the rich (e.g. record companies) for the benefit of the poor (Rojek, 2005: 363). Such views, perhaps indicative of a wider online 'gift economy' (Burkart and McCourt, 2006; see Mauss, 1990; also Sennett, 2003), are championed by indie-rock bands such as Radiohead, recently freed of their record label contract, who released their latest album themselves online on a 'pay whatever you wish' basis (meaning that, if they wished, people could download it for nothing). Thus some claim digital piracy (or 'Net banditry', Rojek, 2005) represents a sea-change in the music industry, if not through blatant pilfering as resistance (Mason, 2008), then at the very least, by rendering the 'old' industry of selling/consuming music obsolete. Furthermore, each phenomenon – *X Factor* and digital music piracy – appears to support, if superficially, suppositions of leisure as participatory, democratic, grassroots, and active consumption. Yet, as I will contend, these assumptions circulate within narrow definitions of leisure that 'fail[ed] to grasp the relevance of divisions in production and reproduction in capitalism' (Bramham, 2006: 385) that mask power hierarchies and social structures.

The links between these issues crystallised during the run-up to 2009 Christmas season, when a group on the social networking site Facebook agitated to 'Rage against the *X Factor*' by inviting people to download (legally) the 1992 song 'Killing in the name' by the rap-rock band Rage Against the Machine (RATM). The Facebook group's aim was to supplant the *X Factor* at the top of the charts. Featuring a rebellious refrain 'Fuck you! I won't do what you told me!' the RATM song was supported by a massive, grassroots viral campaign that swelled to include over 900,000 Facebook 'friends'. As Christmas approached, 'Killing in the name' was neck-and-neck in the top slots of the charts with the recently-crowned 2009 *X Factor* winner, Joe McElderry, singing a cover of 'The Climb' by the Disney teen character/starlet Hannah Montana (played by Miley Cyrus). On the surface this battle appeared to be an angry, politically-minded rap-rock song versus a saccharine pop confection, musical authenticity versus commodified musical product. The 'Battle for Christmas Number 1' also appeared to draw attention to the power of social-networking campaigns to incite populist invective in acts of consumer resistance (i.e. 'don't buy *that*, buy *this* instead'). However, as Fairchild (2008: 9) cautioned, we mustn't get stuck in 'a debate in which the acts of the consuming subject are imagined as autonomous from the social and economic structures that surround them' and through which leisure devolves into self-contained acts of free will and democratic choices that are representative only of the market's (in)effectiveness. That is to say, perhaps the differences between the contenders in the 'Battle for Christmas Number 1' were less than they initially appeared.

This chapter asks 'what's going on?' (Grossberg, 1998: 67) and 'what is at stake?' when looking at paradoxes of leisure consumption of popular music. I explore the 'complex mixture of attraction and repulsion' (Hoggart, 1970 in Storey, 1997) felt towards these recent music and leisure phenomena. Seemingly caught between the devil (*X Factor*) and the deep blue sea (music piracy), I seek to turn the analysis in a different direction, thus providing a radical alternative to the suffocating dualisms of leisure consumption as either reproduction or resistance. In what follows I offer further observations regarding these in-between relations of popular music and leisure. First, I outline several key theoretical intersections between popular music and leisure. Next, I dig in to the examples outlined above – *X Factor*, digital music piracy, and the recent Facebook campaign to 'Rage Against the X Factor for Christmas Number 1'. A final example – the 2007 release of Radiohead's *In Rainbows* – points toward further debates emerging around the online celestial jukebox. Throughout, the chapter calls into question the collective agency of social subjects to change their everyday lifeworlds through music and leisure (Bennett, 2005).

Leisure and popular music

Bennett (2005) and DeNora (2000) have theorised the importance of popular music, arguing its embeddedness in the economic, social, political and material fabric of everyday life. More broadly, Grossberg (1992) had stated that popular culture, especially music, 'is a significant and effective part of the material reality of history, effectively shaping the possibilities of our existence' (p.69). Upon this often mundane yet powerful terrain, Bennett (2005: 333) noted 'At its most spectacular, popular music has contributed to the spread of alternative ideologies – about politics, gender, race, the environment, and so on'. That is, pop music can potentially offer a means for the expression of resistance to social inequities and injustices, as well as serving as a locus of social movements and community solidarity. Such a view risks romanticising the effects of pop music consumption and the agency of music audiences (see Fiske, 1989). However radical and spectacular it may appear, more often than not popular music contributes to the perpetuation of dominant, conservative ideologies and systems (i.e., the music/cultural industries) that act to preserve the status quo. In leisure studies of music, Lashua (2006) explored the limitations of popular music leisure as resistance to racism, social exclusion and spatial inequality, and noted that the power musicians can wrest and wield is often quite marginal and weak outside of limited contexts. Others have discussed the broader problem of casting all creative cultural leisure as grassroots, democratic or resistant (Fairchild, 2008; Fiske, 1989; Frith, 2002; Negus, 2002; Toynbee, 2000). Fairchild warned that such views produced a:

> neo-liberal model of consumption in which consumers are imagined as rational, self-contained units of freewill acting within a context that is so drained of any animating intent that one's range of choice is only a measure of the market's effectiveness, influence, or social function. (2008: 8)

According to Harvey (2003) neo-liberalism is a theory of political-economic practices that values institutional frameworks typified by strong private property rights, free markets and free trade. Against a backdrop of neo-liberal policy toward intellectual property (e.g. protecting music copyright), Fairchild (2008: 8) questioned how we might better define leisure as beyond being 'more or less free to make up or own minds, make our own choices, ingest our own music and use it how we wish'. This kind of individualised, ubiquitous freedom to consume music – e.g. a 'celestial jukebox' (Burkart and McCourt, 2006) that offers access to any song, anytime, anywhere – comes with costs, of course, not only in terms of the fees associated with such a service, but also in terms of hidden social and political costs. For example, the 'participatory democracy' espoused by *X Factor*, which invites public involvement via phone-in voting for the show's contestants (at rates of as much as 75p per vote),[2] conflates being able to pay for music with the freedom to choose.[3] This kind of neo-liberal model of consumption risks reducing democratic participation to the ability to pay for the privilege of choice between commodified products.

Leisure theorists have argued against neo-liberal, individualised freedoms (Arai and Pedlar, 2003; Bramham, 2006) and instead espoused idealised versions of freedom through active citizenship and democratic leisure (Coalter, 1999; Glover et al., 2005; Hemingway, 1996; Storrman, 1993; Sylvester, 1995). Exploring popular music as a terrain for radical democratic leisure is useful, as Grossberg (1998: 390) cheerfully encouraged us to consider:

> If we are to imagine a different, a better future, we need to consider the different ways people participate in social, cultural, economic, and political life...for it is here...that we can examine how people make history, and articulate what history we would – collectively – hope to make.

However, as noted by Bramham (2006), recent theorisations have positioned leisure as 'increasingly more about shopping around, exploring the seductive delights offered by figural consumer culture' (p.386). Along similar lines, Stewart et al. (2008: 371) summarised the position thus: 'To be clear, opposition to neo-liberalism is present within leisure studies, yet in the end, neo-liberal doctrine pervades in practice'. Again, Bennett's (2005) statement that popular music provides a locus of spectacular ideological resistance appears to miss the mark. More spectacular, if less remarked, are the ways in which programmes such as *X Factor* support the status quo and labour to reproduce existing social relations. In the next section I consider the position of *X Factor* at the convergence of neo-liberal leisure and popular music consumption.

X marks the spot? *X Factor* and the mythic allure of the pop star

X Factor is a reality-style TV singing contest broadcast on ITV during Saturday and Sunday nights. By the end of 2009, the show had aired for five seasons. It has become the most popular television viewing in the UK (Frost, 2010). Its contestants are purportedly ordinary folk,[4] drawn from mass auditions held around the UK in the lead up to the televised live series of programmes featuring 12 final contestants. The show follows what is largely a standard format. *X Factor* supplanted its immediate predecessor and the original version of the format, *Pop Idol,* which aired from 2000–2004, and *Popstars* (2000–2002). The series have been hugely successful. *Popstars,* for example, produced such singing celebrities as Cheryl Cole (*Popstars: The Rivals* contestant, 2002) who is a now member of the group Girls Aloud, a successful solo artist and since 2008 also a judge on *X Factor.* Another judge, the acerbic impresario Simon Cowell is largely the public face of *X Factor.* Cowell (along with *Pop Idol* producer Simon Fuller) is in part responsible for the format that, in various guises (e.g. *American Idol, Australian Idol, Pan-Arab Idol, etc.),* has become a global phenomenon (Dann, 2004).

In the UK, the *X Factor*'s immense viewership is matched only by the massive sales of music by its contestants and guest stars. As noted above, 19.1 million people watched the 2009 *X Factor* final, roughly 62% of the total UK viewing audience. This popularity successfully translates into sales, for example, as the 2008 *X Factor* winner had the biggest single-week sales in UK Chart history for her cover of Leonard Cohen's 'Hallelujah'. As a promotional platform for selling music, the show is remarkably effective. The performance of a song by an *X Factor* contestant can catapult a tune into the UK top ten singles chart, as with the 1981 song 'Don't stop believin' by the group Journey. This followed a performance of it in December 2009 by *X Factor* contestant (and eventual winner) Joe McElderry. Similarly, an *X Factor* performance as a guest (such as Alicia Keys, or the Black Eyed Peas, both in December 2009) or being featured as a songbook artist (such as Queen, or Michael Jackson) whose catalogue provides material for entire episodes of the programme can also lead to a return to chart success for those artists. More significantly, however, the show is a fast-track to fame (however fleeting) and typically leads to a Number 1 song in the weeks immediately following the conclusion of the *X Factor* contest in mid-December (i.e., the Christmas season). Indeed, the show's winner has had the Number 1 UK chart single at Christmas four of the last five years, and in the fifth instance, McElderry's song ('The Climb') this year, topped the charts during the week following Christmas.

Such is the success of X Factor, that at the end of 2009, the UK Album charts were saturated by *X Factor* contestants and guest performers (Table 13.1). As the charts indicate for this week at the turn of the New Year 2010, the *X Factor* had an absolute dominance of the UK music market. Only one group (Soldiers) in the top 21 albums in the UK charts during the week of 27 December 2009 to 3 January 2010 was not linked to an appearance on the *X Factor,* and only three others – Snow Patrol, N-Dubz and Paolo Nutini – did not feature prominently on the televised show, either as performers or as guests. While relationships between *X Factor* performance and UK chart success are not always causal (e.g. Michael Jackson's music had already surged into the charts immediately following his death), in other cases (e.g. Burke, Lewis, Cole, Young, and Boyle) the artists are the direct progeny of the format.

Table 13.1 *X Factor* dominance of the UK Album charts, week of 27 December 2009 to 3 January 2010

Chart position	Artist	Link to *X Factor*
1	Michael Bublé	Featured artist and guest performance on 2009 *X Factor*
2	Susan Boyle	Winner 2009 *Britain's Got Talent*, guest performance on 2009 *X Factor*
3	Black Eyed Peas	Guest performance on 2009 *X Factor*
4	Lady Gaga	Guest performance on 2009 *X Factor*
5	JLS	Runners-up 2008 *X Factor* and guest performance on 2009 *X Factor*
6	Robbie Williams	Featured artist and guest performance on 2009 *X Factor*
7	Take That	Featured artist on 2009 *X Factor*
8	Snow Patrol	Hit single 'Run' covered by 2007 *X Factor* winner Leona Lewis
9	Will Young	2002 *Pop Idol* winner, guest performance on 2008 *X Factor*, guest judge on 2009 *X Factor*
10	Cheryl Cole	2002 *Popstars: The Rivals* winner, *X Factor* judge, and guest performance 2009 *X Factor*
11	Westlife	Managed by *X Factor* judge Louis Walsh, guest performance 2009 *X Factor*
12	Leona Lewis	Winner 2006 *X Factor* and guest performance on 2008 and 2009 *X Factor*
13	Rod Stewart	Featured artist and guest performance on 2009 *X Factor*
14	Queen	Featured artist and guest performance on 2009 *X Factor*
15	Paolo Nutini	Nutini's song 'Last request' sung by contestant Lloyd Daniels, 2009 *X Factor*
16	Michael Jackson	Featured artist on 2009 *X Factor*
17	Alicia Keys	Guest performance on 2009 *X Factor*
18	Rihanna	Guest performance on 2009 *X Factor*
19	N-Dubz	Guests on 2009 *Xtra-Factor*
20	Soldiers	–
21	Alexandra Burke	2008 *X Factor* winner and guest performance on 2009 *X Factor*

Source: The Official Charts Company (www.theofficalcharts.com)

What is the secret of the show's success? Fairchild (2007) suggested that programmes such as *X Factor* and *Pop Idol* had become phenomenally successful through their abilities to establish and maintain what have been defined as 'media rituals', or social relationships created and made possible symbolically:

> Our experience of the world, our formed sense of the common and the practical everyday knowledge that sustains these implicit social connections to the larger world are given support and confirmation by a whole range of symbolic expression. 'Idol' is one such semiotic security blanket. (Fairchild, 2007: 357)

The archetypal personae and melodramatic story arcs woven into such a 'semiotic security blanket' as *X Factor* reinforce the show's 'assumed role as transparent, earnest, and benevolent facilitator of the best undiscovered talent it can find, and through this giving us all the drama, tears, pleasure, and pain we can stand' (Fairchild, 2007: 356). In other words, and ironically, shows such as *X Factor* are compelling to audiences precisely because almost everything about them conforms to media rituals (Couldry, 2003) that reinforce dualistic 'good' and 'bad' stereotypes (Reijnders et al., 2006). Cowell (2003: para. 2) suggested that these rituals and stereotypes are reinforced further through the 'large body of metatexts in circulation – websites, electronic newsletters, television and magazine interviews – that underpin, and, in part, constitute the popularity of the show itself'. The 'reality' of programmes like *X Factor* and its siblings, according to Holmes (2004), is an oxymoron as viewers are presented less with reality and more with its antithesis: 'performativity, construction, and the most contrived and artificial situations' (p.148). This artifice appears in direct contrast to the show's 'constant emphasis on not simply specialness but also individuality' (Holmes, 2004: 155). Here Holmes further pointed to the work of Adorno (originally 1941, in Frith and Goodwin, 1990), who noted the efforts of the pop music industry to produce a standardised, yet 'pseudoindividualized' product. This product is the quintessence of the supposedly mysterious and unquantifiable 'X Factor' –the 'whatever' quality that makes someone a popular sensation. Despite this supposedly mysterious quality, the social personae in the show – a cast of increasingly known characters, heroes, villains, and tricksters – play well-established roles in reproducing and reinforcing deep-rooted social relations. For example, each season there is typically the underdog contestant (e.g. the single mother, the older adult, the oddball, the comeback, etc.) who is repeatedly belittled by the villain Simon Cowell who seems to take devilish pleasure in bluntly bashing their dream of singing mega-stardom. Other characters include the good-girl-next-door (such as 2006 winner Leona Lewis)[5] and the good kid (e.g. McElderry, 2007 winner Leon Jackson, or 2005 winner Shayne Ward). Situating these personae within the broader economic, political, social and cultural landscape, the *X Factor* represents a hugely popular media phenomenon. This phenomenon is more broadly indicative of issues (and opportunities) of leisure in late capitalism, within which the *X Factor* exemplifies processes akin to McDonaldisation[6] (Ritzer, 1993) and Disneyisation[7] (Bryman, 2004) in pop music.

A world that is still very much at sea: Leisure, music, piracy and popular culture

In part a contrast to the *X Factor* phenomenon, this section will discuss the issues surrounding the practices of illegal music file-sharing or digital piracy. Opitz (2010: 1) described the continuing relevance of piracy as a means to grasp 'the socially uneven, violent and unstable world we live in – a world that is still very much at sea'. From online digital media, hackers and cyberpunks to Somali hijackers in the Gulf of Aden, piracy is not a practice relegated to the pages of history. Myths continue to circulate about pirates that shift between heroic do-gooders fighting against injustice and villainous criminals. Opitz referred to pirates as historically-situated characters who are simultaneously 'capitalist marauders and militant workers fighting for restoration of the commons' (2010: 1). The impetus for piracy hinges between two concepts – what is legal, and what is profitable. Rojek (2005) added further explanation – piracy as leisure and social resistance. In this regard, others have approached music piracy as a subcultural activity (Cooper and Harrison, 2001; Mann, 2000; Rutter and Bryce, 2008).

Rojek (2005) argued the legality or P2P (peer-to-peer) file sharing as a 'novel' form of resistant leisure. According to Rojek,

> it is possible to interpret P2P as a leisure form that fosters social inclusion, empowerment and distributive justice. This leaves the recording industry in the awkward position of seeking both to impose limitations on the functionality of computer exchange technology and to regulate participation in a mass popular cultural and leisure form. (2005: 357)

P2P file sharing involves the direct transfer of digital media content between users without the use of central transfer sites (e.g. songs in MP3 format, but also images, software, documents, and videos; most notably pornography). P2P file sharing works when users make a location folder available for others to 'see' across a network and from which other users may select content to copy to their own machines. For example, Brett's computer might be able to access the contents of a folder containing MP3 music files on another's computer, and he may choose to copy or download certain files to his own machine. The MP3 format is important to digital piracy because its advent in the early 1990s allowed for easier (e.g. quicker) transferability *via* the limited capabilities of the internet. According to Mann (2000: 44), as soon as MP3 technology was available, 'an active digital-music subculture' was converting songs from CDs to MP3s and sharing these uncopyrighted files across the Internet.

Clearly, when copyrighted files are shared directly, no materials are paid for, and the exchange is illegal. The legality of music file-sharing has dominated music retail since the late 1990s. At that time, a file sharing programme called *Napster* (introduced in 1999) became extremely popular for music piracy. The programme also had broader effects, in the politics of leisure. Rojek (2005) summarised the impact of *Napster*:

> In terms of the commercialization of leisure, *Napster* challenged commodification and rationalization by re-positioning the consumer in relation to ownership, decentralized power to the consumer and extending choice and flexibility. But it also blatantly infringed copyright law, depriving artists and multinationals of the protection of copyright, apart from through the expensive, controversial and haphazard process of litigation. At its peak, *Napster* provided access to as many as one billion music files.

After defeat in the courts in a lawsuit brought by the Recording Industry Association of America, *Napster* was shut down in 2001, only to be purchased by Roxio and relaunched as a legal, for-fee download site in 2003. By this time, however, Apple had launched its iTunes online music store, which has dominated the legal music download market ever since, as it is coupled with the hugely successful iPod mobile music player (Bull, 2005). Despite the ever-increasing regulation of the Internet, online piracy remains a significant practice. RIAA notoriously filed lawsuits directly against individuals who had made as few as a couple of dozen illegal downloads. Furthermore – and in what is widely perceived as an incursion into civil liberties – RIAA has been challenging Internet Service Providers (ISPs) to reveal who illegal file-sharers are. Given these attempts to curtail music piracy, it is all too easy to vilify the music industry and their regulatory bodies (RIAA). Correspondingly, consumers mustn't believe that, once online, their actions exist in a regulatory vacuum, or that they somehow escape the earthly, material web of social structures and institutions – such as the corporate worlds of ISPs, record companies, and copyright laws.

According to the recording industry's trade body, the International Federation of the Phonographic Industry (IFPI) – an anachronistic name if ever there was one – 95% of all music downloads worldwide are illegal (Allen, 2010). The IFPI claims that a country, such as Spain where piracy is reportedly the most extensive in Europe, 'runs the risk turning into a cultural desert' (Allen, 2010: 25). Such analyses contend, for example, that Spanish musicians do not get into the charts because of the loss of huge numbers of sales due to piracy. While this may be the case, these analyses miss the bigger point. Indeed, musical cultures may be flourishing and livelier than ever, online (whether illegal or not), yet are simply not commercially accounted for through the traditional industry index – the sales charts. On this point, it wasn't until 2006 that legally downloaded songs counted toward the calculation of the UK Singles charts (Gibson, 2006). Therefore, where is the risk, and to whose culture? Some, following Adorno, might argue that the *X Factor*, now in the vanguard of the music/cultural industries that churn out carbon-copy artists already poses a greater risk of generating a 'cultural desert' than the online channels of pirated music ever could.

Xmas marks the spot? Raging against the machine

In four of the last five years, the winning contestant on the *X Factor* – which not coincidentally has its finale in mid-December – has been the Number 1 song in the UK singles charts at Christmas. In the 2009 season, Cowell's presence seemed more pervasive than ever and by December, groups associated with the programme (judges, contestants, guest performers, original artists of songs covered during the show) dominated the UK albums and singles charts. Amidst such saturation, a public backlash was perhaps inevitable. As noted earlier, in early December 2009, a campaign was initiated on the social networking site Facebook, to 'Rage against the X Factor' by attempting to hijack the charts via a 'grassroots' campaign to download enough copies of a song called 'Killing in the name of' by an American rap-rock group, Rage Against the Machine. When first visiting the group's Facebook site in mid-December as one of 976,714 'friends', I was greeted:

> Welcome to what is currently the UK's most trafficked facebook group. Are you getting fed up about the possibility of ANOTHER X Factor Christmas No.1?...us too...so we're going to do something about it! We are all buying a download of 'KILLING IN THE NAME' by RAGE AGAINST THE MACHINE right NOW! until the end of Saturday 19th December (23:59pm). ** PLEASE GET INVOLVED AND BUY IT **

To many members of this group, the *X Factor* represents the 'machine' (derisively referred to at times as the '*X Factory*') that churns out pre-packaged pop music, and the show's abilities to produce hits must be 'killed' in the name of breaking the show's stranglehold on the UK charts. Originally an anti-racist, anti-authority tirade, the lyrics of 'Killing in the name' fume about police officers who are also members of the Ku Klux Klan, including the line 'some of those who work forces are the same who burn crosses.' The song repeatedly breaks down into forceful, punctuated, percussive blasts, over which the lead singer Zak de la Rocha shouts the refrain 'And now you do what they told ya/ Now you're under control' which is altered in the final chorus to defiantly proclaim 'Fuck you! We won't do what ya told us!'

In contrast, the song chosen for McElderry to sing in battle for the Number 1 chart spot was 'The Climb', a tune originally written for and performed by the Disney character Hannah Montana (played by Miley Cyrus) for the 2009 film/soundtrack *Hannah Montana: The Movie*. The song peaked for Cyrus at Number 4 on the US singles charts (it achieved Number 11 in the UK charts), and was the eighth best-selling single in the US in 2009, selling over two million copies. Once joined, 'the Battle for Christmas Number 1' between RATM and McElderry saw combined sales/downloads of both songs surpass figures for similar battles in previous years. In the 2008 Christmas season, a battle had also been waged over competing cover versions of Leonard Cohen's song 'Hallelujah'.[8] The 2008 *X Factor* winner Alexandra Burke's rendition was squared off against a popular version by the late Jeff Buckley. Burke's recording sold 576,000 copies, dwarfing Buckley's version

Table 13.2 The 'Battle' for 2009 Christmas Number 1. UK Singles Chart Top 10, week of 20 December 2009 to 26 December 2009

Chart position	Artist	Song	Link to *X Factor*
1	Rage Against the Machine	Killing in the name	Facebook campaign as the 'Anti-*X Factor* song' for Christmas Number 1
2	Joe McElderry	The climb	Winner 2009 *X Factor*, performed this song on *X Factor*
3	Lady Gaga	Bad romance	Guest performance of this song 2009 *X Factor*
4	Peter Kay's Animated All-Stars	The official BBC Children in Need medley	–
5	3OH3 Featuring Katy Perry	Starstrukk	–
6	Robbie Williams	You know me	Guest performance of this song 2009 *X Factor*
7	Cheryl Cole	3 Words	Guest performance of this song 2009 *X Factor*
8	Rihanna	Russian Roulette	–
9	Journey	Don't stop believin'	Song performed on 2009 *X Factor* by Joe McElderry
10	Black Eyed Peas	Meet me halfway	Guest performance of this song 2009 *X Factor*

Source: The Official Charts Company (www.theofficalcharts.com)

with sales of only 81,000 (BBC News, 22 December 2008). In 2009, combined sales neared one million units, as RATM edged McElderry by only 50,000, reaching 500,000 downloads, and also making history as the first Number 1 available only as digital download (NME, 20 December 2009).

The battlefield for the UK Christmas Number 1 chart song is perhaps odd musical terrain. It is marked as much by tradition (e.g. traditional seasonal tunes by Cliff Richard or The Pogues) as by links to children's television characters (such as the cartoon theme 'Can we fix it?' by Bob the Builder, Christmas 2000) as by quirkiness ('Somethin' stupid' sung by Robbie Williams and actress Nicole Kidman, 2001). Such terrain begs questions of what is really at stake in the battle for Number 1, given that the UK charts are more than a measure of the record industry's ability to promote, influence and capitalise on consumers' tastes. The social fabric of the 'semiotic security blanket' (Fairchild, 2007) and associated media rituals (Couldry, 2003) noted earlier are surely the stuff of making seasonal tradition and familiar songs into Christmas chart hits. Some of the controversy regarding the 2009 battle for the Christmas Number 1 song centred upon not only the unseasonal lyrics of 'Killing in the name' and the song's overt politics, but also the overall anti-consumerism stance of the Facebook campaign that launched the battle.[9] Songs such as 'Killing in the name' thus appear to tear at the fabric of deeply-rooted and ritualistically symbolic leisure practices. Massive consumption, arguably, is largely what the Christmas season, pop music and leisure are all about (Rosen, 2002).

In terms of leisure, music consumption and social resistance, the 2009 battle for Christmas Number 1 held an ironic twist. Despite RATM's politically charged music, they are nevertheless hugely successful American musicians signed to a major music corporation, Interscope, which is a subsidiary of Sony BMG. Sony BMG is the same record label that owns the rights to 'The Climb'. Fans may thus purchase and attach whatever set of symbolic meanings to these songs they wish – resistance or reproduction. In either case, profits go to Sony BMG, and the cultural victory of the 'Rage against X Factor' campaign echoes fairly hollow. The Facebook campaign implored fans to purchase 'Killing in the name' *via* official channels that would allow the song to chart successfully, such as the iTunes online music store. Yet as Burkart (2010) noted, outlets such as iTunes retain 70% of the price of song downloads. Again I ask how much mileage, beyond mere symbolism, such campaigns offer as resistance or opportunities for radical change.

Fans, as cultural studies theorists such as Fiske (1989) and Grossberg (1992) remind us, are not merely passive dupes who do what they are told (as the RATM lyrics also remind us), but are actively involved in the production of popular culture. Others, however, have argued that the power of active consumers to generate their own cultural meanings is inevitably quite limited. Ang (1990: 247) noted that 'it would be utterly out of perspective to cheerfully equate "active" with "powerful", in the sense of "taking control" at an enduring, structural, or institutional level' because of the power imbalances between groups such as cultural producers and fans. While the Facebook campaign did keep *X Factor* out of the top spot in the charts, it only did so for one week (McElderry's 'The Climb' then became Number 1), and the entire battle occurred on the field that is representative of the industry: the charts. Therefore, the impact of the active consumption of the *X Factor* is limited, and 'the extent to which "people make the popular" has yet to be resolved' (Harrington and Bielby, 2001: 9).

The example of the Facebook campaign to 'Rage Against the *X Factor*' opens up questions of leisure consumption and audience interactivity as inherently democratic. Fairchild (2008) referred to cultural democracy and cultural resistance as the 'strawmen' of consumerism. In terms of cultural democracy in *X Factor*, viewers appear to have choices and influence, through opting to pay to vote (via a phone-in system, or sending a text message) for contestants. Earlier in the 2009 *X Factor* season, such democracy (assuming that one would wish to reduce the concept to the mere ability to vote for a fee) had included voicing resistance to the seriousness of the show by voting, for example, to keep the oddball 'anti-contestant'[10] twins John and Edward Grimes in the show. For several weeks of the programme 'Jedward' (as they came to be called in the tabloid press) were voted to stay in the show, supposedly to Cowell's chagrin. However, such issues elide the broader context of the show and the point of the contest overall. Whether viewers voted for the Grimes or for the show's eventual winner, McElderry, is almost entirely beside the point: any vote at all – even a seemingly resistant vote for 'Jedward' – was in general a vote for the show, and above all else, a vote for the system of producing, promoting, and selling music/musicians that it represents. Leisure in such cases is reduced to the freedom to pay for a choice between commodified products (Hemingway, 1999). In sum, as Habermas (1991: 171) counselled, 'the world fashioned by the mass media is a public sphere in appearance only'.

The next battle is up in the air: The 'heavenly jukebox' versus *In Rainbows*?

In a mass-mediated cultural environment, it is little wonder that many people viewed Cowell's imperious influence on the UK charts through the *X Factor* as inspired devilry. Some have turned to illegal music downloads or broader social campaigns to rage against 'the machine' of corporate power (Klein, 2000). However, as the example of 'Rage Against the *X Factor*' illustrated, the depth and efficacy of such resistance is often limited and in this case, highly paradoxical. If the campaign to make RATM the Christmas Number 1 – to 'get involved and buy it' as the Facebook group exhorted its 'friends' – was not radically resistant to the pop hegemony of *X Factor*, then what alternatives for opposition are there?

In 2007, Radiohead, a successful art-rock band, recently freed from their major label contract, took what many believed to be a bold anti-industry step and released their seventh full-length album *In Rainbows* through the band's website, on a 'pay what you want' basis – free if one wished to pay nothing.[11] Yet, despite the option to download the album for free, during the two-month period the album was available on the band's website, 2.3 million people chose instead to obtain the album *via* illegal file-sharing networks (Theissen, 2008). This continued pirate activity raises questions of why fans illegally download the album when it is freely available on the group's official website. Theissen (2008) put forward two hypotheses. First, people may have been frustrated simply with early glitches in Radiohead's website that prevented authorised download – simply put, the band's site was overwhelmed by demand and crashed. Second, and more interestingly, Thiessen offered a 'venue hypothesis' that argued illegal download-savvy computer users simply stuck to the download channels that they were most accustomed to, such as various peer-to-peer sites. In addition, these 'off limits' channels for downloads may have been more attractive (i.e. symbolically resistant) to today's 'youth market' (Theissen, 2008: para. 3). This suggests that illegal downloading is a deeply entrenched cultural phenomenon that not even the offer of freely available materials can radically transform. Although *In Rainbows* was available free, I am reminded by Mary Douglas (in her Foreword to Mauss, 1990) that there are 'no free gifts... the whole idea of a free gift is based on a misunderstanding... A gift that does nothing to enhance solidarity is a contradiction' (Mauss, 1999: 145). The ritualistic, rebellious, and communitarian appeal of free music (i.e. piracy) does not bode well for the divinely-named 'celestial jukebox' (Burkart and McCourt, 2006), an emerging license-model (for fee) service for access to any music, anytime, anywhere. According to Burkart (2010) the early days of P2P music piracy served only as a '*cause célèbre*' that helped ignite resistance to the music industry, and larger battles over the emerging celestial jukebox (also called the 'heavenly' jukebox, Mann, 2000) are yet to follow. It would appear that further opportunities to 'rage against the machine' lie imminently ahead.

Conclusions

The phenomena discussed in this chapter represent the major changes that have occurred in popular music consumption in the last decade, in-between the devil (Cowell and the *X Factor*) and the deep blue sea (music piracy). This chapter raises questions about popular music, leisure, culture and consumption, legitimacy and illegality. Through examples primarily based around a reality TV singing contest and Internet use (e.g. Napster, Facebook), I have examined some of the relations between digital music piracy, *X Factor* and social networking campaigns to 'Rage Against the *X Factor*'. Seemingly mundane, everyday leisure practices (such as downloading music files or watching an interactive television programme) are at the heart of debates surrounding these phenomena. As such, understanding ordinary leisure is critically important to understanding the unequal power effects that these practices produce.

As Holmes (2004: 149) has said, it is precisely because an 'emphasis on the ordinary' (i.e. 'ordinary people' as contestants and as voting audiences) that programmes like the *X Factor* obscure the ways that such 'ordinariness' sustains existing social institutions and relations. Perhaps because of this ubiquity and familiarity, the *X Factor* has seeped deeply into contemporary popular culture. Although writing more generally than in terms of shows like *X Factor*, During's (2005: 193) comments on popular culture are *apropos*:

> It's the mirror in which the culture recognizes itself. It peoples the world: for many, celebrities and fictional characters are like distant acquaintances. It draws national – and international – communities together dotting conversations and private and communal memories. For some elements of popular culture become an obsession or help form an identity. What would society be without television, sport and pop music? Different hardly catches it.

While it may be difficult to imagine a world without *X Factor* and its sibling contests, as Grossberg (1998: 390) suggested, many people do imagine a different and 'better future' through leisure and popular cultural activities. For some, this progressive outlook holds the promise of resisting the perceived tyranny of the *X Factor* through grassroots campaigns on social networking sites. For others, resistance may take the form of digital music piracy. In both instances, such opposition has produced some limited changes so far, and, to some extent, this resistance has appeared to support the very same systems that it was organised to act against. While unresolved and seemingly caught between the devil and the blue sea, these issues of popular music are nevertheless important to understanding the economic, social, and political forces at work in the wider 'leisure project' of the twenty-first century. There are many ways to study the relations between popular music and leisure, and I have covered only a little of this contested terrain. Whether 'killing in the name of' social resistance or joining 'The climb' of *X Factor* ascendancy, whether on the airwaves, *via* pirate-infested online channels, or through more 'celestial' contexts, phenomena such as the *X Factor* and the annual battle for Christmas Number 1 will likely continue to rouse debate and stir campaigns that, at heart, are issues of leisure.

Acknowledgments

The author would like to thank Beccy Watson and students in the 'Mass Media' module at Leeds Metropolitan University who listened patiently and offered thoughtful comments to an earlier version of this chapter. Thanks also to Sara Cohen, at the University of Liverpool's Institute of Popular Music for her critical support and suggestions for improvements.

Notes

1. With apologies to Ella Fitzgerald (1993) and Marcus Rediker (1987).
2. Phone-in voting became controversial after scandals involving other ITV and BBC reality TV and audience participation programmes emerged in 2009. Incidents included fee overcharging and charging fees after the voting had closed. Programmes including BBC's *Strictly Come Dancing*, and *Blue Peter* were involved, as well as *Ant and Dec's Saturday Night Takeaway* on ITV. In the latter instance, as many as ten million calls were overcharged, resulting in repayments by the network of over £7.8 million.
3. Dann (2004: 20) reported: 'Rupert Murdoch...has enjoyed his ride with *American Idol* so much that he is using its format for *American Candidate*. Airing this summer, the show is the ultimate blending of politics and media. One hundred political hopefuls will compete in a game/talent show, broadcast by *Fox TV*. Using elimination rounds and viewer vote-ins to determine the winner, the final show will be broadcast live from the Mall in Washington D.C.'
4. While some contestants clearly come onto the programme with little performance experience, I qualify this statement about 'ordinariness' because other contestants, including winners Joe McElderry and Leona Lewis have come to the show via 'Fame'-style performing arts schools, such as the Brit School.
5. There also are gendered and racialised overtones here. Leona Lewis' mother, for example, claimed that Black women couldn't win the X Factor because of the majority of White male viewers who voted along racial and gendered lines (McGarry, 2006; see also Lee, 2006).
6. McDonaldisation refers to 'the process by which the principles of the fast-food restaurant are coming to dominate more and more sectors of American society as well as of the rest of the world' (Ritzer, 1993: 1).
7. Parallel to McDonaldisation, Disneyisation refers to 'the process by which the principles of the Disney theme parks are coming to dominate more and more sectors of American society as well as the rest of the world' (Bryman, 2004: 10).
8. There is a precedent here as well, for example, as *Popstars: The Rivals* in 2003 set up two vocal groups to battle for the Christmas Number 1.
9. Incidentally, the song 'Don't stop believin' by Journey also was supported by a less successful Facebook campaign to make it the Christmas Number 1, in response to the Facebook campaign to take RATM to the top spot.
10. Again, there is much precedent here, such as efforts to irk Cowell by keeping the oddball duo 'Same Difference' in the 2007 season, or the eccentric singer 'Chico' in the 2005 season.
11. The band later released the album in a boxed CD format.

References

Adorno, T. W. (1990[1941]) 'On popular music', in S. Frith and A. Goodwin (eds) *On Record: Rock, Pop, and the Written Word*, pp.22–38 (London: Routledge).

Adorno, T. W. and M. Horkheimer, (2000[1944]) 'The culture industry: Enlightenment as mass deception', in J. Schor and D. Holt (eds) *The Consumer Society Reader*, pp.3–19 (New York: The New Press).

Allen, K. (2010, 21 January) 'Piracy continues to cripple music industry as sales fall 10%'. *The Guardian* [Online] http://www.guardian.co.uk/business/2010/jan/21/music-industry-piracy-hits-sales [Accessed 21 January 2010].
Ang, I. (1990) 'Culture and communication: Toward an ethnographic critique of media consumption in the transnational media system', *European Journal of Communication*, 5(2–3): 239–60.
Arai, S. and A. Pedlar (2003) 'Moving beyond individualism in leisure theory: A critical analysis of concepts of community and social engagement', *Leisure Studies*, 22: 185–202.
BBC News (2008, 22 December) 'Christmas double for Hallelujah', *BBC News* [Online]. http://news.bbc.co.uk/1/hi/7794709.stm [Accessed 1 February 2010].
Bennett, A. (2005) 'Editorial: Popular music and leisure', *Leisure Studies*, 24(4): 333–42.
Bramham, P. (2006) 'Hard and disappearing work: Making sense of the leisure project', *Leisure Studies*, 25(4), October, pp.379–90.
Bryman, A. E. (2004). *The Disneyization of Society* (London: Sage).
Bull, M. (2005) 'No dead air! The iPod and the culture of mobile listening'. *Leisure Studies*, 24(4): 343–55.
Burkart, P. (2010) *Music and Cyberliberties* (Middletown CT: Wesleyan University Press).
Burkart, P. and T. McCourt, (2006) *Digital Music Wars: Ownership and Control of the Celestial Jukebox* (Lanham MD: Rowman & Littlefield).
Coalter, F. (1999) 'Leisure sciences and leisure studies: The challenge of meaning', in E. Jackson and T. Burton (eds) *Leisure Studies: Prospects for the 21st Century*, pp.507–22 (State College, PA: Venture).
Cooper, J. and D. M. Harrison (2001) 'The social organization of audio piracy on the Internet', *Media, Culture & Society*, 23(1): 71–89.
Couldry, N. (2003) *Media Rituals: A Critical Approach* (London: Routledge).
Cowell, S. (2003) 'All together now! Publics and participation in *American Idol*', *Invisible Culture: An Electronic Journal for Visual Culture*, 6 [Online]. http://www. rochester.edu: 8011/in_visible_culture/Issue_6/cowell/cowell.html [Accessed 28 December 2009].
Dann, G. (2004) 'American Idol: From the selling of a dream to the selling of a nation', *Mediations*, 1(1): 15–21.
DeNora, T. (2000) *Music in Everyday Life* (Cambridge: Cambridge University Press).
During, S. (2005) *Cultural Studies: A Critical Introduction* (London: Routledge).
Fairchild, C. (2007) 'Building the authentic celebrity: The "Idol" phenomenon in the attention economy', *Popular Music and Society*, 30(3): 355–75.
Fairchild, C. (2008) *Pop Idols and Pirates: Mechanisms of Consumption and the Global Circulation of Popular Music* (Aldershot: Ashgate).
Fiske, J. (1989) *Understanding Popular Culture* (London: Routledge).
Fitzgerald, E. (1993) *Between the Devil and the Deep Blue Sea*, The best of the song book (The Verve Music Group: UMG Recordings, Inc.).
Frith, S. (2002) 'Fragments of a sociology of rock criticism', in S. Jones (ed.), *Pop Music and the Press*, pp.235–46. (Philadelphia, PA: Temple University Press).
Frost, V. (2010, 10 January) 'Jedward steal the show as X Factor triumphs at National Television Awards', *The Guardian* [Online]. http://www.guardian.co.uk/tv-and-radio/2010/jan/20/national-television-awards-jedward [Accessed 15 February 2010].
Gibson, O. (2006, 29 December) 'Oldies but goldies benefit in digital revamp of charts', *The Guardian* [Online] http://www.guardian.co.uk/media/2006/dec/29/radio.music-news [Accessed 20 February 2010].

Glover, T. D., K. J. Shinew and D. C. Parry (2005) 'Association, sociability, and civic culture: The democratic effect of community gardening', *Leisure Sciences,* 27(1): 75–92.

Goodman, N. (2009) 'Are you voting?, BCS: The Chartered Institute for IT' [Online]. http://www.bcs.org/server.php?show=ConWebDoc.25741&changeNav=5664 [Accessed 1 February 2010].

Grossberg, L. (1998) 'The cultural studies' crossroads blues', *European Journal of Cultural Studies,* 1(1): 65–82.

Grossberg, L. (1992) *We Gotta Get Out of this Place: Popular Conservatism and Post-modern Culture* (London: Routledge).

Habermas, J. (1991) *The Structural Transformation of the Public Sphere: An Inquiry into a Category of Bourgeois Society* (Cambridge, Massachusetts: MIT Press).

Harrington, C. L. and D. D. Bielby (2001) 'Constructing the popular: Cultural production and consumption', in C. L. Harrington and D. D. Bielby (eds) *Popular Culture: Production and Consumption,* pp.1–16 (Oxford: Blackwell).

Harvey, D. (2003) *The New Imperialism* (Oxford: Oxford University Press).

Hemingway, J. (1996) 'Emancipating leisure: The recovery of freedom in leisure', *Journal of Leisure Research,* 28(1): 27–43.

Hemingway, J. L. (1999) 'Critique and emancipation: Toward a critical theory of leisure', in E. L. Jackson and T. L. Burton (eds) *Leisure Studies: Prospects for the Twenty-First Century,* pp.487–506 (State College, PA.: Venture Publishing Inc.).

Holmes, S. (2004) 'Reality goes Pop!': Reality TV, popular music, and narratives of stardom in *Pop Idol', Television New Media,* 5(2): 147–72.

Klein, N. (2000) *No Logo: Taking Aim at the Brand Bullies* (New York: Picador).

Lashua, B. D. (2006) '"Just another native"? Soundscapes, chorasters, and borderlands in Edmonton, Alberta, Canada', *Cultural Studies ↔ Critical Methodologies,* 6(3): 391–410.

Lee, J. (2006) 'American Idol: Evidence of same-race preferences?', *Institute for the Study of Labor,* discussion paper No. 1974, February. [Online]. http://www.iza.org/ [Accessed 1 August 2009].

Mason, M. (2008) *The Pirate's Dilemma: How Hackers, Punk Capitalists, Graffiti Millionaires and Other Youth Movements are Remixing Our Culture and Changing Our World.* (London: Penguin Books Ltd).

Mann, C. (2000) 'The heavenly jukebox', *Atlantic Monthly,* 286(3), September, pp.39–59.

Mauss, M. (1990) *The Gift: The Form and Reason for Exchange in Archaic Societies.* (London: Routledge Classics).

McGarry, L. (2006) 'Is X Factor racist?' [Online]. http://www.unrealitytv.co.uk/X Factor/is-X Factor-racist/ [Accessed 20 October 2009].

Negus, K. (2002) 'The work of cultural intermediaries and the enduring difference between production and consumption', *Cultural Studies,* 16(4): 501–15.

NME (20 December 2009) 'Rage Against the Machine beat "X Factor"'s Joe McElderry to Christmas Number One', *NME* [Online]. http://www.nme.com/news/rage-against-the-machine/48971 [Accessed 1 February 2010].

Opitz, A. (2010) Editorial notes: 'Pirates and piracy – Material realities and cultural myths', *Darkmatter Journal,* 5: 1–2. [Online]. http://www.darkmatter101.org/site/category/journal/issues/5-pirates-and-piracy/ [Accessed 23 February 2010].

Ritzer, G. (1993) *The McDonaldization of Society* (London: Pine Forge Press).

Rediker, M. (1987) *Between the Devil and the Deep Blue Sea: Merchant Seamen, Pirates, and the Anglo-American Maritime World, 1700–1750* (Cambridge: Cambridge University Press).

Reijnders, S., G. Rooijakkers and L. van Zoonen (2006) 'Idols: An eruption of festive pleasures', *European Journal of Cultural Studies*, 9(2): 131–48.
Rosen, J. (2002) *White Christmas: The Story of an American Song* (New York: Scribner).
Rutter, J., and J. Bryce (2008) 'The consumption of counterfeit goods: "Here be pirates?"', *Sociology*, 42(6): 1146–64.
Rojek, C. (2005) 'P2P leisure exchange: Net banditry and the policing of intellectual property', *Leisure Studies*, 24(4), October, pp.357–69.
Sennett, R. (2003) *Respect: The Formation of Character in an Age of Inequality* (London: Allen Lane).
Stewart, B., D. C. Parry and T. Glover (2008) 'Writing leisure: Values and ideologies of research', *Journal of Leisure Research*, 40(3): 360–84.
Storey, J. (1997) *An Introduction to Cultural Theory and Popular Culture* (London: Prentice Hall/Harvester Wheatsheaf).
Storrmann, W. (1993) 'The recreation profession, capital, and democracy', *Leisure Sciences*, 15: 49–66.
Sweney, M. (2006, 26 May) 'American Idol outvotes the President', *The Guardian* [Online]. http://www.guardian.co.uk/media/2006/may/26/realitytv.usnews [Accessed 21 September 2009]
Sylvester, C. (1995) 'Relevance and rationality in leisure studies: A plea for good reason', *Leisure Sciences*, 77(2): 125–31.
Thiessen, B. (2008) 'Radiohead's *In Rainbows* illegally downloaded by 2.3 million', Exclaim.ca [Online] http://exclaim.ca/articles/generalarticlesynopsfullart.aspx?csid1=0&csid2=844&fid1=32800 [Accessed 20 October 2009].
Toynbee, J. (2000) *Making Popular Music: Musicians, Creativity and Institutions* (London: Routledge).
Ytreberg, E. (2009) 'Extended liveness and eventfulness in multi-platform reality formats', *New Media Society*, 11: 467–85.

14

Doublethink:[1] 'Deregulation', Censure and 'Adult-Sex' on Television

Julian Petley

> *The power of holding two contradictory beliefs in one's mind simultaneously, and accepting both of them.* George Orwell, 1984

***Chapter 14** is a polemical account of the **politics of leisure and pleasure** that surround **censorship** in Britain at the turn of the twenty first century. It argues that these politics are wholly contradictory and, in doing so, it points up once again the central purpose of this book – to explore the **tension** between **freedom and constraint in contemporary leisure worlds**. The deregulation of the mass media in various societies and by various governments – the US presidency of Ronald Reagan (1981–1989) or the British Conservative administrations of Margaret Thatcher (1979–1990) – exalted personal freedom and responsibility. One might therefore have expected that, in contemporary Britain, the public might be able to watch what they pleased. However, as this chapter shows, the **rhetoric of emancipation** has gone hand in hand with continued **restriction and censorship**.*

'Deregulation'

In December 2000, the British government published its White Paper, *A New Future for Communications*, which paved the way for what would eventually become the Communications Act 2003. Very much in the same spirit as the previous Conservative government's Broadcasting Act 1990, 'New' Labour proudly proclaimed itself an apostle of 'deregulation', arguing that 'it is important that communications can flourish with the minimum of regulatory intervention' (DTI/DCMS, 2000: 59) and making it clear that the new communications regulator Ofcom 'would have a duty to keep markets or sectors under review and roll back regulation promptly when regulation becomes unnecessary' (ibid.: 13). The same free market philosophy was also very much apparent in the document which followed in May 2002, *The Draft Communications Bill – The Policy*, which trumpeted that 'unnecessary regulations need to be removed wherever possible' (DTI/DCMS, 2002: 3), with the result that 'for viewers and listeners there will be wider access, better quality services, more choice and improved value for money' (ibid.: 5). Ofcom, it stated, would be 'required to ensure that regulation is kept to the minimum necessary' and would be 'subject to a duty to secure light touch regulation, requiring it to carry out regular reviews of its functions to identify any areas where regulation is no longer necessary or appropriate' (ibid.: 27).

Regulation

However, whilst keen to liberalise certain aspects of broadcasting, such as the regulations governing media ownership and the obligations placed on the commercially funded public-service broadcasters, 'New' Labour was no more prepared than its Tory predecessors to lift the restrictions on certain kinds of broadcast content. Thus *A New Future for Communications* made it clear that 'we seek to combine a lighter touch in many aspects with tough protection of the genuine public interest in others' (DTI/DCMS, 2000: 4) and that 'the freedoms which are at the heart of our arrangements for communications bring with them responsibilities and we want to ensure that the growth of multi-channel, multi-media services serves society and the interests of citizens and does not harm them' (ibid.: 59). The newly-created Ofcom was given the task of protecting 'the interests of citizens by maintaining accepted community standards in content, balancing freedom of speech against the need to protect against potentially offensive or harmful material' (ibid.: 79). In a neat ideological sleight of hand, the maintenance of broadcasting censorship became thus equated with serving the public or citizen interest, and the notion of 'accepted community standards', which at this time appeared nowhere in statute law and carries distinct overtones of US pro-censorship discourse, made an early official appearance as unwelcome as it was unnoticed. We were, however, assured that the new framework would be 'flexible enough to recognise the differences between different services and to respond to rapid changes in technologies, services or public expectations' (ibid.: 62). This, as we shall see, would turn out to be very far from the truth.

Harm and offence

Clause 3 (2) (e) of the Communications Act 2003 requires Ofcom to apply to all broadcast services 'standards that provide adequate protection to members of the public from the inclusion of offensive and harmful material'. This too represents a significant change in terminology, as previous broadcasting legislation had referred not to harm and offence but to taste and decency. This was welcomed by Richard Hooper, then chair of the Ofcom Content Board, who stated that the new wording 'supports a move away from the more subjective approach of the past, based on an assessment of taste and decency in television and radio programmes, to a more objective analysis of the extent of harm to audiences' (quoted in Millwood Hargrave and Livingstone, 2009: 27). Section 3 (4) (h) of the Act lays down that in performing its duties, Ofcom must also have regard to 'the vulnerability of children and of others whose circumstances appear to Ofcom to put them in need of special protection'. Ofcom's 'standards objectives' require 'that persons under the age of eighteen are protected' (Section 319 (2) (a)) and 'that generally accepted standards are applied to the contents of television and radio services so as to provide adequate protection for members of the public from the inclusion in such services of offensive and harmful material' (Section 319 (2) (f)). Finally, in drawing up its standards, Ofcom is instructed to have particular regard, *inter alia*, to 'the degree of harm or offence likely to be caused by the inclusion of any particular sort of material in programmes generally, or in programmes of a particular description' (Section 319 (4) (a)).

Pace Hooper, however, the notion of harm, in this context, is no more objective than are the notions of taste and decency. In 1979, the Williams Committee argued that 'there is often a real difficulty in identifying what the harmful effect of the material is supposed to be' and 'the causal concept of obscenity, in terms of doing harm, has in legal practice proved very resistant to being given the precise application, and submitting to the canons of proof, required in general by the law' (1979: 59, 60). These words are as true now as when they were first written, and the subsequent enshrining of such a dubious notion in laws such as the Communications Act 2003 and the Video Recordings Act 1984 does not make it any more intellectually coherent or philosophically cogent, it merely gives it legal force and, in so doing, paves the way for acts of official censure and censorship on highly questionable grounds, as this chapter will illustrate.

The Broadcasting Code

Ofcom was also required by the Communications Act to draw up a Broadcasting Code covering all aspects of programme content. This is periodically revised, and the most recent edition was published in December 2009. This clarified and tightened the rules governing the broadcasting of material of a sexual nature, and it is because these rules so clearly illustrate Ofcom's ongoing role as a censor – its deregulatory rhetoric notwithstanding – that I wish to concentrate on them in this chapter.

Section One of the Code is headed 'Protecting the Under-Eighteens', and lays down a number of basic rules (Ofcom, 2009a: 7–8). These include a ban on the broadcasting of 'material that might seriously impair the physical, mental or moral development of people under eighteen', and the requirements that 'broadcasters must take all reasonable steps to protect people under eighteen' and that 'the transition to more adult material must not be unduly abrupt at the watershed (in the case of television) or after the time when children are particularly likely to be listening (in the case of radio). For television, the strongest material should appear later in the schedule'. The Code also reminds television broadcasters that these rules are in addition to their obligations resulting from the Audiovisual Media Services Directive of the European Union, Article 22 of which states that: 'Member States shall take appropriate measures to ensure that television broadcasts by broadcasters under their jurisdiction do not include any programmes which might seriously impair the physical, mental or moral development of minors, in particular programmes that involve pornography or gratuitous violence' (quoted in ibid.: 70).

This section of the Code also deals specifically with sexual material (ibid.: 11–12), making it clear that 'material equivalent to the British Board of Film Classification (BBFC) R18-rating must not be broadcast at any time'. Under the previous version of the Code, Rule 1.25 laid down that 'BBFC R18-rated films or their equivalent must not be broadcast'. Ofcom allows the broadcasting of what it calls 'adult-sex material', but effectively this is simply bowdlerised hard-core – what is known as 'vanilla porn'. Ofcom defines 'adult-sex material' as containing 'images and/or language of a strong, sexual nature which is broadcast for the primary purpose of sexual arousal or stimulation', but even this feeble concoction, Ofcom dictates:

> must not be broadcast at any time other than between 22:00 and 05:30 on premium subscription services and pay per view/night services which operate with mandatory restricted access. In addition, measures must be in place to ensure that the subscriber is an adult.

'Mandatory restricted access' means simply that there is a PIN-protected system in operation which cannot be removed by the user and which restricts access solely to those authorised to view. Broadcasters are also required to ensure that material broadcast after the watershed which contains images and/or language of a strong or explicit sexual nature, but is not 'adult-sex material', as defined above, is justified by its context. What is meant here by context is explained below.

Section Two of the Code is entitled 'Harm and Offence' (ibid.: 15–19), and is aimed at protecting adults as well as those aged under 18. It thus has to be read in conjunction with Section One, but it also directs readers to Article 10 of the European Convention on Human Rights, Article 10 (1) of which states that:

> everyone has the right to freedom of expression. This right shall include freedom to hold opinions and to receive and impart information and ideas without interference by public authority and regardless of frontiers.

However, it is all too often forgotten that Article 10 (2) seriously qualifies this ringing declaration of principle and contains numerous useful get-out clauses for those wishing to indulge in censorship. It states that:

> the exercise of these freedoms, since it carries with it duties and responsibilities, may be subject to such formalities, conditions, restrictions or penalties as are prescribed by law and are necessary in a democratic society, in the interests of national security, territorial integrity or public safety, for the prevention of disorder or crime, for the protection of health or morals, for the protection of the reputation or rights of others, for preventing the disclosure of information received in confidence, or for maintaining the authority and impartiality of the judiciary (quoted in ibid.: 72–3).

The principle governing this section of the Code is 'to ensure that generally accepted standards are applied to the content of television and radio services so as to provide adequate protection for members of the public from the inclusion in such services of harmful and/or offensive material'.

In applying these 'generally accepted standards' broadcasters are required to ensure that:

> material which may cause offence is justified by the context...Such material may include, but is not limited to, offensive language, violence, sex, sexual violence, humiliation, distress, violation of human dignity, discriminatory treatment or language (for example on the grounds of age, disability, gender, race, religion, beliefs and sexual orientation).

The Code explains that:

> context includes (but is not limited to):
> - the editorial content of the programme, programmes or series;
> - the service on which the material is broadcast;
> - the time of broadcast;
> - what other programmes are scheduled before and after the programme or programmes concerned;
> - the degree of harm or offence likely to be caused by the inclusion of any particular sort of material in programmes generally or programmes of a particular description;
> - the likely size and composition of the potential audience and likely expectation of the audience;
> - the extent to which the nature of the content can be brought to the attention of the potential audience for example by giving information; and

- the effect of the material on viewers or listeners who may come across it unawares.

As noted earlier, the version of the Code published in December 2009 tightened the rules governing the broadcasting of material of a sexual nature. Ofcom had explained the reasons for its actions in this respect in its *Broadcasting Code Review*, which was published in June 2009. According to Ofcom, the revisions were necessitated by a

> significant number of compliance failures in this area. During 2007 and 2008 there were six sanction decisions against licensees and 22 published findings regarding the broadcast of material of a strong sexual nature, including findings in relation to 'adult-sex' material which was transmitted without a mandatory access restriction. Ofcom has found that recent material which has been transmitted without any access restrictions has featured nudity of a strong sexual nature, and sustained sex scenes and sexual language that has not, in some cases, been justified by the context in which the material was transmitted (2009b: 22).

Ofcom went on to explain that it had previously investigated and adjudicated on much of this material under Section Two of the Code then in force, most notably Rule 2.1 which stated that 'generally accepted standards must be applied to the contents of television and radio services so as to provide adequate protection for members of the public from the inclusion in such services of harmful and/or offensive material' and Rule 2.3 which laid down that 'in applying generally accepted standards broadcasters must ensure that material which may cause offence is justified by the context'. Ofcom explained that the latter rule had been applied to material of a sexual nature that was considered to be strong but not as being broadcast for the primary purpose of sexual arousal. However, Ofcom had now decided that this rule did not provide sufficient information to broadcasters regarding the requirements to protect those under 18, namely that material of a strong sexual nature requires strong contextual justification.

Ofcom also explained that it had hitherto drawn on Rule 1.24 of the then current Code which stated that premium subscription services and pay per view/night services may broadcast 'adult-sex' material between 22:00 and 05:30 provided that mechanisms restricting access to adults are in place. However Ofcom admitted that this rule did not clearly state that this material must not be broadcast *outside* these specific services and restrictions.

It is for these reasons, then, that the revised Code defines more specifically what Ofcom means by 'adult-sex' material and, in particular, distinguishes this more clearly from what it calls 'strong' sexual material. Furthermore, the rules regarding the latter now make it clearer that the inclusion of such material on general entertainment channels (which is anyway permitted only after the 21:00 watershed) must be strongly justified by context. Factors which broadcasters now need to consider in determining whether such material is acceptable include:

- the explicitness of any sexual material and/or sexual language used;
- the purpose of any sexual material in a programme;
- whether any plot or narrative provides sufficient editorial context for its transmission; and/or
- whether there is any other strong editorial justification for its transmission (ibid.: 24).

The limits of the possible

At the time of writing, the revised Code has only just come into force, and Ofcom has thus not had the opportunity to censure any programmes for breaches of its provisions. However, analysis of a representative sample of Ofcom's judgements on a number of programmes which it censured during 2009, some of which gave rise to the Code revisions discussed above, gives a very clear indication of the limits of the possible on British television in the case of material of a sexual nature.

Let us start by looking at Ofcom's verdict on a programme on a general entertainment channel, Virgin 1: *Sin Cities*, broadcast at 22.00 on 8 November 2008. This was a documentary about two actresses in the US porn industry; it was preceded by a warning about 'strong language and sexual scenes' and was the subject of one complaint from a viewer. In its *Broadcast Bulletin*, 23 March 2009 (26–8), Ofcom argued that although the programme 'did have some editorial content and purpose', was shown after the watershed, was preceded by information about its sexual content and was to some extent scheduled around complementary programming, 'these factors taken together did not ensure that the potentially offensive material was justified by the context'. In particular, in Ofcom's view it did not:

> provide adequate editorial context for, or analysis of, what the broadcaster described as 'the moral dilemmas of being married to a porn star'. Instead, at times, the programme lacked editorial distance and a considerable amount of the content concentrated on the detail of the sexual acts the actresses were undertaking rather than a serious analysis of the subject matter. More importantly, some of the sexual content shown did not appear directly relevant to the subject matter of the programme – in particular, a scene where De'Bella removed an anal plug and placed it in her mouth in a sexual manner, and a sequence in which the narrator made reference to bleeding from an anal tear De'Bella had suffered, as a result of the prolonged anal sex

> she was engaged in with three pornography actors and the programme showed her wiping herself.

Ofcom concluded that it did not consider *Sin Cities* 'to be a work of sufficient seriousness or rigorous enquiry to attract special latitude in the strength of material it can properly contain'. This material included scenes of oral, anal and vaginal intercourse, whose frequency and explicitness had, in Ofcom's opinion, 'the potential to cause considerable offence, especially to viewers who might come across such content unawares'. The fact that these images were cropped and masked cut no ice with Ofcom, and indeed caused it to point out that 'licensees should consider carefully whether the need to obscure images of sexual activity or intrusive nudity is in fact an indication that the material as a whole is unsuitable for broadcast'. Virgin was found guilty of failing to apply 'generally accepted standards' and thus of breaching Rule 2.3 of the version of the Code in force at the time that the programme was broadcast.

Moving on to a more niche broadcaster, the music specialist Channel AKA, Ofcom censured its late night slot *XXX Channel AKA* for broadcasting 'Playtime Two', a music video by Giggs Featuring Kyze, at 22:45 approximately on 25 June 2009. Although the channel is available without any access restrictions, *XXX Channel AKA*, which is broadcast between 22:00 and 05:30, features music videos of a more adult nature. One viewer complained about the sexual nature of the images in 'Playtime Two'.

Ofcom argued in its *Broadcast Bulletin*, 14 September 2009 (22–3), that the nature and strength of the sexual imagery broadcast in this particular music video 'had the clear potential to cause offence. Therefore the broadcaster was required to ensure that the material was justified by the context in order to provide adequate protection for viewers and compliance with the Code'. In spite of the fact that 'there would have been a certain amount of audience expectation for the broadcast of more challenging material during this particular programme', Ofcom was particularly concerned by the 'frequent shots of naked breasts; women touching their breasts and genital area in a sexual manner; women licking whipped cream off each other's breasts; and a man simulating sexual stimulation on a woman'. In Ofcom's view, the strength of the material 'would have exceeded audience expectations for a music programme of this nature broadcast at 22.45 without any access restrictions on a music channel'. The regulator thus concluded that 'on balance, the broadcaster did not apply generally accepted standards to this content and the material was not justified by the context', and so had breached Rules 2.1 and 2.3 of the Code.

Advancing into the realm of the specialist adult broadcasters, let us first of all consider a number of Ofcom's judgements on adult sex chat channels. A repeat offender here, in Ofcom's view, was the programme *Bang Babes*, which was censured on three occasions by Ofcom in 2009 (*Broadcast Bulletin*, 6 July: 12–16; *Broadcast Bulletin*, 26 October: 47–9). One of these concerned an episode which was broadcast immediately after the 21.00 watershed, and thus complicated matters by raising issues concerning scheduling, but the other two were broadcast well after midnight, and thus Ofcom's comments on these episodes illustrate its view about the acceptability of the material *per se* on channels of this type.

Bang Babes is an adult sex chat service, available freely without access restrictions on the channels Tease Me and Tease Me 3, both of which are situated in the 'adult' section of the Sky electronic programme guide. The channels broadcast programmes after the 21:00 watershed which are based on interactive adult sex chat services: viewers are invited to contact via premium rate telephone services scantily dressed on-screen female presenters who talk and behave in an overtly sexual fashion. Regarding the programme broadcast on Tease Me 3 from 01:45 to 02:30 approximately on 20 June 2009, a viewer complained that the presenter had mimed vaginal and anal masturbation, and that overall the sexual content was in excess of the material generally available on a channel without restricted access at that time of night. In the case of a programme broadcast on Tease Me from 01:00 to 03:00 approximately on 20 June 2009, a viewer complained that the presenter spanked her buttocks and that close-up shots of her vaginal and anal areas were shown while she was wearing only a thong. The complainant argued that the sexual content included in this programme was excessive.

Ofcom concluded in its *Broadcast Bulletin*, 26 October 2009, that although neither programme consisted of 'adult-sex' material, which would have necessitated the channel utilising a mandatory PIN protection system, the behaviour of the presenters on both Tease Me and Tease Me 3 was 'highly sexualised and sexually provocative, and a number of the images were filmed in a prolonged and intrusive manner'. In both cases:

> given the strength of the material, Ofcom considered that this content clearly had the potential to cause offence. Therefore its treatment by the broadcaster required justification by the context to provide adequate protection for viewers. Ofcom took into account all the relevant contextual factors including, for example, the explicit sexual content, the nature of the channel, and the time of broadcast. In Ofcom's opinion, given the strength of the material shown, it would have exceeded the likely expectation of viewers watching a channel without access restrictions. Ofcom was also concerned by the degree of offence likely to be caused to viewers watching at this time and the significant effect this material would have had on those who may have come across it unaware. There was no sufficient editorial justification for the broadcast of these strong sexual images. Further, the nature and location of the channel in the 'adult' section of the EPG and the time of broadcast are not sufficient in Ofcom's view to

> justify broadcast of such content. The broadcast was therefore not justified by the context and breached Rules 2.1 and 2.3.

Indeed, so concerned was Ofcom about the 'serious' and 'repeated' breaches of its Code by adult sex channels that it inserted a special note into its 6 July 2009 *Bulletin* reminding such broadcasters that they

> must take all reasonable steps to protect people under eighteen and ensure that generally accepted standards are applied to their material... Material of a sexual nature broadcast after the 21:00 watershed must be appropriately limited and justified by the context to ensure compliance with generally accepted standards. For instance, broadcasters operating in the free-to-air 'adult' sex chat sector should take great care not to include physically invasive shots, in particular images of anal or genital areas for example, or of any real or simulated sex acts including masturbation or intercourse, or inappropriate shots of simulated oral sex.

A further problem encountered by Ofcom with two such channels, Lucky Star and Babeworld TV, was that at 21.00 on 21 May 2009 and 23.30 on 28 May 2009 respectively, they included in their programmes on-screen references to their websites. These, Ofcom discovered, contained sexually explicit material equivalent to that classified by the British Board of Film Classification at R18, which may only be sold in licensed sex shops to those over 18, and which, as already noted, is prohibited outright by Ofcom on television; furthermore, access to these sites required no mandatory age-verification checks. In its *Broadcast Bulletin*, 20 July 2009 (18–20), Ofcom thus reminded Lucky Star of a statement in its *Broadcasting Bulletin*, 21 July 2008, to the effect that 'while the content of websites is not in itself broadcast material, and therefore not subject to the requirements of the Code, any on-air references to websites are clearly broadcast content. Such references must therefore comply with the Code'. This judgement concerned website references in a trailer broadcast on 13 February 2008 on the encrypted adult service Red Hot TV, and made it abundantly clear that whilst Ofcom does not regulate broadcasters' websites, 'in no circumstances may such websites contain R18 material if they are promoted on a licensed service'. In Ofcom's view, the fact that the promotional references to the website URL were broadcast on a channel in the 'adult' section of the EPG did not entail that the broadcasting of these references was justified by the context in which they appeared. Lucky Star was thus judged to have breached generally accepted standards.

When approached by Ofcom, Babeworld had the temerity to query Ofcom's power to regulate on-air promotional references to websites.

Ofcom sharply reminded them in its *Broadcast Bulletin*, 9 November 2009 (8–11), that it possessed both the duty and the power to regulate such references under the Communications Act 2003, and quoted the passage from the legislative background to the Broadcasting Code, which states that:

> although a link included in the service may lead to features outside of that service which are not regulated by Ofcom, the provision of access to those features by, for instance, the inclusion of a link, is within the control of the broadcaster and so within Ofcom's remit. Ofcom may therefore require such a link or facility to be removed where Ofcom has concerns, in the light of its statutory duties and, in particular, the standards objectives set out in Section 319 of the Act [cited above], about the material to which it leads.

Rather testily, it also noted that it had 'made its position regarding this matter very clear to date', expressed its concern that Babeworld was unfamiliar with its earlier judgements on this subject, and drew the company's attention to their decision, published on 18 May 2009, to impose a fine of £25,000 on repeat-offender RHF Productions Limited for broadcasting on several of its Red Hot channels between 21 July and 28 August 2008 URLs which gave access to R18-type content. Indeed, Ofcom concluded its judgement by stating that:

> in view of the serious and repeated nature of these contraventions of the Code, Ofcom reviewed carefully whether they should be considered for referral to the Content Sanctions Committee. On balance Ofcom decided not to do so on this occasion. However, Ofcom will seriously consider further regulatory action should Babeworld breach the Code in the future.

This brings us on to the final part of this survey of Ofcom's censure of sexual material in 2009, namely sex films.

In its *Broadcast Bulletin*, 6 April 2009 (5), Ofcom announced that Playboy TV UK/Benelux Limited (another repeat offender) had been fined £22,500 for broadcasting sexually explicit content that was unsuitable on a free-to-air unencrypted channel, thereby failing to protect the under-18s and breaching generally accepted standards. The films in question were

Jenna's American Sex Star (26 September 2007, 23:35)
Adult Stars Close-up (27 September 2007, 00:35)
Blue Collar Babes (27 September 2007, 01:05)
Sexy Girls Next Door (27 September 2007, 02:00)
Sexy Urban Legends (29 November 2007, 23:00)

Sex House (30 November 2007, 00:35)
Sex Guides (9 December 2007, 03:30)

The Ofcom Content Sanctions Committee explained that the reasons why the above breached the Code were that:

> each of the programmes featured sexual material which – depending on the individual programme – was a combination of content so explicit, strong and/or sustained and strong sexual language that it was unacceptable when shown free-to-air and unencrypted. Its primary purpose was to arouse the audience sexually. None of this material was in the opinion of the Executive editorially justified. This content included for example depictions of a sustained and sexually explicit all-girl group sex 'gang-bang' featuring naked women performing sex acts, including oral sex, on each other; sexual intercourse (whether simulated or real); oral sex; masturbation, both with and without dildos; and full nudity, in some instances showing labial detail. Further, a number of these sequences were of considerable duration, and with little or no qualifying narrative to justify their inclusion on a free-to-air service. Some for example featured female porn stars stripping to camera and touching themselves explicitly in scenes which included full and frequent nudity including instances of clear 'open-legged' labial detail. The Executive was also concerned by the use of frequent explicit sexual language, for example (from *Jenna's American Sex Star*), 'I would stick a big fucking dildo in your fucking twat, I'd fucking lick it all up and I'd taste all your fucking pussy juices 'cos I'd make you come harder than you've ever come before'.

The Committee concluded that such material, shown unencrypted 'is totally unacceptable. It has the potential to cause offence to the audience and harm to under-eighteens, and children in particular, especially those who come across such material unawares'.

It also gave due warning that:

> 'adult' channels generally and 'adult chat' channels should be in no doubt of Ofcom's concerns about the broadcast of sexual material which is too explicit. Should further such cases be considered for sanction in future, the Committee will continue to regard them very seriously. If highly graphic sexual material is broadcast without editorial justification on a free-to-air channel even on a single occasion it can be a very serious breach of the Code. (http://www.Ofcom.org.uk/tv/obb/ocsc_adjud/playboytv.pdf)

The *Broadcast Bulletin*, 26 May 2009 (11–13), revealed that a further seven films shown on Playboy One had been found to breach the same rules of the Code as the above. These were:

Hollywood Sins (10 September 2007, 00:00)
Sex Games Cancun (30 November 2007, 00:05)

Sexy Urban Legends (2 January 2008, 23:05)
Confessions of a Porn Star (2 January 2008, 23:35)
Sex Games Vegas (3 January 2008, 00:05)
Sex Court (11 January 2008)
Girl for Girl (11 January 2008)

In Ofcom's view:

> the sexual nature of the material in each of the programmes conflicted with the standards viewers generally expect on channels that broadcast free-to-air without encryption. Further, the potential for the material to cause offence was not justified sufficiently.
> In reaching this decision Ofcom took into consideration: the positioning of the channel in the EPG; the nature of the channel, including the recognised character of the Playboy brand; the programme titles; and the time of broadcast.
> Whilst these factors would have signalled to any potential viewer that the content was likely to be 'adult' in nature, Ofcom considered the material (in particular the focus on, and the frequency, duration and explicitness of, the sexual scenes) was not consistent with viewer expectations of an unencrypted free-to-air channel. The material was therefore likely to cause offence, particularly to those who were able to come across it unaware because it was transmitted without encryption. In assessing the Programmes where sexual content featured as part of a drama or film, Ofcom concluded that whilst they contained clear storylines they appeared to be constructed primarily to facilitate sexual encounters which did not provide sufficient justification for the emphasis on the sex scenes, their frequency, duration and explicitness (and, in particular, the emphasis on female genitalia). In assessing the programmes that were in a documentary and/or reality TV style, Ofcom judged the explicitness of the sexual discussions and portrayal of sex acts went beyond what is acceptable for an unencrypted service. The strength of the material transmitted was contrary to general viewer expectations and was not justified by the context in which it was shown.

These programmes were thus found to be in breach of Rules 2.1 and 2.3, but not so seriously as to warrant any sanction.

However, as already noted, even on encrypted services Ofcom absolutely refuses to allow material of R18 strength to be shown. Thus when TVX2 (which is owned by the Portland Media Group, which is also the owner of RHF, the licensee for the above-mentioned Red Hot Channels) showed *Bathroom Bitches* at 21.53 on 4 September 2008, it was fined £27,500 for including scenes which included 'prolonged and explicit scenes of a woman masturbating, some of which were shown in close-up and depicted vaginal penetration using a dildo'. (http://www.Ofcom.org.uk/tv/obb/ocsc_adjud/rhfportland. pdf).

Ofcom was clearly irked by the fact that this particular breach occurred shortly after the publication, on 23 July 2008, of its Content Sanctions

Committee's decision to fine Portland £25,000 for showing R18-strength material on its encrypted TVX channel from 22.10–22.40, as well as an unencrypted trail for the programme, on 8 June 2007. As the Sanctions Committee put it in its judgement on the programme, the encrypted material featured

> two naked presenters/actresses engaging in very explicit sexual acts including frequent and prolonged masturbation, shown in close-up; explicit scenes of oral sex; and explicitly depicted scenes of vaginal penetration by fingers and dildos. There is an absolute prohibition on the transmission of such material under Rule 1.25 of the Code (http://www. Ofcom.org.uk/tv/obb/ocsc_adjud/portland.pdf).

Effects: 'A severe lack of research'

As the above examples demonstrate, one of the main reasons why Ofcom is so censorious regarding the broadcasting of sexual material is that it regards it as harmful to those under 18. Whether or not exposure to pornography harms the young is, like the question of harm-by-media in general, an extremely controversial issue, and beyond our scope here.[2] However, what is most certainly worth noting in this respect is that a review of the research on this topic commissioned by Ofcom noted quite correctly that, 'due to ethical restrictions, there is a severe lack of research regarding the effects of exposure of minors to R18 pornography which contributes to the evidence being inconclusive'. And whilst admitting that 'there is some evidence that indicates that sexual material influences the moral development of young people under the age of 18. In other words, that through exposure to pornography young people become more cynical towards traditional relationships such as marriage and become sexually active at a younger age', the review also concludes 'that there is no empirical research that proves beyond doubt that exposure to R18 material seriously impairs the mental or physical development of minors' (2005a: 4).

Double vision

Whether such extremely tentative conclusions justify Ofcom's strictures about certain kinds of broadcast sexual material is left to the reader to judge. What is absolutely certain, however, is that these strictures cannot be simply laid at the door of the door of Article 22 of the Audiovisual Media Services Directive, nor of Article 10 (2) of the European Convention on Human Rights, since the kind of material to which it takes such strenuous objection is broadcast in every other EU country except Ireland and Poland without any form of regulatory hullabaloo. Indeed, both Ofcom and the British government need forcefully to be challenged on why it is that they appear to believe that the British public is so vulnerable, feeble-minded or wicked that its members need a far greater degree of protection from certain kinds of sexual material than do almost all of their continental counterparts.

It should also be noted in this respect that Ofcom is highly selective in those aspects of the Directive which it chooses to enforce. For example, Article 4 lays down that:

> member States shall ensure where practicable and by appropriate means, that broadcasters reserve for European works, a majority proportion of their transmission time, excluding the time appointed to news, sports events, games, advertising, teletext services and teleshopping. This proportion, having regard to the broadcaster's informational, educational, cultural and entertainment responsibilities to its viewing public, should be achieved progressively, on the basis of suitable criteria (http://ec.europa.eu/ avpolicy/docs/reg/avmsd/ recitals_en.pdf).

Not only has Ofcom remained resolutely deaf to those who have repeatedly pointed out that it is both practicable and desirable to apply Article 4 to BSkyB, but, in the starkest possible example of 'regulatory capture', from the moment that proposals for the Directive were announced in 2005, it actually joined forces with those whom it is supposed to be regulating in aggressively lobbying against any extension of EU regulation to new modes of delivering content. Thus in a position paper on the Directive, published in April 2006, it argued that it was

> particularly concerned with the extension of scope of the Directive to cover all 'audiovisual media services' (AVMS). The claimed target of this revision is video on demand (VOD) services that are competing directly with broadcasting. Yet, the definition of AVMS advanced by the Commission is very broad, vague and ambiguous, going well beyond VOD. It potentially, and perhaps inadvertently, catches a significant number of new third generation mobile and web-based services, including videoblogs, online video games, webcams, online newspapers or magazines which carry significant amount of video content, and even individual websites that host user-generated content.

In Ofcom's view:

> these services are a major source of potential future EU creativity and competitiveness. They are also uniquely vulnerable at this stage of development to regulatory risk and uncertainty. The imposition of additional regulatory layers could lead to market distortions and to a net outflow of jobs and businesses in the ICT and media industries from the EU area.

It thus proposed that 'a revised Directive should be limited in scope to cover *only* those services which "look and feel" like television broadcasting (e.g. television delivered over IP networks – IPTV)'. One might also note how, exactly as in the passages quoted at the start of this chapter, 'deregulation' and its alleged benefits are conceived in purely economic terms.[3]

The viewers' views

Returning to the subject of sexual material on television, the examples quoted from Ofcom's *Broadcast Bulletins* demonstrate that the regulator also justifies prohibitions by arguing that it has to protect all audience members, and not simply those under 18, from material which may be harmful and/or offensive. As noted earlier, there is actually no consensus over what constitutes harmful media content; indeed, there is a good deal of controversy over whether any media content at all can meaningfully be categorised in this fashion – except, of course, content whose production involved harm to the participants, such as death, injury or non-consensual sexual activity. Meanwhile, the idea that offensiveness can be defined in terms of breaching 'generally accepted standards' simply denies the basic fact that what is regarded as offensive is a highly subjective matter, particularly in a society as diverse and heterogeneous as the contemporary United Kingdom. Consideration of this topic is beyond our scope here, but what is certainly relevant is the research which Ofcom itself has either commissioned or carried out into viewers' attitudes to sexual material on television.

In *Language and Sexual Imagery in Broadcasting: A Contextual Investigation*, research carried out for Ofcom by The Fuse Group with focus groups in 2004 found that:

> sexual imagery was less of an issue for people than offensive language, particularly in their day-to-day lives...Parents in particular tended to worry about the degree of sexual imagery which surrounded their children, largely because they felt that it could lead to premature sexualisation of their children. However, many other participants in this piece of research felt that the growth in sexual imagery in public life generally indicated a more liberal and tolerant attitude towards sexual matters, and this was thought to be a good thing (Ofcom 2005b: 49).

The great majority of those who participated in the research thought that there was more sexual imagery on television than before, that it was more explicit, and that it started earlier in the evening. However, not everyone was equally offended nor concerned by such imagery, and, not altogether surprisingly, it was older people who were most likely to be offended. Teenagers did not think that there was too much sex on television overall, although some girls did feel that broadcasters showed more explicit content than they would like. Both girls and boys were embarrassed by sexual scenes if watching programmes with their parents or younger siblings; this coloured their attitudes towards the most appropriate scheduling times for such programmes. In terms of rules and guidelines: 'most think that general entertainment satellite, cable and DTT channels should adhere to a similar regulatory code as the five terrestrial channels. Premium rate subscription channels should have more freedom to broadcast a range of content at all times of the day/night' (ibid.: 77).

Turning to the Ofcom report, *The Communications Market 2006*, we find that 36% of viewers felt there was too much sex on television. This compares to 56% who thought there was too much violence, 55% (swearing) and 56% (intrusion). These figures were down by six, three, two and three percentage points respectively compared to the previous 12 months. 52% thought there was 'about the right amount' of sex on television compared to 38% (violence), 39% (swearing) and 35% (intrusion) (2006: 268).

In 2009, Opinion Leader produced for Ofcom the qualitative research report *Attitudes Towards Sexual Material on Television*. Again, television was not the greatest concern of participants – this time it was the Internet. Whilst they were concerned about sexual content on television, this was not their area of greatest concern, with violence, sexism and racism also cited as examples as equally or more unacceptable. Protection of the under-18s was the main concern with respect to sexual material, with participants worried that younger children might come across it by accident and older ones might actively look for it. According to the report:

> a wide range of factors appeared to inform participants' levels of concern, with clear differences according to demographics (particularly age but also gender), life-stage (particularly whether they have children at home) and attitudes (liberal or conservative). In particular, older participants (those aged 35–54 and 55+, and particularly older women) and those with children at home (especially those with older children) were more likely to be concerned about sexual material on television. Older participants tended to be more concerned from the perspective of personal offence. Those with children (particularly older children, who may actively seek out stronger sexual content), were more concerned about the protection of under-18s.
>
> Most participants believed in general that there is a place for sexual material on television for those adults who would choose to watch it. However, participants voiced the need for mandatory access restrictions where appropriate, depending on the type and strength of sexual material, and highlighted the importance of other contextual considerations such as: the channel, time of broadcast and pretransmission announcements. This was a common finding across demographic categories, although there were differences in the strength of the material participants would consider to be in need of access restrictions (2009c: 5).

In the context of protecting the under-18s, Ofcom and the Youth Research Group undertook research in 2005 into the effectiveness of PIN protection systems. From this they discovered that:

> some children and young people say they know their parents'/guardian's PIN numbers, even though their parents think they do not. Furthermore, a significant minority of children say they have, at least occasionally, used the PIN number without permission, even though their parents think they have not done so (2005c: 6).

More specifically, 'slightly less than half of children who had awareness of the PPV PIN number claimed to have used it, at least occasionally – without parental permission, but 90 per cent of parents believed their children had NOT done so' (ibid.: 7). On this basis, the report concluded that 'the evidence does not demonstrate that such security measures are sufficient to protect under eighteens from "R18" material and its equivalent in the *current* PIN environment' (ibid.: 6). Ofcom noted the possibility that parental attitudes and behaviour regarding PIN numbers might change if they felt there was a possibility that children might access, either accidentally or intentionally, R18 or R18-equivalent material, but stated that this was a 'hypothetical issue that is not – and cannot – be accurately covered by this piece of research' (ibid.: 7).

But it could, of course, have been the subject of further research – namely into devising more effective forms of parental control, whether by the PIN system or other means. However, this has not been forthcoming, and it is not exactly difficult to see why. Such is the disjunction between the tentative conclusions reached by Ofcom's research and the strength of its strictures against certain kinds of broadcast sexual material that the only conclusion to be drawn is that the regulator, no doubt with the government's blessing,[4] has from the start been absolutely determined to interpret its obligations under the Communications Act 2003 in such a way as to outlaw material of below R18 strength on unencrypted channels, even late at night on niche channels, and material of R18 strength on encrypted channels at any time. Such an attitude, virtually unique in the EU, has deep political and cultural roots.

Putting a stopper on the poison

Although it is now more or less accepted that the written word will not be prosecuted for obscenity, the situation is very different when it comes to images, moving or still. Hard-core pornography may be sold legally only in licensed sex shops, but even here the range allowed by police and prosecution practice is far more restricted than in virtually every other EU country, and in this respect it's particularly worth noting that the BBFC cuts nearly 30% of the DVDs which it passes even at R18. In legal terms, then, 'very little sex please, we're British' stubbornly remains the order of the day. And although television was brought under the Obscene Publications Act in 1990, this was largely an act of spite by the Thatcher government, since television, before and since, has always been so tightly regulated by its own codes that nothing remotely obscene has ever been broadcast – at least on the mainstream terrestrial channels.

It was thus a double affront – to UK television regulation and to the laws on obscenity and indecency – when technological developments made it possible to beam pornography into British television sets by satellite. When Margaret Thatcher was first alerted to this possibility her first, and characteristic, reaction was that such satellites should simply be shot down. When the possibility became an actuality with the arrival of Red Hot Dutch in

1992, the official response, though less extreme, was no less hysterical. Thus William Rees Mogg, who then headed the Broadcasting Standards Council, created by the Thatcher government to ensure that its 'deregulation' of British television did not extend to matters of content, was quoted in the *Independent*, 21 January 1993, as stating that the channel's programming 'is as unpleasant as being taken into an abattoir to view a series of butcheries'. Inevitably the *Daily Mail* embarked on one of its many censorship campaigns, alleging that Britain was under 'electronic attack from the forces of decadence' and 'besieged by degrading and violent images' (25 February 1993). The fact that the images were of continental origin also enabled the *Mail* to embroil its anti-porn campaign with its ongoing crusade against the hated 'Europe'. Thus as a particularly laborious (not to mention inaccurate) headline spelled it out on 18 January 1993: 'Hard-core pornography is now being beamed into thousands of British homes. Sold over the counter, it would be illegal. But thanks to an EC directive signed by the Government, there is nothing at all to stop it appearing on our television screens'. Its message to that government in its 25 February edition was unequivocal: 'put a stopper on this poison'. This it rapidly did by making it illegal for British shops to sell the decoders necessary for receiving the channel, and for British firms to advertise on it or supply it with programmes. In so doing, it invoked exactly the same article of the European Commission's Television Without Frontiers Directive of 1989 which is now replicated as Article 22 of the Audiovisual Media Services Directive. Subsequent satellite 'invaders' were met with the same chorus of outrage from the *Mail* and other moral campaigners, and were dealt with in exactly the same fashion. That the 'enemy' is now within simply makes it that much easier to deal with, technological advances notwithstanding, and any government which failed to do so would know *exactly* what to expect from the *Mail* and its ilk.

From censure to self-censorship

The advent of a form of economic and structural 'deregulation' has not been accompanied by a concomitant loosening of the reins when it comes to all aspects of television content. In this respect, the strictures by Samuel Brittan, a member of the Peacock Committee whose 1986 report led eventually to the 1990 Broadcasting Act, are as applicable to 'New' Labour as they were to Thatcherism:

> in putting forward the idea of a free broadcasting market without censorship, Peacock exposed many of the contradictions in the Thatcherite espousal of market forces. In principle, Mrs Thatcher and her supporters are all in favour of deregulation, competition and consumer choice. But they are also even more distrustful than traditionalist Tories of policies that allow people to listen and watch what they like, subject only to the law of the land. They

> espouse the market system but dislike the libertarian value-judgements involved in its operation (1989: 40).

One can agree with Brittan without necessarily arguing for the abolition of certain *positive* forms of regulation which consist of regulating *into* the broadcasting system elements which most people find desirable, such as impartiality, honesty, diversity, indigenous production and so on. Indeed, it is perfectly consistent to argue for retaining these whilst arguing against retaining absolute prohibitions on adult viewing, particularly when these cannot be justified by anything other than subjective judgements about alleged harm and offence – however fancily these may be dressed up semantically in order to make them appear a great deal more authoritative than they actually are. Of course it should be made as difficult as possible for viewers, of any age, to come across unawares material which might offend them, and for children to circumvent technical devices which have been put there expressly to protect them from such material, but this should not prevent soft-core material being shown, late at night and with due warning on unencrypted general or niche channels, and full-on, hard-core R18 material being shown late at night on encrypted channels. As Geoffrey Robertson and Andrew Nicol argue:

> it is absurd for Ofcom to imagine that its role is to stop teenagers from getting their hands on pornography – that is impossible. Its role is to make the exercise difficult, in relation to television, and to impress the unsuitability of such films for the young whilst permitting some access – if only in the early hours of the morning – for adults. Using teenage familiarity with PIN systems as an excuse for a blanket ban is disproportionate (894–5).

The alternative is to continue with the increasingly anomalous situation in which adults are legally entitled to buy R18 material in licensed sex shops and watch it on pay-per-view services in hotel rooms, but not to view it on broadcast television in their own home. It is important to remember that this is not material which has been found by a court to be indecent or obscene, nor which has been explicitly proscribed by statute law; it is simply material which Ofcom has deemed to convene the terms of its own Broadcasting Code and the provisions of the Communications Act 2003. Of course, Ofcom does not have the power of pre-broadcast censorship, but, as we have seen, it most certainly has the power of post-broadcast censure, and the sanctions with which to back up its judgements. As these include not only hefty fines but the ability to revoke a station's licence to broadcast, it is but a very short step from censure to self-censorship, which is precisely the form of governance which Ofcom is designed to instil into the broadcasting system as a whole.

The 'context' and 'editorial content' of pornography

This is also a situation in which the regulator in all seriousness searches for 'editorial content and purpose' in a soft-core programme such as *Sin Cities*, whose 'purpose', self-evidently, is purely titillatory, and appears to believe that viewers will unable immediately to understand the 'context' in which programmes are broadcast on an adult sex chat channel entitled Babeworld TV. The absurdity and fatuity into which this leads is nicely illustrated by Ofcom's judgement of an item broadcast on that channel at 23.30 on 28 May 2009, in which, as its *Broadcast Bulletin*, 9 November 2009 (8–11), inimitably put it:

> the female presenter was shown to move her left hand down to her crotch and then pull the string of her G-string to one side to show viewers clear, unambiguous and close-up shots of her vagina and anus. She repeated this action five times in just under three minutes. In Ofcom's view it is a breach of generally accepted standards and offensive to broadcast such images on a service without mandatory access restrictions unless they are justified by the context.

Ruminating on the 'context' of these actions, Ofcom noted that:

> the programme was broadcast well after the 21:00 watershed. It judged however that the repeated and seemingly deliberate actions by the presenter to show her vagina and anus had either no, or a completely insufficient, justification in the context. Ofcom considered that the actions of the presenter were clearly not a 'one-off accident' because she needed to act with deliberation to locate the string with her fingers and then move it to the side. Also the action was repeated five times in total. Ofcom questions the speed with which the onsite compliance officer reacted to the presenter's behaviour, given that the presenter only appeared to be informed of the incident while she was on screen once she had changed position and the G-string was back in place.

One would have thought that the 'justification' for the presenter's actions was all too glaringly apparent, but seemingly not to Ofcom, which decided that:

> even though viewers of 'adult-sex' channels are used to a great extent to the type of material they show, the degree of offence capable of being caused by the broadcast of the very explicit images shown in this case was likely to be considerable. In Ofcom's view, this material would have exceeded the likely expectation of the audience, especially for viewers who may have come across it unaware. For all these reasons, this

> content was offensive and not justified by context, and so breached generally accepted standards. It was therefore in breach of Rules 2.1 and 2.3 of the Code.

With a sex chat performer's actions minutely scrutinised by a compliance officer and members of a quango, we appear to have plunged straight into a world imagined by television satirist and prankster Chris Morris. The fact that the quango in question lists one of its core activities in its *Annual Report 2008/9* as 'reducing regulation and minimising administrative burdens' (Ofcom, 2009c: 29) adds only a further touch of the bizarre. However as the quango was created by a government which equates censorship with 'tough protection of the genuine public interest', and which in its *Digital Britain* report and through the appointment of a 'Digital Inclusion Champion' gives the distinct impression that it is every citizen's moral duty to connect to broadband forthwith, but which in the Criminal Justice and Immigration Act 2008 threatens them with imprisonment and being placed on the Sex Offenders Register if they use that connection to access the 'wrong' kind of porn (Petley, 2009), then really we shouldn't be too surprised.

Notes

1. It's extremely revealing of the continuity between Thatcherite and 'New' Labour broadcasting policies that in 1989 I wrote an article with exactly the same title, and making many similar points, about the White Paper which preceded the Broadcasting Act 1990. It's also extremely depressing.
2. For a useful review of this issue see Cumberbatch and Howitt (1989).
3. For a good example of Ofcom at work on lobbying to 'liberalise' the directive, see the speech by then Ofcom Chairman David Currie to the 2005 Liverpool conference on the Television Without Frontiers Directive (http://www.Ofcom.org.uk/media/speeches/2005/09/liverpool_conf). For a detailed analysis of this process, and in particular of the alliances which Ofcom formed with industry groups which shared the same 'liberal' objectives, see Granville Williams (2007).
4. Having decisively lost a ferocious, but very much behind-the scenes, battle with the BBFC over the latter's relaxation of the R18 guidelines in the late 1990s (for a detailed account of which see Petley, 2000), it is hardly likely that the government would have permitted R18 material to be shown on television.

References

Brittan, S. (1989) 'The case for the consumer market', in C. Veljanovski (ed.) *Freedom in Broadcasting*, pp.25–50 (London: The Institute of Economic Affairs).

Cumberbatch, G. and D. Howitt (1989) *A Measure of Uncertainty: The Effects of the Mass Media* (London: John Libbey).

Department of Trade and Industry/Department for Culture, Media and Sport (2000) *A New Future for Communications* (London: TSO).

Department of Trade and Industry/Department for Culture, Media and Sport (2002) *The Draft Communications Bill – The Policy* (London: TSO).

Millwood H. and Livingstone, Hargrave, A. S. (2009, second edition) *Harm and Offence in Media Content: A Review of the Evidence* (Bristol: Intellect).

Ofcom (2005a) *R18 Material: Its Potential Impact on People Under 18: An Overview of the Literature* (London: Ofcom).

Ofcom/The Fuse Group (2005b) *Language and Sexual Imagery in Broadcasting: A Contextual Investigation* (London: Ofcom).

Ofcom/Youth Research Group (2005c) *Research into the Effectiveness of PIN Protection Systems in the UK* (London: Ofcom).

Ofcom (2006) *The Communications Market 2006* (London: Ofcom).

Ofcom (2009a) *The Ofcom Broadcasting Code* (London: Ofcom).

Ofcom (2009b) *Broadcasting Code Review: Proposals on Revising the Broadcasting Code* (London: Ofcom).

Ofcom (2009c) *Annual Report 2008/9* (London: Ofcom).

Opinion Leader (2009) *Attitudes Towards Sexual Material on Television* (London: Opinion Leader).

Petley, J. (1989) 'Doublethink', *IPPA Bulletin*, Summer, pp.4–5.

Petley, J. (2000) 'The censor and the state: Or why *Horny Catbabe* matters', *Journal of Popular British Cinema*, 3, pp.93–103.

Petley, J. (2009), 'Pornography, panopticism and the Criminal Justice Act 2008', *Sociology Compass*, 3(3): 417–32.

Robertson, G. and A. Nicol (2008, fifth edition) *Media Law* (London: Penguin).

Williams, B. (1979) *Report of the Committee on Obscenity and Film Censorship* (London: HMSO).

Williams, G. (2007) 'From isolation to consensus: The UK's role in the revision of the *Television Without Frontiers* directive', *Westminster Papers in Communication and Culture*, 4(3): 26–45. http://www.wmin.ac.uk/mad/pdf/WPCC-VolFour-NoThree-Granville–Williams.pdf.

15

Afterword: Closing Reflections on the New Politics of Leisure and Pleasure

Peter Bramham and Stephen Wagg

When this book was first proposed, a number of reviewers commented on its viability. It was interesting, said one, 'but a lot of what they talk about isn't political'. This was mistaken, on two counts, both of which we hope are clear from reading the book. First, leisure has always been political. Leisure time had to be fought for, in a time of unremitting toil, and then defended. Successive factory and employment acts across a range of countries created space within the working week variously for men, women and children but a succession of legislative interventions governed what people, particularly the working class, could do with their leisure time – what they could see or read, what they might ingest or otherwise do with their bodies, whether or not they could place a bet, where they might walk, what they might hear a comedian say from the stage, and so on. In essence, as this book shows, this remains the case – albeit in ways that are undeniably complex. Secondly, the word 'politics' has, since the 1960s, acquired a diverse set of meanings which extends far beyond the doings of professional politicians. Indeed, one can draw a useful distinction between a narrow (P1) and broader definitions of politics (P2). In the UK, Politics 1 refers to what happens in the Westminster village, what is discussed daily in the politics pages of the press, during weekly political programmes on the television and so. Politics 2 refers more broadly to all social relations underpinned by the exercise of power and constraint. So even the most intimate personal relationships, take for instance in the family, are grounded in the exercise of power. One only needs to remember recent debates as to whether parents should be allowed to smack their own children, before considering moral panics about child abuse by parents and relatives, domestic violence between partners, neglect of the elderly by carers and so on. Leisure relations are, therefore, in essence political, grounded in the recursive tension between human freedom and constraint.

A growing number of people in what some commentators refer to as post-modern society may feel increasingly remote from affairs of state, parliamentary debates and the like. But they are probably more aware than previous generations of the politics of identity – that is, they are conscious of being male or female, straight or gay, black or white and so on, and of the political and social consequences of these identities. They are also more likely than earlier generations to have some kind of relationship to the politics of pleasure. They may not vote in a parliamentary election, for instance, but will likely rush to the phone to do so for *The X Factor* or *Strictly Come Dancing*. They will recognise few politicians but many celebrities. They will hold a view on the ban on smoking in pubs, perhaps, or pornography or censorship. They may take 'recreational' drugs. They may, like the 'lovely girls' observed by the novelist Alan Sillitoe one recent Friday night in his native Nottingham, begin their weekends by 'queuing up at cashpoints to get money to go to the clubs and get stinking'.[1] They will find some comedians funny and take strong exception to others. They may even appreciate that previous generations fought, to paraphrase the Beastie Boys, for the 'right to party'.[2]

The emergence of these new cultural politics, the attendant tensions contradictions between freedom and control and the specific issues arising in specific cultural realms have been the subject of this book. As such we hope that the book has been, and will be, read not only by academics and their students but by any interested party who participates, knowingly or not, in whatever form and to whatever extent, in these politics.

Notes

1 James Walker, 'I've always strongly believed in a meritocracy' [Interview with Alan Sillitoe] *The Independent Viewspaper*, 27th April 2010, p.10.

2 '(You gotta) fight for your right (to party!)' was written by Adam Horovitz, Rick Rubin and Adam Yauch and was first heard on the Beastie Boys' album *Licensed to Ill* (Def Jam/ Columbia CK-40238) in 1986.

Index